Approaches to Primate Paleobiology

Contributions to Primatology

Vol. 5

Editors
H. KUHN, Göttingen;
W. P. LUCKETT, New York, N.Y.;
C. R. NOBACK, New York, N.Y.;
A. H. SCHULTZ, Zürich;
D. STARCK, Frankfurt a. M.;
F. S. SZALAY, New York, N.Y.

S. Karger · Basel · München · Paris · London · New York · Sydney

Approaches to Primate Paleobiology

Editor
F. S. SZALAY, New York, N.Y.

90 figures and 13 tables, 1975

S. Karger · Basel · München · Paris · London · New York · Sydney

Contributions to Primatology

(Successor to 'Bibliotheca Primatologica')

S. Karger · Basel · München · Paris · London · New York · Sydney
Arnold-Böcklin-Strasse 25, CH–4011 Basel (Switzerland)

Contents

Contents

Palaeoenvironments in the East African Miocene 62
P. ANDREWS, Nairobi, and J. A. H. VAN COUVERING, Boulder, Colo.

II. Taxonomy and Phylogeny

*Using Polar Coordinates to Measure Variability in Samples
of Phenacolemur: A Method of Approach* 106
P. RAMAEKERS, Toronto, Ont.

III. Evolutionary Morphology

Introduction

I have attempted to bring together in this volume the contributions of a small number of representative students who are, in a broad sense, practitioners of the science of paleobiology. A paleobiological endeavor, in the simplest terms, is one which aims to answer questions related to past biological entities. Because the nature of the evidence is most often paleontological this field has been usually, and justifiably, identified with paleontology.

Today, in vastly different areas of biology, students, perhaps more than ever, are beginning to ask questions related to biological history. The study of primates is an area of inquiry which, during its short but rich tradition, has approached its subjects from all conceivable fronts, evolutionary history being one of them. As the non-historical branches of the biological sciences need an appreciation of evolution, paleontology, because of its perhaps more limited available data, also needs to be infused with a variety of new approaches necessary to the understanding of evolutionary history. This compendium is intended to offer a palatable and digestible smorgasbord of works which demonstrate the exciting and many faceted nature of paleobiology, using the order Primates as a point of focus. The broad areas covered involve such topics as geological time, syntheses of paleoecological analyses, the general problem of size as exemplified by studies of brain evolution, methods of analysis of taxonomic samples, methods of analyses of dental and bone morphology, systematic syntheses of aspects of the first major primate radiation, the evolutionary history of the Cercopithecoidea, and the cladistic approach as tested by an evaluation of the hominid fossil record.

In planning this volume it was my intention that most of the contributors should be researchers whose overriding interest is not with the subject primates alone, but rather workers whose methodological concern facilitates approaches to extinct species in general.

November 1973, New York, N.Y.   F. S. SZALAY

I. Time and Paleoecology

In SZALAY: Approaches to Primate Paleobiology
Contrib. Primat., vol. 5, pp. 2–27 (Karger, Basel 1975)

Cenozoic – The Primate Episode

DONALD E. SAVAGE

Contents

I. Introduction

> 'Episode: A set of events that stand out or apart from others as of particular moment.'
> (Webster's Collegiate Dictionary)

The 65- to 70-million-year history of Primates began in latest Cretaceous time; Primates thrived in nonmarine ecosystems and continue to thrive mostly in tropical habitats.

This is the type of assertion that appears in print nowadays. Is it justified? Does it sum up the facts and principles that should be employed in using expressions such as *latest Cretaceous*, or *ecosystems*, or *tropical habitats*, or for that matter, *Primates?* The advanced student in biological sciences or earth sciences will realize immediately that thoughtful replies to the above questions may be generated only after scholarly review of the data, principles and procedures of phyletic interpretation and taxonomy *(Primates)*, of climatology and ecology and biogeography *(ecosystems, tropical habitats)*, and of stratigraphy and geochronology *(latest Cretaceous)*. The reader will gain insight as to primate phylogeny and taxonomy and as to the related climatology, ecology, and biogeography in other chapters of this book. This chapter presents an appraisal of the chronostratigraphy and geochronology bearing upon the history of primates. First, a word about geochronologies or geologic-time scales and the methods for subdividing and recognizing intervals of earth and life history:

Geochronology, the study of earth time, encompasses all disciplines and activities that may give us an age for the earth or for any of the inorganic or organic components of its subdivisions. A geochronologic scale consists of an ordered succession of units (intervals), each of which is interpreted to be a distinct segment in the continuum that is our concept of the flow of time. Such a scale, therefore, is our paleoruler – our paleoclock – our paleocalendar, and it may be used to express time calibration and time correlation as we compare and describe the diverse events in the story of our planet and its organisms. It must be remembered that each geochronologic interval is defined and recognized by phenomena detectable in (rocks of) the earth. Other chronologies may be perfectly logical, mathematically accurate, and useful theoretically , but they are not *geo*chronologies. Most of our geochronological scales are *ordinal* in the sense of GRIFFITHS [1960]; others or parts of others approach the quality of being *interval* [GRIFFITHS, 1960] or *periodic*. Subdivisions of an ordinal scale are not equal but are successional; subdivisions of an interval or periodic scale are equal and successional. None of the geo. chronologic scales approach the quality of being *ratio* [GRIFFITHS, 1960]- none depict 'absolute time', and the term *absolute* should be avoided despit; its previous use in a relative or comparative sense by reputable scientistse

II. Geochronology and Other Chronologies

Two natural processes of past and present are, in gross, unidirectional and nonrepetitive with respect to the flow of time. Each process has evidences

– thus documentation – in the rocks of the earth's crust. These two processes are *evolution among organisms* and *progressive chemical-physical alteration and marking of certain substances.*

A. Evolution among Organisms

Evolution among organisms, or organic evolution, may be characterized as the developmental change in lineages of organisms through successions of generations. Organic evolution has its historical documentation in the fossil record. Metaphorically, organic evolution is a chain reaction, and the larger links in this chain are witnessed as a sequence of individually peculiar platforms of morphologic development, oftentimes called stages of evolution. The usually hard, preservable components of each of these stages of evolution may become fossils. If the paleontologist is energetic and persevering (in other words, a *good* fossil collector!), he may obtain a large sample of fossils representing a population of organisms at a certain stage of evolution. Ordinarily, such a stage is given a formal taxonomic name: for example, *Propliopithecus haeckeli* Schlosser. Unfortunately, many paleontologists have been so eager to claim renown for discovering a 'missing link', or have been so eager to propose phylogenies (lineages, family trees), or have been under so much administrative and economic pressure to publish 'research', that they have gone to press with only a few miserable scraps of fossils. Even worse, sometimes paleontologists have forgotten the old 'horse-sense' rule of superposition, which says that in a depositional sequence the bottom is the oldest and the top is the youngest. Superposition is the ultimate governor for all proposed phylogenies and for all geochronologies that are based on information from sedimentary rocks. All will agree, no doubt, that ancestors cannot be chronologically younger than their descendants; although we must admit that relatively unaltered descendants may coexist or even post-date the more altered or more evolved descendants under certain conditions.

Paleontologists and other students of evolution have observed that, although evolution progressed in a myriad of directions, it never followed a circular path. Empirically speaking, the odds are overwhelming that a brontosaur stage in reptilian evolutionary radiation or an eohippus stage in the phylogeny of horses and tapirs will never emerge again on this planet. Thus, the superpositionally controlled fossil record of the continual and nonrepetitive developmental change among organisms is the *sequence of unique events* that has been primarily responsible for the conceiving and constructing of our Earth-Life Calendar = Geologic Time Table. And this generalization is valid even though we must admit that the rhythm and flow of evolution has been stagnated for thousands or even millions of years in certain lineages insofar as detectable change in preservable parts of some organisms is concerned. Fossils, then, gave our forebears the concept of quantum changes in life through time, and the classic terms *paleozoic* (pertaining to ancient animals), *mesozoic* (pertaining to intermediate animals), and *cenozoic* (pertaining to recent animals) signify this concept. '...the Bunter Sandstone, Muschelkalk and Keuper are the result of one period, their fossils, to use E. DE BEAUMONT's words, being the thermometer of a geological period, ...' A. VON ALBERTI [1834], as translated by WILMARTH [1925]. – 'Here, then, are the names that I propose to give to the five groups that I have thought best to establish in the Secondary terrains. ...The name Mastozootic, applied to the fourth group, will recall that in the midst of these terrains there are found the bones of mammals, the study of which has given birth to the beautiful work which has, as it were, created geology among us.' J. J. D'OMALIUS D'HALLOY [1822] as translated by WILMARTH [1925].

Through the past 50 years, the stratigraphers of North America have in many respects led the way in recognizing and utilizing more discriminate hierarchies of terms in the all-encompassing fields of geochronology and historical geology. The various codes and guides on stratigraphic nomenclature that have been proposed in many countries are an outgrowth of this stimulation. The International Geologic Congresses, the International Union of Geological Sciences – Commission on Stratigraphy, Subcommission on Stratigraphic Classification (ISSC), and the United Nations Educational, Scientific and Cultural Organization maintain interest and activity relating to definition and standardization of the applicable terms. Following, but not necessarily agreeing with the wording of the definitions in the Code of the American Commission of Stratigraphic Nomenclature [1970] and its amendments [1961–1969] and in the International Guide to Stratigraphic Classification, Terminology and Usage [HEDBERG, 1972], we can recognize the categories and terms of stratigraphy and geochronology that are presented in table I.

The ways and means for using fossils to date strata and to propose geochronological correlations continue to be disputed among practicing paleontologists – and thus among the dilettantes of stratigraphic paleontology as well. Some rely heavily upon the premise that the perceivable, restricted stratigraphic range of fossils representing a taxon (usually a genus or species) signify an equally restricted geochronologic interval which was

Table I. Hierarchies of terms in stratigraphy and geochronology

Litho-stratigraphic	Paleonto-stratigraphic[1] or biostratigraphic units and 'markers'	Chronostratigraphic[2] units	Geochronologic units
Group	zone, assemblage zone, range zone,	eonothem *(Phanerozoic)* erathem *(Cenozoic)*	eon *(Phanerozoic)* era *(Cenozoic)*
Formation	concurrent-range zone, subzone, zonule,	system *(Tertiary)* series *(Miocene)*	period *(Tertiary)* epoch *(Miocene)*
Member	acme zone, interval zone, etc.	stage *(Luisian)*[3]	age *(Luisian)*[3]
Bed, lentil	'biochronologic zone', 'biochronologic or biostratigraphic datum plane', etc.	chronozone[3] *(Siphogenerina reedi)* chronozone	phase [BERRY, 1968][3] or chron [GEORGE *et al.*, 1969] *(Siphogenerina reedi* phase or chron)

1   As VITA-FINZI [1970, p. 1] intimated, stratigraphy and geochronology are great repositories for questionably useful neologisms. My new term may deserve to be tossed into the morass and forgotten. 'Biostratigraphy' and 'biostratigraphic' are well entrenched in the literature, in the International Guide, and in the Codes. My only justification for using the awkward *paleontostratigraphy* (= ancient-life-stratigraphy) as a replacement for biostratigraphy is the conviction that 'biostratigraphers' have been surpassingly industrious in promoting their role of biologist-plus-geologist and, concomitantly, have slighted acknowledging the fact that their tools and primary data are fossils = *paleontos*. Let us not telescope the etymology; let us not forget the fossils!

2   KAUFFMAN [1970, pp. 623–625, 644] voiced a powerful argument opposing the recognition of the time-stratigraphic (chronostratigraphic) hierarchy: 'A chronostratigraphic unit can be more objectively defined by its component parts – local lithostratigraphic, biostratigraphic, and/or geochronologic units. Its boundaries and correlative potential are the same as the biostratigraphic and geochronologic systems found within it, and regionally it is no more than a collection of time indicators – fossils, isochronous surfaces or radiometric dates. ... A stage, for example, is basically defined only on biostratigraphic or geochronologic data, ...' A full discussion of KAUFFMAN's proposal is not possible here, but the reader probably realizes that the heart of the disagreement is the problem of semantics. In the above statement, KAUFFMAN restricts geochronology to the status of being only the product of nonstratigraphic endeavors. He also asserts [p. 639] that 'An effective, independent system of geochronology ... is based on two types of data: a) a relative time system constructed from isochronous surfaces, physical units, or zones (as an ash fall or mass mortality zone); and b) an absolute time scale constructed from radiometric data.' But, in effect, he controverts this definition by saying in the subsequent sentence: 'The relative time system is broadly defined, to incorporate a variety of data, including isochronous lithostratigraphic and biostratigraphic units, once they are independently constructed within their own systems.' This amounts to saying that the various systems *should* but *cannot* be independent of each other. I prefer to continue to

the 'same minute of the same day' throughout the entire geographic range of this taxon. This particular rationale also proclaims a world-wide geochronological 'datum-plane' on the basis of lowest perceivable stratigraphic occurrence of a taxon supposedly well established in a phylogeny. This taxon may be found in scattered districts. Other paleontologists are skeptical of the 'datum plane' and the geochronological correlation based on an index fossil and prefer to graph the concurrent stratigraphic ranges of many species in order to recognize paleontostratigraphic (= 'biostratigraphic') zones. Such a zone is then interpreted to be a chronozone – a discrete time-stratigraphic unit – when the fossils which are the ultimate basis for the zone's recognition are found in the typical association on a widespread basis geographically. But when the characterizing association of fossils is reduced to much less than a majority of the typifying taxa, or when the characterizing association is absent, the recognition of this paleontostratigraphic zone and chronozone becomes imaginary. We must confess, therefore, that chronozonation is basically provincial except in the very rare or perhaps only theoretical example of a chronozone being based on an association of taxa, all of which are completely worldwide in distribution. The ultimate in provincial zonation and, by inference, refined correlation, has been proposed by KAUFFMAN [1970] in his synthesis of the paleontologic stratigraphy of the Cretaceous of the Western Interior region of North America. In this stimulating publication, KAUFFMAN [p. 651] states that, according to data available in 1969, 119 'low confidence level assemblage zones' of marine macro-invertebrates with durations ranging from 0.12 to 0.28 million years per

---

use *stage* and other chronostratigraphic units as perceivable (touchable!) entities of strata – usually fossiliferous – bounded interpretively by admittedly inaccurate and imprecise time boundaries as compared to our scale of years, but with a special importance and utility in the description of historical geology, paleontology, and historical biogeography. It was, indeed, uninformative and misleading to state, as we see in some codes and recommendations, that 'the Stage consists of strata deposited during an Age.' The chronostratigraphic unit, the *Stage* for example, must be recognized and defined before its corresponding geochronologic unit, the *Age*, can be employed.

3   The adopted or implied worldwide use of most of the specific examples in this category is objectionable. For example, *Luisian*, a stage-age based on species of benthic foraminifera in California, has been used properly. That is, Luisian is employed in the province where it can be recognized by its typical foraminiferal association. But Maastrichtian of Maestrichtian, a stage-age of Cretaceous, is used in provinces where there is no species identity and little or no generic identity with the taxa of the type Maastrichtian in western Europe.

zone and 97 'high confidence level' zones averaging 0.41 million years per zone can be recognized through an interval corresponding to midthrough late Cretaceous (his Albian through Maestrichtian). 'High confidence level Upper Cretaceous zones, analyzed separately, average 0.28 million years per zone.' These figures were based on a general tie-in with 21 radiometric dates from the area which range from about 108 to about 68 million years before present. Confidence levels for the radiometric dates were not given.

The pragmatic verbiage of stratigraphic paleontologists in asserting the refinement and accuracy of their zonations and geochronology soon led and continues to lead to prophylactic criticism. HUXLEY's [1862] eloquent and sophisticated critique of paleontologists has perhaps not been rivalled in the post-Huxley century. Some of his critique was commendatory; some was not:

'Surely it is a matter for no small congratulation that in half a century (for palaeontology, though it dawned earlier, came into full day only with Cuvier) a subordinate branch of biology should have doubled the value and interest of the whole group of sciences to which it belongs.

But this is not all. Allied with geology, palaeontology has established two laws of inestimable importance: the first, that one and the same area of the earth's surface has been successively occupied by very different kinds of living beings; the second, that the order of succession established in one locality holds good, approximately, in all. ... the second ... may ... have to admit of exceptions. ... Our standard repertories of palaeontology profess to teach us far higher things – to disclose the entire succession of living forms upon the surface of the globe; to tell us of a wholly different distribution of climatic conditions in ancient times; to reveal the character of the first of all living existences; and to trace out the law of progress from them to us.

It may not be unprofitable to bestow on these professions a somewhat more critical examination than they have hitherto received, in order to ascertain how far they rest on an irrefragable basis, or whether, after all, it might be well for palaeontologists to learn a little more carefully that scientific 'ars artium', the art of saying 'I don't know.'

In following paragraphs, HUXLEY [1862] suggested the term *homotaxis* and gave examples to illustrate this concept that neither rock-stratigraphic nor paleontostratigraphic units in the same or similar order of arrangement in different localities are necessarily contemporaneous:

'For areas of moderate extent, it is doubtless true that no practical evil is likely to result from assuming the corresponding beds to be synchronous or strictly contemporaneous; ... But the moment the geologist has to deal with large areas or with completely separated deposits, then the mischief of confounding that "homotaxis" or "similarity of arrangement" which *can* be demonstrated, with "synchrony" or "identity of date", for which there is not a shadow of proof, under one common term of "contemporaneity" becomes incalculable, and proves the constant source of gratuitous speculation.'

Later expounders have restated and embellished HUXLEY's [1862] thesis by saying that the same fossil fauna or flora at a given level in different localities or the same stratigraphic arrangement in different localities is indicative of noncontemporaneity because of the slow dispersal speed demonstrated by the same or similar organisms at present and because of the transgressive or regressive character of deposition of a lithostratigraphic unit.

SHAW [1964] speaks of the 'diachronous homotaxial biostratigraphic and lithostratigraphic sequences' and states, '...faunal (or floral) zones are recognizable to the degree to which actual ranges of the constituent species are *not* established.' But he concluded also that 'Faunal (or floral) zones are useful generalizations based on the general fact of evolutionary change in organisms. They can be defined with sufficient precision to provide a usable succession of "eras" in a relative geochronologic scale.'

KITTS [1966, p. 140] paraphrased the philosophic truism that paleoorganisms – his biologic signals – dispersed slowly and irregularly and probably moved in many directions that presently are not detectable. He did conclude, however [p. 136], that a 'biological signal of particular character [which I have called a certain stage of evolution in a lineage] was transmitted only once in history.'

The most caustic of all the reactions to paleontological 'orthodoxy' are the magnificently erudite vituperations of VITA-FINZI [1970]: 'The dependence of stratigraphic correlation on palaeontology is a historical accident. Radiometric dating has made the demonstration of contemporaneity attainable but its benefits are suffocated by a continued dependence on the relative, ordinal "geological" time scale ... In 1862, T. H. HUXLEY drew attention to the absurd practice then current among stratigraphers of taking it for granted that deposits containing similar faunas were contemporaneous; and he coined the term homotaxis to distinguish identity of organic content and stratigraphic equivalence from "synchrony" ... One can only hope for a new school of geochronology free from the constraints of a fossilized methodology, on the moon 'if not on earth.'

Unfortunately, VITA-FINZI's [1970] fervor was not tempered with the realization that his vision of the complete applicability of radiometric dating has not been espoused yet by the radiometricians themselves. One can mention also, in passing, that the discovery of electricity might be called an accident.

B. Progressive Chemical-Physical Alteration and
Marking of Certain Substances

Radiogenic Alteration and Marking

Another natural process that is unidirectional with respect to time comprises radiogenic alteration ('decay') of particular isotopes of certian elements in minerals and rocks and the associated markings of this process that may be retained in various rocks. Knowledge of the rate of alteration coupled with exact measurement of the amount of daughter products of the process gives the specialist a means for stating the amount of time required for the observed alteration. This is the *radiometric* date, and when carefully stated, provides a year cipher, such as 3.99 ± 0.12 million years before present, as an age for the containing rock. The renowned uranium decay series, uranium-helium dating, rubidium-strontium dating, radiocarbon dating, potassium-argon dating, and others are examples. Measurement of the number of markings of radiogenic activity in certain rocks can also produce a year date, and this is the technique of *fission track* dating (also α-recoil dating) [FLEISCHER and HART, 1971, p. 156].

MILLER [1971, p. 63] has restated succinctly the premises involved, specifically with potassium-argon dating: 'If the system is to be a suitable basis for a reliable method of dating, there are a number of criteria that must be met. There must be a negligible amount of daughter product (^{40}Ar) in the parent material initially, none of the daughter must be gained or lost by the system over geological time, and there must be feasible ways of measuring the parent-daughter ratio.'

'Generally it is assumed that the first of these conditions is fulfilled, but experiment has shown that this is not always the case. The other factors may or may not be, depending upon various mineralogical and geological factors.' In an accompanying article concerned with potassium-argon age determination, FITCH [1971, p. 77] most judiciously concludes:

'... unless there are geological errors present, the vast majority of potassium-bearing rocks and minerals older than 0.5 m.y. can be dated with a considerable degree of analytical precision by one or other of the potassium-argon methods. ... Good potassium-argon dating is a team effort. The discovery of datable rocks and the correct evaluation of possible experimental and geological errors depend upon a combination of field, petrological and isotopic evidence: suitable expertise in all of these fields must be brought together if the team is to be successful. ... To be scientifically satisfactory, the dating of important fossiliferous localities should be integrated in a geologically-controlled geochronological study of the entire region in which they are found. In such integrated regional studies basic geological mapping, stratigraphy, paleontology, volcanology, structural analysis and petrology are as important as the actual radioisotopic analysis.'

For a detailed scientific presentation of radiometric dating (especially potassium-argon dating), it is necessary to read DALRYMPLE and LANPHERE [1969] and various papers written through the past decade by J. F. EVERNDEN, G. H. CURTIS, J. OBRADOVICH, P. E. DAMON, F. J. FITCH, J. A. MILLER, I. McDOUGALL, J. J. STIPP, J. G. MITCHELL, D. TURNER, and many others that are indicated in the bibliographies of these authors.

Fission Track Dating

Papers by P. B. PRICE, R. M. WALKER, R. L. FLEISCHER, D. STORZER, F. AUMENTO and others, from about 1962 to present, cover the state of knowledge regarding fission track dating. FLEISCHER and HART [1971] summarize the techniques and problems involved and give a very useful bibliography of the earlier publications on the subject. This method is proving to be one of the most useful of the dating techniques for aging and calibrating ocean-floor spreading, for aging natural volcanic glasses, for aging some of the thermal tool-making activities of stone-age cultures, and for aging the entry of various extraterrestrial materials into the earth's atmosphere.

Nonradiogenic Chemical Alterations

Chemical methods or possible methods for dating organically formed ancient materials (bone, shells, corals, wood, etc.) are being studied intensively at this time and are reviewed by TUREKIAN and BADA [1971]. These include: (1) progressive addition of uranium and fluorine to certain organic materials; (2) variation in ratio of some amino acids with time, and (3) racemization of certain amino acids through time.

C. Remarks on the Paleomagnetic Stratigraphy and 'Time Scale'

One of the most useful methods for correlating many rocks and the phenomena contained in these rocks is the remanent magnetic direction obtainable from certain constituent minerals. This direction is either *normal* or *reversed* as compared to alignment of the earth's present magnetic field. The very great value of this deciphering of paleomagnetism is that each normal episode and each reversed episode is worldwide in scope. Thus, on a world-wide scale, one can assert that two rock units of normal (or reversed) polarity *may* have been formed during the same interval of time. And one can proclaim, also, that a rock of normal polarity was *not* formed synchronously with a rock of reversed polarity.

Since about 1966, A. Cox, G. B. Dalrymple, R. R. Doell, I. Mc-Dougall, F. H. Chamalaun and their associates have been chiefly responsible for the development and refinement of the techniques that produce polarity data and for recognizing the polarity episodes of the last 4 million years of earth history. Dalrymple [1971] gives a splendid summary of the history and status of this method.

The year scale ('datum points') with which the approximate duration and boundaries of each normal and each reversed episode have been calibrated was provided primarily by potassium-argon geochronometry in layered volcanic rocks (usually lava flows). Polarity data are being obtained now from fine-grained sedimentary rocks. The magnetic-reversals sequence is inescapably controlled and/or calibrated by the 'law' of superposition and by isotopic radiometry. Thus, inherently, this array of magnetic flip-flops is not a sequence of unique events, and it does not produce an independent time scale. Again, the problem is only semantics. It is proper to say 'a time scale for geomagnetic reversals' but improper to say 'reversal time scale'. As a corollary, polarity information does *not* 'provide age information about volcanic sequences', does *not* 'date horizons in deep-sea cores', and does *not* 'determine the rates of sea-floor spreading' – in the absence of a control matrix of radiometric dates (sometimes supplemented to a minor degree by ordinal data from the paleontostratigraphy). This criticism of the terminology of some of our most eminent colleagues in no way detracts from the established value of polarity data as tools for worldwide correlation of volcanic sequences, 'horizons in deep sea cores', etc.

D. Remarks on the Age Estimates of Episodes of Organic Evolution Made from Studies of Phenomena in Living Organisms

Sarich [1971] has summarized the premises and procedures that molecular biochemists and their colleagues use in estimating (1) the degree of phyletic affinity among taxa of living organisms and (2) the length of time required for a given taxon to attain its present phyletic distance from other taxa. His provocative conclusions are now well known to all students of evolution among vertebrates. An adequate résumé of the technique and philosophy of 'dating by molecules rather than by fossils' is beyond my knowledge and beyond the scope of this chapter; so a terse summation from SARICH [1971, p. 64] is appropriate:

'Thus, to measure species, in Nuttall's terms, we need only to measure the extent of DNA and/or protein sequence difference between them. Several techniques of measurement are available, none of which is technically simple, but three – DNA hybridization, protein sequencing, and immunology – have been widely used with the results showing comforting concordance to the student of human evolution.'

SARICH and WILSON [1967] have worked primarily with immunology in albumins and derive a unit termed *immunological distance* (ID) as a means for measuring dissimilarity between recognized species. Their computations are based on the conviction that the rate of evolution of albumin molecules or components has not varied significantly. And their original year calibration [SARICH and WILSON, 1967, p. 1202] was established as follows: 'Although the primate fossil record is fragmentary, it does, in combination with the available immunological evidence, provide sufficient evidence to suggest that the lineages leading to the living hominoids and Old World monkeys split about 30 million years ago.' With an ID of 2.3 units for the split specified above, SARICH and WILSON derived a constant (k = 0.012) to use in their formula for calculating the number of years before present when other lineages of living animals began to depart from a common ancestry. Their immunological scale, which is in harmony with the other biomolecular scales, gives antiquities for certain vertebrate evolutionary events that are much less than those previously estimated by most workers.

SARICH's [1971, pp. 75–76] enthusiasm regarding his discipline is climaxed by his reply to E. SIMONS' unjustifiably and unrealistically vitriolic comments on the immunology method:

'We require that the various lines of evidence be used to place more and more marked constraints upon interpretations derived from other areas, and in this cybernetic fashion to more closely approach evolutionary reality. No single way leads to perfection; at present the protein picture lacks something in resolution, the fossil record is necessarily rather incomplete, and the anatomical data are often difficult to interpret. ...I have yet to see any suggestion as to how a twenty million year date for the origin of the hominid line can possibly be used to explain the molecular evidence. To put it as bluntly as possible, I now feel that the body of molecular evidence on the *Homo-Pan* relationship is sufficiently extensive so that one no longer has the option of considering a fossil specimen older than about eight million years as a hominid *no matter what it looks like.*' Sarich is to be admired for making this straightforward assertion.

His 'blunt' statement shows us that we are faced with severe differences of opinion in the field of systematics. What is a hominid (Hominidae)? Is it a stage of evolution of the components of a molecule or is it defined and recognized by other criteria? Must a hominid be a bipedal post-brachiator? Why must this family exclude pre-brachiators? Would cyberneticist Sarich's trauma be fatal if I paraphrased part of his statement to say that paleontologists can see the 'origin of the hominid line' in certain groups of tetrapods that lived about 250 million years ago? – My wild statement actually can be substantiated if one of the many possible definitions of 'origin of the hominid line' is conceived.

I believe that the solution to differences of opinion on a subject such as that which produced Simons versus Sarich is neither to 'place more and more marked constraints' on one or the other of the interpretations nor to let the computers or robots average out the interpretive differences, (hopefully) 'to more closely approach evolutionary reality'. Rather, let each discipline continue to clean its own house by gathering more data and by frankly admitting the uncertainties and assumptions. Then, as regards the final interpretation, 'let the chips fall where they may' ...just as Sarich has. And this clean-up would most certainly *not* include an attempt to 'explain the molecular evidence' on the basis of the paleontostratigraphic record or on the basis of associated isotopic geochronometry.

Thus, during the past decade, geochronology has been jet-propelled into an era of 'control-matrix' activity (cf. Kauffman, 1970]; the control matrix being geochemically and/or geophysically derived year dates for select minerals in select rocks at select localities in the crust of the earth. A corollary to this stage of evolution of geochronology is the certainty that the beginning student or beginning researcher or intelligent consumer in this field, no matter what his special subdiscipline may be, must now, more than ever, be thoroughly grounded in mathematics, physics and chemistry.

The evolving geochronological control matrix should provide, eventually, enough dates with small analytical and geological error to substantiate a truly interval scale of time (still not a ratio or 'absolute' scale, however!) for correlating and describing microepisodes in the history of the earth and its life. When all rocks in all localities become susceptible to a more precise aging by sophisticated geochemical-geophysical methods, paleontologists must then withdraw their fossil record and its ordinal geochronological scale from the field of action and concentrate activity in the equally broad field of paleobiology. And on that occasion, the time-honored 'orthodox' Earth-Life Calendar with its subdivisions such as Paleozoic, Cenozoic, Ordovician, Tertiary, Eocene and the like will have become inutile also.

III. Cenozoic Erathem and Era

'As many systems or combination of organic forms as are clearly traceable in the stratified crust of the globe, so many corresponding terms (as Palaeozoic, Mesozoic, Kainozoic, etc.) may be made, nor will these necessarily require change upon every new discovery.' J. PHILLIPS, Penny Cyclopedia, vol. 17, pp. 153–154, as cited by WILMARTH [1925].

Following PHILLIPS' first use of the term, Kainozoic or Cainozoic or *Cenozoic* was adopted by practically all earth scientists of the world as an integral component of the (ordinal) geologic time table. Cenozoic can be recognized by its paleontostratigraphy and one can assign to the Cenozoic Era the geochemical-geophysical year dates ranging from 65 ± 2 million years [or 65 ± 2 megennia, *vide* WICKMAN, 1968] ago to the Present. The year 1950 AD is used by some as the zero point or 'Present' datum on the 'before Present' year scale [GEORGE *et al.*, 1969, p. 142]. The difference between 1950 and 1973 may not be significant to us in considering micro-episodes of earth history, but a 'Present' extending through a millenium might be very troublesome to our progeny in 2950 AD.

Many generations of earth scientists have been convinced that the Cenozoic era was much shorter than the Paleozoic of Mesozoic eras. Radiometric dates have substantiated this conviction. The inequality of eras is logical and is in keeping with the fact that eras are major parts of our (ordinal) geochronological scale. Conceptual boundaries of the Cenozoic were implied but not specified by PHILLIPS, and the base of the Cenozoic erathem, for example, is regarded usually as coincident with the base of the Tertiary system and the base of the Paleocene series. One may ponder, however, the false concept of accuracy that is generated in the mind of the beginning student in earth science or in paleobiology as she or he examines and memorizes a textbook table in which, say, the lower or early boundary of a chronozone-phase, a stage-age, a series-epoch, a system-period, and an erathem-era coincide graphically in one line. Such a boundary, as graphed, would be conceived to have zero magnitude stratigraphically and in time. In this way, we have falsely indoctrinated tyros as to the great accuracy, the great precision, and the absoluteness of our chronostratigraphic-geochronologic scales. With the rise to near dominance of geochemical-geophysical dating for rocks, stratigraphers must reappraise their graphic presentations as well as their nomenclature, and they must better express the inequalities and inaccuracies of their ordinal tables.

IV. Subdivision of the Cenozoic

There are an infinite number of ways to subdivide the Cenozoic. Sometimes it seems that most of these ways have been tried, for there is a tremendous mass of literature in which sundry methods are acclaimed. The exhaustive reviews of the literature on the subject by BERGGREN [1971] and HAYS and BERGGREN [1971] are necessary reading for those who would familiarize themselves with the complexities. Somewhere in this maelstrom we can hope to find and secure a judicious balance between the extremes of 'down with the old; if it's new, it's true!' and 'hold on to the old for the sake of priority and stability, even though it may be no more than ambiguous and stereotyped rigidity.' There is strong impetus at present, as exemplified by GEORGE et al. [1969], to abandon the classic, formal twofold subdivision of Cenozoic into Tertiary and Quaternary, substituting other terms. Paleogene and Neogene would replace Tertiary in this scheme. This change is supported in part because of the long-recognized overly simplistic concept of the organization of rocks in the earth's outer shell which led to 'Tertiary'

Erathem-era	Systems-periods	Series-epochs	Informal, descriptive units
Cenozoic	Quaternary	Recent (Holocene) Pleistocene	
	Tertiary	Pliocene Miocene	Neogene
		Oligocene Eocene Paleocene	Paleogene

and 'Quaternary'. Some people have argued for the change, also, because of the unequalness in gross number of years represented by Tertiary and Quaternary respectively. Tertiary is 20–30 times longer. Such inequality is not troublesome to the geochronologist who remembers that Tertiary and Quaternary belong in a chronostratigraphic and/or geochronologic table. Continued use of the following traditional subdivisions of the Cenozoic is recommended:

A. Recognizing Series-Epochs of the Cenozoic; Boundary Problems

Many reviews, symposia, committee, reports and the like have been published in recent years, detailing the problems of content, subdivisions, and boundary problems of each of the series-epochs of the Cenozoic. Inevitably, each of these publications reconsiders the original statements of LYELL as he presented the concept of Eocene, Miocene, Older Pliocene, and Newer Pliocene to the world. LYELL clearly explained his basis for recognizing these subdivisions and reemphasized his method in later publications. He devised this method as a means for emphasizing the continuity of species through the Tertiary – '... the connexion of the geological remains of deposits formed at successive periods.' You may remember that Eocene *et cetera* were based upon percent of Linnaean species of living testacea in the fossil assemblages that had been collected by various paleontologists from marine sedimentary rocks in Western Europe. It must be reemphasized that the Lyellian percentage was derived from evaluation of an aggregate of a large number of species that had been collected from a stratigraphic interval of essentially unspecified boundaries. Regarding Eocene, for example, he said, 'To this era the formations first called Tertiary, of the Paris and London basins, are referrible... The total number of fossil shells of this period already known is one thousand two hundred and thirty-eight, of which number forty-two only are living species, ... In the Paris basin alone, 1122 species have been found fossil, of which thirty-eight only are still living.' And as late as 1863, LYELL wrote, 'But the reader must bear in mind that the terms Eocene, Miocene and Pliocene were originally invented with reference purely to conchological data, and in that sense have always been and are still used by me.'

Thus, the series-epoch subdivisions of the Tertiary should be recognized and the names of the subdivisions applied ultimately on the basis of pert centage of living species of marine molluscs in a bulk aggregate (i.e., abou-

1,000 species) of fossils from a bulk section of strata if the resultant chronostratigraphic and geochronologic nomenclature is Lyellian. Delineation of
a boundary for Eocene, Miocene, or Pliocene (as well as Paleocene, Oligocene and Pleistocene) is not resolvable by the Lyellian method. The search
(and voting!) for 'more refined' boundaries, neostrato-types and the like for
Lyellian series-epochs, and for the still-larger categories in our chronostratigraphic and geochronologic hierarchies has been an exercise in futility, in
my opinion. I propose that we leave the erathems-eras, systems-periods, an
series-epochs as the vaguely-bounded bulk units that we inherited – units
that serve essentially for international communication of ideas – and that
we desist from trying to 'refine' the entities that were unrefined typologically.
So much of this 'refinement' activity seems to have been the unreality of
'putting ideas into the minds of our predecessors!' Our international com-
missions, our society committees, and such can properly concentrate on
stratotypes, neostratotypes, better zonations with radiometric 'control
matrices', tying-in to polarity episodes, and other such refinements within the
true paleogeologic and paleobiogeographic provinces and in the lower levels
of our hierarchies of stratigraphic geochronologic classification. Then, perhaps, VITA-FINZI's [1970] dream of a 'new school of geochronology free from
the constraints of fossilized methodology' can become a reality. The Cenozoic
Primate Calendar, which follows in this chapter, attempts to treat the series-
epochs of the Cenozoic in the above-recommended manner.

B. Plant Ages

J. A. Wolfe and his colleagues [WOLFE, 1966, 1972; WOLFE et al., 1966]
have defined fossil-plant provincial stages-ages for use in the northwestern
part of North America. A summary of these units is given by WOLFE
[1972]:

The Seldovian stage, for example, was assigned to Lower Miocene and
was typified in the plant-bearing strata of Seldovia Point, Alaska. A stratigraphic top was designated and the overlying stage was recognized; no base
was defined. Of the 76 species identified from leaves and 24 'forms' identified
from pollen and spores, 12 'significant megafossil species' were believed by
WOLFE to be restricted to this stage (plus 2 restricted in Alaska only), and
there is a characterizing joint occurrence of species for the stage. This is the
first time that paleobotanists who study Cenozoic plants have systematically
organized and described an independent system for recognizing subdivisions

Pliocene	Clamgulchian	
Miocene	{ Homerian Seldovian	{ late early
	{ Angoonian	{ late early
Oligocene	{ unnamed Kummerian	
Eocene	{ Ravenian	{ late middle early
	{ Fultonian Franklinian	
Paleocene	unnamed	

of the Tertiary. The work of WOLFE and colleagues evidently results from the conclusion, completely rejectable 25 years ago (!), that the fossilized parts of plants do depict evolution and a concomitant succession of species through time.

C. Mammal Ages

Mammal age is used here particularly in reference to the 'North American Provincial Age' of WOOD *et al.* [1941] and to units of equivalent concept in Europe. The construct of the mammal age was summarized by EVERNDEN *et al.* [1964] and has been reviewed critically by TEDFORD [1970]. Widespread use of mammal age is based on the opinion that fossils of mammals are excellent tools for paleontostratigraphic geochronology:

(1) Many genera and probably many species have and did have transoceanic and trans-world geographic distribution.

(2) The often-used taxa are noteworthy for their restricted chronostratigraphic range as compared to many of the taxa of other Cenozoic organisms (a claim that has been surprizingly well verified by radiometric dating in recent years).

(3) Phylogenies of the often-used taxa are relatively well understood and are easily demonstrable, in contrast to many taxa in other groups.

Insofar as utility is concerned, these usable attributes of mammals

Table II. Cenozoic primate calendar

European 'standard' ages	European mammal-ages and faunas	Megennia B.P.	Primate records	North American mammal-ages	Megennia B.P.	
Tyrrhenian	Taubach			Rancholabrean 0.008		Pleistocene
Milazzian				0.03		
Sicilian			Homo Faunas	0.5		
Emilian	Mosbach	1	Lemuridae (Madagas.)	1.0	1	
				Irvingtonian		
Calabrian				1.3, 1.5		
		2	Australopithecus	2.1, 2.3	2	
Villafranchian	Villafranchian 1.9 2.5		Faunas	Blancan		
	3.4		(Af., As.)			
	3.8			3.3, 3.5		Pliocene
Astian		4		4.1	4	
Piacenzian	Ruscinian					
Tabianian				5.2		
Zanclian						
		6		Hemphillian	6	
Andalusian			Cercopithecidae	6.4		
	(7.4)		(EuAs., Af.)			
Pontian	Pikermian= 8.5 to	8		9.2	8	
Messinian	'Turolian' 9.3			9.1		
		10	Ramapithecus	10.1, 10.0	10	
			Faunas	Clarendonian 10.7		
Sarmatian	Vallesian		(As., Af.)	11.2 10.8, 10.9, 11.1		Miocene
				11.7, 11.4, 11.5		
Tortonian	12.4	12	Dryopithecus and		12	
Vindobonian			Oreopithecus Faunas	13.2, 13.3		
			(EuAs., Af.)	13.4, 13.6		
Helvetian	Maremmian (14.0)	14	Hylobatinae (Eu.)	(14.2, 14.4)	14	
	LaGrive ?(14.6)		Ceboidea (S.A.)	Barstovian		
	Sansan			15.0, 15.1, 15.2		
Serravallian				15.9 15.4, 15.6		
Langhian	LaRomieu	16		16.1	16	
Burdigalian	Wintershof-West		Lorisidae (Af.)	17.1		
				17.4		
Girondian	Laugniac	18	Dryopithecus Faunas (Af.)	Hemingfordian	18	
	Paulhiac	20			20	
				21.3		
Aquitanian	St-Gerand-le-Puy	22	Hylobatinae (Af.)	22.3	22	
				Arikareean		
			Ceboidea (S.A.)			
		24			24	
Bormidian			Omomyidae (N.A.)	24.9		
				25.3		
Chattian	Coderet, Rickenbach	26		25.6	26	
	Cournon			26.8		
	Bumbach			to		Oligocene
Stampian				28.0		
		28			28	
Rupelian	Montalban, LeSauvetat			Whitneyan		
			Aegyptopithecus-	Orellan		
		30	Propliopithecus Faunas (Af.)		30	
				31.1		
Sannoisian	Ronzon, Hoogbutsel	32		31.5?, 31.6?	32	
Lattorfian			Oligopithecus (Af.)	32.0		

Table II. Continuation

Stage	Locality	Ma	Taxa	Ma (taxa)	N.A. land mammal age	Ma	Epoch
				31.6, 32.4, 32.6, 32.9 / 33.0 / 33.9			
Tongrian		–34–	Omomyidae (N.A.)		Chadronian	–34–	
			Ceboidea (S.A.)	34.3			
				34.7 34.8			
				35.2 35.7			
				36.5			
Ludian	LaDébruge, Mont-martre Euzet	–36–	Adapinae, Tarsiidae (Eu.)	37.5 36.8		–36–	
			Amphipithecus (As.)	37.5			
Priabonian		–38–	Paromomyidae, Microsyopinae, Omomyidae, Uintasoricinae (N.A.)			–38–	Eocene
		–40–	Omomyidae, Adapidae (As.)		*Uintan sensu lato*	–40–	
Bartonian				41.2			
Marinesian	Robiac	–42–				–42–	
Ledian	Lissieu			?42.7			
Auversian	Egerkingen						
	Buchsweiler						
Lutetian		–44–	Adapinae, Tarsiidae, Microsyopinae (Eu.)			–44–	
Wemmelian	Messel, Upper Geiseltal	–46–		45.0, 45.4?		–46–	
Cuisian			Omomyidae,	f and k			
Ilerdian	Grauves, Cuis	–48–	Notharctinae, Microsyopinae, Uintasoricinae (N.A.)	[34.0] / 48.5	*Bridgerian*	–48–	
	Condé-en-Brie		Plesiadapidae, Paromomyidae, Notharctinae, Adapinae, Omomyidae (Eu.)	49.0 / 49.2			
Ypresian		–50–				–50–	
Sparnacian	Mutigny, Avenay		Paromomyidae, Notharctinae, Omomyidae (N.A., Eu.)				
Landenian		–52–	Microsyopinae, Uintasoricinae (N.A.)		*Wasatchian*	–52–	
		–54–				–54–	
Thanetian	Cernaysian		Plesiadapidae, Paromomyi-(Eu.) dae?				Paleocene
		–56–	Plesiadapidae, Paromomyidae, Carpolestidae, Picrodontidae (N.A.)		*Tiffanian sensu lato*	–56–	
		–58–				–58–	
Montian		–60–	Plesiadapidae, Paromomyidae, Carpolestidae, Picrodonti-dae (N.A.)		*Torrejonian*	–60–	
		–62–				–62–	
Danian			Purgatorius (N.A.)		*Puercan*		
		–64–		64.8		–64–	
		–66–	? *Purgatorius* ?? (N.A.)			–66–	
		–68–			*Late Cretaceous*	–68–	Cretaceous
		–70–				–70–	

remain counterbalanced by the relative scarcity of mammalian fossils. Nevertheless, mammals have been the principal means for constructing a geochronologic trellis for the nonmarine strata of the Cenozoic.

Each mammal age is based on a characterizing aggregate of genera, none of which is necessarily confined to the age it helps to identify. Joint occurrence of genera is the significant criterion. Each age is further characterized by particular species within genera that may be long-ranging. Thus, species phylogenies and stage-of-evolution continue to play a strong role in mammal-age determinations.

Historically and lamentably, mammal ages ('North American Provincial Ages') were not paleontostratigraphically constructed – there has been little or no typification and description of the stratigraphic discreteness of the characterizing faunas (although this can be done now with most of the ages) – so, we do not give them the title of stage-ages. Most Cenozoic mammal ages of North America, however, are as big geochronologically and geographically as the recognized marine stages-ages. The boundary problems between mammal ages, as would be displayed on the radiometric year-scale, are not severe to date.

Recognized mammal ages of North America and Europe are based on fossils of nonmarine mammals and are shown on the Cenozoic Primate Calendar above. Especially from the Miocene on, however, fossils of the marine Cetacea (whales), pinnipeds (marine carnivores) and Sirenia (sea cows) are widespread and are becoming exceptionally useful for intercontinental correlation and for geochronologic tie-in with the dominantly shallow-water deposited marine strata of European stages.

V. Cenozoic Primate Calendar

The calendar that is shown on pages 20–21 offers a technique for more objective presentation of the *lack of resolution* in correlating subdivisions of the Cenozoic around the world. It is proclaimed to be, therefore, a frank depiction of the geochronologic scale that is our present state of knowledge, but the reader is sophisticated enough to know that probably some of my isolated judgements (biases?) have crept into the compilation! Absence of boundaries between some of the age names and other units indicates possible or probable overlap of adjacent units or other uncertainties. For example, certain stratigraphic or paleontostratigraphic entities that have been assigned to *Messinian* by some workers may be, in part, the same age as things assigned to *Astian* by other workers. No one worker is able to judge each and every age assignment by other specialists.

Although the presentation on this chart at first may appear to be a derogation of the marine stages-ages and their proposed zonations, it is not intended as such. Rather, there is necessary concentration upon the non-marine phenomena and related radiometric datings. VAN COUVERING [1971] and BERGGREN [1971, 1972] are some of the latest attempts to formulate a detailed correlation within a radiometric control-matrix of marine and non-marine Cenozoic units in Europe, Africa and North America.

Names in quotes may have been used widely but their use is not recommended. The year dates given on the Calendar or on the list of sources that follows the Calendar are based on the potassium-argon method, supplemented by fission-track dating, except within the Rancholabrean and (a few) within the early Cenozoic of Europe. These dates have direct stratigraphic tie-in with the mammalian fauna or faunas by which the geochronologic age is recognized. Dates associated with correlated faunas whose taxa are unpublished or are unknown to me and dates correlated from interdigitating marine strata are shown in parentheses. Space on the Calendar does not permit giving the amount of possible analytical error for each radiometric date; these possible errors may or may not be stated by the original authors of the dates, as indicated in the list of sources. Radiometric dates listed within one geochronologic age are not necessarily verified stratigraphically; for example, within Chadronian, 36.8 actually may be higher stratigraphically than 35.7.

A. Sources for the Radiometric Dates Given on Calendar

a = BLACK [1969]
b = BODELLE *et al.* [1969]
c = BONHOMME *et al.* [1968]
d = CURTIS [personal commun., 1970–1974]
e = EVERNDEN *et al.* [1964]
f = McKENNA [personal commun., 1972], see remark opposite SMEDES and PROSTKA [1972]
g = Odin *et al.*, [1970]
h = Odin *et al.* [1969a]
i = ODIN *et al.* [1969b]
j = SAVAGE and CURTIS [1970]
k = SMEDES and PROSTKA [1972], work in Yelllowstone Park area indicates that base of Uintan may be much earlier than previously estimated

l = TURNER [1970]
m = VAN COUVERING [1971]
n = VAN COUVERING and MILLER [1971]
o = LIPPOLT *et al.* [1963]
p = WILSON *et al.* [1968]
q = EVERNDEN and JAMES [1964]
r = OBRADOVICH [personal commun., 1971]
s = NAESER *et al.* [1971]
t = OBRADOVICH *et al.* [1973]

Arikareean: 22.3 – d, 21.3 – e, 24.9 – e, 25.3 – e, 25.6 – e, 26,8 ± 2.5 – t, 27.0 ±
    0.6 – t, 27.8 ± 0.7 – t, 28.0 ± 0.7 – t

Barstovian: 13.2 – d, 13.3 – d, 13.4 – d, 13.6 – d, 12.3 – e, 14.6 – e, 15.0 – e, 15.1
    – e, 15.2 – e, 15.4 – e, 15.6 – e, 16.1 – e, (14.2) – 1, (14.4) – 1, 15.9 – q

Bartonian: 41.8 ± 3 – b, 45.2 – c, 43.1 ± 2 – g, 42.0 ± 2 – h, 43.1 ± – h

Bridgerian: [34.0] – q, date from CLARNO 'Nut Bed' [Bridgerian assignment is based
    on personal communication from BRUCE HANSON, University of California,
    Department of Paleontology]

Blancan: 2.1 – e, 2.3 – e, 3.3 – e, 3.5 – e

Bridgerian-Wasatchian: 49.0 – e, 49.2 – e

Burdigalian: (19 ± 1) – m

Chadronian: 32.6 – e, 32.9 – e, 34.3 – e, 34.8 – e, 35.7 – e, 36.8 – e, ? 36.5 – e, 37.5
    – e, 31.6 – e, 32.4 ± 1.7 – p, 33.0 ± 1.1 – p, 33.9 ± 1.8 – p, 34.2 ± 3.0 – p,
    34.7 ± 2.0 – p, 35.2 ± 2.3 – p, 36.5 ± 1.2 – p

'Chattian' ('Coderet'): 22.8 ± 0.6 – m

Chattian: 28.8 ± 1.5 – g, c, 29.5 ± 1.5 – g, c, 30.1 ± 1.5 – g, c, 31.2 ± 1.5 – g, c

Clarendonian: 9.9 – e, 10.0 – e, 10.7 – e, 10.8 – e, 10.9 – e, 11.1 – e, 11,4 – e, 11.5
    – e, 11.2 – e, 11.7 – e, 10.1 – q

Cuisian: 48.9 – b, 49.4 – b, 51.2 ± 2.5 – b, ? 55.3 – b, ? 56.8 – b, 51.3 – c

Hemingfordian: 17.1 – e, 17.4 – d

Hemphillian: 4.1 – e, 5.2 – e, 6.4 – e, 8.1 – e, 9.1 – e, 9.2 – e, 10.0 – e

Irvingtonian: ± 1.0 – r, 1.3 – e, ± 1.5 – r

Lutetian: 43.0 ± 2.5 – b, 44.7 ± 2.5 – b, 45.2 ± 3 – b, 47.4 ± 2.5 – b, 47.7 ± 2.5
    – b, 48.7 – c, 44.0 – h, 48.6 ± 3 – h, 49.4 ± 3 – h

Maremmian: (14.0) – m, ? (14.6) – m

Pikermian: (7.4) – m, 8.5 ± 0.8 – n, 9.3 ± 0.7 – n

Puercan: 64.8 – e, post-Puercan 58.7 – e

Rancholabrean: miscellaneous radiocarbon dates

Thanetian: ± 54 (Bracheux Sand) – r, 53.6 – c, 68.1 ± 4 – h, 57.9 – e

Uintan: 41.2 ± 1.4 – a, 37.5 – e, ? 42.7 – e, 45.0 – e, ? 45.4 – e

Vallesian: Vallesian? – 12.4 – m, o

Villafranchian: 1.9 – j, 2.5 – j, 3.4 – j, 3.8 – j

Vindobonian: (16 ± 1) – m, Vindobonian? – (16 ± 1) – m, Vindobonian? – (16.2
    ± 0.4) – m

Wemmelian: 43.1 ± 2 – c, 45.2 ± 2 – c, 45.6 ± 2 – c, 43.6 ± 2 – h, 44.0 ± 2 – h,
    45.2 ± 2 – h, 45.6 ± 2 – h, 48.6 ± 2 – h

Whitneyan-Orellan: ? 31.1 – e, ? 31.5 – e, 31.6? – e, 32.0? – e

VI. References

American Commission of Stratigraphic Nomenclature: Code of stratigraphic nomenclature (Amer. Ass. Petrol. Geol., Tulsa 1970).

BERGGREN, W. A.: Tertiary boundaries and correlations; in FUNNELL and RIEDEL Micropaleontology of oceans, pp. 669–692 (Cambridge Univ. Press, London 1971).

BERGGREN, W. A.: A Cenozoic time scale; some implications for regional geology and paleobiogeography. Lethaia 5: 195–215 (1972).

BERRY, W. B. N.: Growth of a prehistoric time scale based on organic evolution (Freeman, San Francisco 1968).

BLACK, C. C.: Fossil vertebrates from the late Eocene and Oligocene; Badwater Creek area, Wyoming, and some regional correlations. Symp. on Tertiary Rocks of Wyoming, Wyo. Geol. Ass. Guidebook, 21st Field Conf., 1969.

BODELLE, J.; LAY, C. et ODIN, G. S.: Détermination d'âge par la méthode géochronologique 'potassium-argon' de glauconies du Bassin de Paris. C. R. Acad. Sci. 268: 1474–1477 (1969).

BONHOMME, M.; ODIN, G. S. et POMEROL, C.: Age des formations glauconieuses de l'Albien et de l'Eocène du Bassin de Paris. Colloque sur l'Eocène. Mém. Bur. Rech. géol. min. 58: 339–346 (1968).

COUVERING, J. A. VAN: Radiometric calibration of the European Neogene; in BISHOP and MILLER Calibration of hominoid evolution, pp. 247–272 (Scottish Academic Press, 1971).

COUVERING, J. A. VAN and MILLER, J. A.: Late Miocene marine and non-marine time scale in Europe. Nature, Lond. 230: 559–563 (1971).

DALRYMPLE, G. B.: Potassium-argon dating of geomagnetic reversals and North American glaciations; in BISHOP and MILLER Calibration of hominoid evolution, pp. 107–134 (Scottish Academic Press, 1971).

DALRYMPLE, G. B. and LANPHERE, M. A.: Potassium-argon dating: principles, techniques and applications to geochronology (Freeman, San Francisco 1969).

EVERNDEN, J. F. and JAMES, G. T.: Potassium-argon dates and the Tertiary floras of North America. Amer. J. Sci. 262: 945–974 (1964).

EVERNDEN, J. F.; SAVAGE, D. E.; CURTIS, G. H., and JAMES, G. T.: Potassium-argon dates and the Cenozoic mammalian chronology of North America. Amer. J. Sci. 262: 145–198 (1964).

FITCH, F. J.: Selection of suitable material for dating and the assessment of geological error in potassium-argon age determination; in BISHOP and MILLER Calibration of hominoid evolution. pp. 77–92 (Scottish Academic Press, 1971).

FLEISCHER, R. L. and HART, H. R., jr.: Fission track dating: techniques and problems; in BISHOP and MILLER Calibration of hominoid evolution, pp. 135–170 (Scottish Academic Press, 1971).

GEORGE, T. N. et al.: Recommendations on stratigraphical usage. Proc. Geol. Soc., No. 1656, pp. 139–166 (1969).

GRIFFITHS, J. C.: Aspects of measurement in geosciences. Mineral Industries, January (1960).

HAYS, J. D. and BERGGREN, W. A.: Quaternary boundaries and correlations; in FUNNELL and RIEDEL Micropaleontology of oceans, pp. 669–692 (Cambridge Univ. Press, London 1971).

HEDBERG, H. D. (ed.): International subcommission on stratigraphic classification. Rep. 7a: Introduction to international guide to stratigraphic classification, terminology and usage. Rep. 7b: Summary of international guide to stratigraphic classification, terminology and usage. Lethaia 5: 283–324. (1972).

HUXLEY, T. H.: The anniversary address. Proc. Geol. Soc. xl–liv (1862).

KAUFFMAN, E. G.: POPULATION systematics, radiometrics and zonation; a new biostratigraphy. Proc. N. amer. paleont. Conv. 1: 612–666 (1970).

KITTS, D. B.: Geologic time. J. Geol. 74: 127–146 (1966).

LIPPOLT, H.-J.; GENTNER, W. und WIMMENAUER, W.: Altersbestimmungen nach der Kalium-Argon-Methode im tertiären Eruptivgestein Südwestdeutschlands. Jb. Geol. Landes Baden-Württemberg 6: 507–538 (1963).

LYELL, C.: Principles of geology, vol. 3 (1833).

LYELL, C.: Antiquity of man, p. 5 (1863).

MILLER, J. A.: Dating Pliocene and Pleistocene strata using the potassium-argon and argon-40/argon-39 methods; in BISHOP and MILLER Calibration of hominoid evolution, pp. 63–76 (Scottish Academic Press, 1971).

NAESER, C. W.; IZETT, G. A., and WILSON, R. E.: Zircon fission-track ages of Pearlette-like volcanic ash beds in the great plains. Abstract Geol. Soc. Amer. Annu. Meet., Washington (1971).

OBRADOVICH, J. S. et al.: Radiometric ages of volcanic ash and pumice beds in the Gering Sandstone (earliest Miocene) of the Arikaree group, southwestern Nebraska. Abstracts with programs. Geol. Soc. Amer. 5: 499–500 (1973).

ODIN, G. S.; BODELLE, J.; LAY, C. et POMEROL, C.: Géochronologie de niveaux glauconieux paléogènes d'Allemagne du Nord (méthode potassium-argon). Résultats préliminaires. C. R. somm. Soc. géol. France 6: 220 (1970).

ODIN, G. S.; CURRY, D.; BODELLE, J.; LAY, C. et POMEROL, C.: Géochronologie de niveaux glauconieux tertiaires des bassins de Londres et du Hampshire (méthode potassium-argon). C. R. somm. Soc. géol. France 8: 309 (1969a).

ODIN, G. S.; GULINCK, M.; BODELLE, J. et LAY, C.: Géochronologie de niveaux glauconies tertiaires du bassin de Belgique (méthode potassium-argon). C. R. somm. Soc. géol. France 1969: 198–199 (1969b).

SARICH, V. M.: A molecular approach to the question of human origins; in DOLHINOW and SARICH Background for man, pp. 60–81 (Little, Brown, Boston 1971).

SARICH, V. M. and WILSON, A. C.: Immunological time scale for hominid evolution. Science 158: 1200–1203 (1967).

SAVAGE, D. E. and CURTIS, G. H.: The Villafranchian stage-age and its radiometric dating, Geol. Soc. Amer., suppl. 124, pp. 207–232 (1970).

SHAW, A. B.: Time in stratigraphy (McGraw-Hill, Maidenhead 1964).

SMEDES, H. W. and PROSTKA, H. J.: Stratigraphic framework of the Absaroka volcanic supergroup in the Yellowstone National Park region. US Geol. Surv. Prof. Paper, 729-C (1972).

TEDFORD, R. H.: Principles and practices of mammalian geochronology in North America. Proc. N. amer. paleont. Conv., part F 666–703 (1970).

TUREKIAN, K. K. and BADA, J. L.: The dating of fossil bones; in BISHOP and MILLER Calibration of hominoid evolution, pp. 171–186 (Scottish Academic Press, 1971).

TURNER, D. L.: Potassium-argon dating of Pacific coast Miocene foraminiferal stages. Geol. Soc. Amer., spec. paper 124: 91–130 (1970).

VITA-FINZI, C.: Time, stratigraphy and the Quaternary. Scientia, CV N. DCIII–DCCIV, 12 pp. (1970).
WICKMAN, F. E.: How to express time in geology. Amer. J. Science 266: 316–318 (1968).
WILMARTH, M. G.: The geologic time classification of the United States geological survey compared with other classifications. US geol. Surv. Bull. 769: 1–138 (1925).
WILSON, J. A.; TWISS, P. C.; DeFORD, R. K., and CLADBAUGH, S. E.: Stratigraphic succession, potassium-argon dates, and vertebrate fauna, Vieja Group, Rim Rock country, Trans-Pecos Texas. Amer. J. Sci. 266: 590–604 (1968).
WOLFE, J. A.: Tertiary plants from the Cook inlet region, Alaska. US geol. Surv. Prof. Paper 398/B: B1–B32 (1966).
WOLFE, J. A.: An interpretation of Alaskan Tertiary floras; in Floristics and paleofloristics of Asia and eastern North America, chap. 13 (Elsevier, Amsterdam 1972).
WOLFE, J. A.; HOPKINS, D. M. and LEOPOLD, E. B.: Tertiary stratigraphy and paleobotany of the Cook Inlet region, Alaska. US geol. Surv. Prof. Paper 398/A: A1–A29 (1966).
WOOD, H. E. et al.: Nomenclature and correlation of the North American continental Tertiary. Geol. Soc. Amer. Bull. 52: 1–48 (1941).

Author's address: Dr. DONALD E. SAVAGE, University of California, Museum of Paleontology, Berkeley, CA 94720 (USA)

In SZALAY: Approaches to Primate Paleobiology
Contrib. Primat., vol. 5, pp. 28–61 (Karger, Basel 1975)

Paleoecology of the Paleocene-Eocene Transition in Europe

DONALD E. RUSSELL

Contents

I. Introduction

Paleocene mammalian faunas occupy a special place in evolutionary history and are as discrete, in many ways, as those that developed in the late Cretaceous. Most of the mammals of the Paleocene can be characterized as primitive or generalized, and at the same time their relationships to modern orders are often unclear. They represent, in fact, more the culmination of Mesozoic forms than the initial development of the fauna living today. This heyday of archaic mammals came to a dramatic ending with the sudden appearance of a new fauna at the beginning of the Eocene. It is obvious, however, that the new forms descended from Paleocene ancestors, even if the latter cannot always be positively identified and, in the western United States where deposition was continuous in passing from one period to the next, the faunal change can be shown to be somewhat less abrupt than was at first supposed. Nevertheless, the transition from the Paleocene and its alien mammalian assemblages to the Eocene with its elements of modernity constituted an event of particular significance in mammalian history.

Prosimian primates, appearing tentatively at the end of the Cretaceous, underwent a development that by the middle Paleocene assured them a major place in the mammalian assemblages of the time. A climax in numbers and diversity was reached in Europe and North America during the Eocene. Their subsequent disappearance on both continents early in the Oligocene left their further evolution to take place in the more southern regions of the globe.

While the early Tertiary history of mammals in the North American record has long been well documented, that of Europe has been more refractory. The striking resemblance existing between early Eocene Wasatchian faunas of North America and those of the European Ypresian (= Sparnacian + Cuisian) has made the absence of information from most of the period preceding the Ypresian particularly vexatious. Although new discoveries and revisions have greatly broadened our knowledge of the European Paleocene in the last decade, it must be admitted that the principal localities still can be counted on the fingers of a single hand: Mons (Belgium), Walbeck (DRG), Menat (France) and the Cernaysian sites centered around the Mont de Berru (France). The earliest Paleocene faunas remain totally unknown, and for vertebrate life of the middle of the period we are afforded only tantalizing glimpses. Because of their rarity, the contribution made by each of the Paleocene localities will be treated individually; the subsequent early Eocene faunas will be grouped by stages.

II. Cernay-lès-Reims

Alone among the few European Paleocene mammal localities, Cernay-lès-Reims offers both a large and relatively well-balanced fauna and the assurance of excellent stratigraphic control. Most of its 34 mammalian species are amply represented in the collections and correlation with the marine Thanetian Sables de Bracheux to the west is established through deposits which become increasingly less brackish in nature and in which remains of the Cernaysian fauna become progressively scarce. The view of

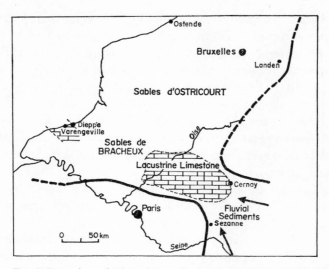

Fig. 1. Extension of Thanetian sediments in the Paris Basin. The episode during which the lacustrine marls and limestones were deposited was posterior to that of the Sables de Bracheux. [Modified from POMEROL, 1967.]

terrestrial life in the Paleocene afforded by the fauna of Cernay, supplemented by contemporaneous floras from neighboring sites, is unique in Europe.

The Mont de Berru, a small erosional Tertiary remnant isolated on a plain of late Cretaceous chalk a few miles east of Reims, actually contains two principal Paleocene localities. The first, worked primarily during the last quarter of the 19th century, is that situated near the village of Cernay-lès-Reims. Most specimens from these deltaic deposits have suffered considerable attrition during transportation. Farther over the hill, on the eastern slope, a large sand quarry exists today near the village of Berru. Commercial exploitation of this quarry began about 1955 and fossils were discovered therein soon after. The fossiliferous sediment represented a river channel and produced quantities of large bones, even a number of skulls, and several associated skeletons. Marine elements, such as shark teeth and molluscs were conspicuously absent. From an analysis of the fauna [RUSSELL, 1964], the Berru deposit appears to be slightly younger than the Cernay sediments. The same species occur in both places, but some of those from Berru show an evolutionary advance over their Cernay relatives. For general purposes, both assemblages can be grouped as the Cernaysian fauna.

The total recovered mammalian material is made up, in very large part, of two species, the primate *Plesiadapis tricuspidens* and the meniscotheriid condylarth *Pleuraspidotherium aumonieri*. There would appear to be little doubt that the latter constituted a major herbivore element in the region, but opinion has been more divided concerning the habits of *Plesiadapis*. This genus has usually been considered a primate [for a minority view, see CHARLES-DOMINIQUE and MARTIN, 1970] and the attendant image of a tree-dwelling habitus has been frequently assumed. However, TEILHARD DE CHARDIN [1922] remarked that, given the abundance of individuals, *Plesiadapis* must have been terrestrial. Arborial animals, as a rule, are rarely found in such numbers. SIMONS [1967] suggested a terrestrial way of life in order to account for its wide geographic distribution, necessitating an independance of forested regions. He further surmised that *Plesiadapis* lived somewhat like a large rat or squirrel and was partly cursorial. In a similar opinion, GINGE-RICH [in press] regarded it as a gregarious, terrestrial herbivore with a stem-cutting adaptation of the incisors, resembling the living *Marmota* in habits and appearance (but see SZALAY, TATTERSALL, and DECKER, pp. 136-166).

An important role in the ecology must have been played by the carnivorous aquatic reptile *Simoedosaurus*, whose remains predominate in the collections. Two genera of crocodiles were also present in and along the river that flowed through Berru, but are represented by fewer individuals. Mammalian carnivores were very few, perhaps restricted to the infrequent mesonychid *Dissacus*. The arctocyonids, of course, are considered to be carnivorous to some extent, but their dentition belies any specialization in this direction. In fact, as was pointed out by VAN VALEN and SLOAN (1966), it is nearly impossible to reconstruct in any detail the diets of Paleocene mammals. Demonstration of this difficulty can be found by confronting the diversity of food habits that occur in recent species possessing dentitions of similar morphology.

Other members of the Cernay fauna include numerous multituberculates and other small forms of insectivorous and herbivorous tendency. Most frequent among the latter are several species of condylarth. Also of small size are two primate genera: *Chiromyoides* is the size of the smallest specimens referred to *Plesiadapis tricuspidens* and from which isolated cheek teeth are practically impossible to distinguish. Several mandibles are known, however, testifying to the validity of the generic disctinction. The other primate, *Berruvius*, is minute with teeth measuring about 1 mm in length. In partial consequence, it is known from very few specimens and, lacking complete dentition, its affinities are uncertain. Relationships to the North American *Navajovius* and *Uintasorex* have been suggested. In contrast to *Plesiadapis*, these small and rare primates probably lead a retiring existence as scattered individuals in the undergrowth by the river.

The garpike, *Lepisosteus*, is extremely rare, while aquatic turtles abound in the Cernaysian sediments and three types can be distinguished. Several lizards and an amphisbaenid are known, as well as three salamanders and two frogs. Snakes do not appear to have lived in the region at that time. About a dozen types of birds are represented in the Cernaysian fauna, many of them only tentatively referred to modern groups.

From the above vertebrate assemblage it seems quite reasonable to imagine a wooded or brushy riparian environment at the Mont de Berru during the late Paleocene, but the great number of probably gregarious forms suggest that the river crossed relatively open country. Rather surprising is the absence of large plains-dwelling mammals, like the pantodonts and uintatheres of North America and Asia. *Arctocyon*, the largest mammal, was the size of a small bear but probably frequented wooded areas. A likely solution could be that of a sparsely wooded region, situated on a flat coastal plain, well watered by streams and lakes. Evidence in support of the latter aspect is found in the presence of amiid fish and elongate salamanders. As noted by ESTES *et al.* [1967], these forms indicate permanent, rather quiet waters.

A little to the south of the Mont de Berru, at Rilly-la-Montagne, a late Paleocene lake existed from which a flora has been described. Still farther to the south, the contemporaneous travertine of Sézanne has yielded an even more complete flora. Predominating forms include *Alnis*, *Laurus*, *Sassafras*, *Salix*, *Magnolia*, and *Vitis*. This, and other evidence, suggests for the area a warm temperate climate.

It should be emphasized that none of the mammalian genera found in the Mont de Berru directly gave rise to modern forms. The Cernaysian fauna presents a picture of a primitive epoch, a last view of the life that reigned before our world of modern mammals began to emerge.

III. Walbeck

Next in faunal completeness after Cernay, but older, is the assemblage collected from a fissure filling near Walbeck (DRG). The sediments containing the late middle Paleocene Walbeck fauna were reworked by a transgressing Oligocene sea and the fossils transported to protected hollows in the Muschelkalk. Nothing, therefore is known of the original deposit, or of the flora and additional fauna it might have contained. Examination of the recovered mammalian fauna reveals curious deficiencies when compared to the closely similar fauna of Cernay. Two of the groups which most typify the Cernaysian assemblage are the archaic multituberculates and the aberrant meniscotheriid condylarths, the latter being present in predominating numbers. Both are lacking in the Walbeck fauna. The cause of their absence may be presumed to have been due to a differing environment since multituberculates were abundant elsewhere (Mongolia and North America, for example) at the time of the original deposition of the Walbeck sediments, as well as in the succeeding European locality of Cernay; it seems unlikely that they had disappeared from Europe during the middle Paleocene. Furthermore, their absence at Walbeck cannot be attributed to their having been overlooked during collection. Although possessing diminutive teeth, equally small or smaller insectivore teeth were found in large numbers during the thorough search of the fissure-filling matrix made by the staff of the 'Geologisch-Paläontologisches Institut' of Halle (DRG). VAN VALEN and SLOAN [1966] suggest that, among other features, 'the decline of the multituberculates was probably initiated by competition with condylarths, increased by primates, and completed by rodents.' Primates in general began expanding at the end of the middle Paleocene, or at about the time when the Walbeckian mammals were thriving. As constituents of that fauna, however, primates form a relatively small part (about 5% of the known population). As such, it appears difficult to attribute the complete absence of multituberculates to their presence and competitive powers. Multituberculates constituted a considerably diversified herbivorous group that occupied a wide variety of ecologic niches. It is peculiar that insectivores and condylarths found in association with them at Cernay are present in closely related forms at Walbeck; the existence of a similar habitat is thus implied. The nature of the environmental differences, then, which may have been responsible for the absence of multituberculates from the Walbeck fauna is not clear. Something or, more likely, some combination of conditions at Walbeck was singularly inhospitable to multituberculate existence. One might add that the habitat was relatively hostile to primates as well; at Cernay, primates rival the meniscotheres in numbers.

While it is difficult not to believe that multituberculates inhabited the European continent continuously from the Jurassic, the absence of meniscotheriid condylarths from Walbeck may be perhaps explained by suggesting a late Paleocene immigration into Europe. Of course, environmental factors could have played a decisive role. The original Walbeck sediments might have been deposited in highlands more subject to continental climatic influences and under conditions quite different from those of the supposed hot, coastal plain at Cernay. However, the presence of a region south of present-day Berlin sufficiently elevated to create a biotope markedly different from that at Cernay is rather difficult to justify paleogeographically. It is likely, but unprovable, that the original Walbeck deposit was also in an area of low relief. The role of gregarious herbivores may have been occupied at Walbeck by the large numbers of a arctocyonid condylarths, which make up over 50% of the recovered fauna. Although arctocyonids were long classified among the primitive, generally predaceous creodonts, their flat, rather pig-like molars indicate an omnivorous diet; their large canines served probably more for combat than for satisfying nutritive needs. But even if they be considered as comprising the dominant herbivore-omnivore element of the fauna, the concept of herds of arctocyonids remains disturbingly unfamiliar. At other fossil localities, they usually occupy a much more modest part of the total.

In definitive, it could be argued that the unusual composition of the Walbeck faunal assemblage (of which arctocyonids and insectivores, in nearly equal proportions, constitute more than 90%) is an example of the faunistic unbalance frequently noted in fissure-filling faunas. But whether or not this is applicable can only be conjectured since the fossils were contained in a reworked sediment. The low number of genera, 13, present in the Walbeck assemblage could also indicate that either specialized local environmental conditions existed, or that the mammals of the area are incompletely represented in the collection. But possible evidence against the latter hypothesis lies in the large number of specimens recovered – nearly 7,000 teeth alone – and in the fact that both large and small size groups are present in ample quantities. The more normally varied Cernay fauna is known from about 4,000 teeth.

As was noted above, the primate element at Walbeck is unusually small for a late middle Paleocene fauna. The dominant form, however, is a plesiadapid curiously more closely related to North American members of the same genus, *Plesiadapis*, than to the species found in the French late Paleocene locality of Cernay. The only other primate at Walbeck is an aberrant carpolestid, *Saxonella*, known from about 50 specimens. North American carpolestids form a homogeneous group and a neat evolutionary sequence in the Paleocene, but *Saxonella*, the only European member of the family, represents a divergent though primitive lateral branch. Both Walbeck primates possess large procumbent and rather gliriform incisors and occupied herbivorous niches in the fauna.

The salamanders found in the Walbeck fauna could indicate a temperate climate and a high continental zone with permanent lakes nearby, even though some newts occur in relatively warm, lowland areas. The presence of crocodiles testifies to a warm climate. BERG [1964] has noted that the mean temperature of the coldest month would have had to be at least 10–15°C for crocodiles to have existed; crocodiles, however, are very rare at Walbeck, being known from only a few dermal scutes. Quite surprisingly, both turtles and the freshwater bowfin, *Amia* (as well as any other fish), are lacking in the Walbeck collection. Representatives of a closely related group of *Amia* species are frequently found in other European localities, as are two or three genera of turtles. Only one lizard, an anguid, is present. The evidence, then, contributed by the lower vertebrates (or by their unusual absence), seems to suggest, like the mammals, the presence of a specialized environment or, on the contrary, that many forms were somehow excluded from the fossil assemblage. Useful information concerning this quandary may be obtained when the rather large number of bird specimens [WEIGELT, 1939] are analyzed.

IV. Menat

In marked contrast to the conditions of deposition and to the fossil material recovered at Walbeck, the Menat locality in the south of France represents a small, isolated lake basin wherein vegetal fossils abound. The most numerous animal remains are those of fish; insects are not rare but they are most often preserved in fragmental form. Also known are a few specimens of arthropods and fresh-water gastropods, three individuals of turtle, one frog, one crocodile, one lizard, and two birds. Mammals, too, are extremely rare; only four individuals are known so far, of which one, identified as closely related to the Walbeck species of *Plesiadapis* [RUSSELL, 1967], first permitted an accurate dating of the deposit. An insectivore was described by GUTH [1962], but the other two specimens lack skulls and provide little information of stratigraphic interest.

Evaluations of the age of the Menat deposit have varied considerably. Paleobotanists first considered it Aquitanian (late Oligocene), but the high degree of similarity to the Paleocene floras of Gelinden and Sézanne and to those of the 'Laramie' in North America, led to the conclusion that it was earliest Tertiary; later, an early Eocene age was proposed, and still later, early Oligocene. PITON [1940], who studied both the fauna and the flora, concluded that the latter indicated an early middle Eocene age. However, he noted that archaic Paleocene (or even Cretaceous) types of plants dominated the flora in numbers of species. PITON explained the mixture of floral elements by supposing that an ancient holoarctic type of flora from the north occupied the highlands surrounding the Menat basin and that a Mesogean flora, more modern in aspect, occurred in the lowlands. More recently, KEDVES [1967] was able to state with certainty that the fossil pollen from Menat indicated a Paleocene age, early in the Thanetian. As suggested by the *Plesiadapis* specimen, this corresponds well with the supposed age of Walbeck. The latter, extrapolated from the stage of evolution of its mammalian fauna, was considered to be late middle or early late Paleocene. Specimens of insects are relatively common at Menat, and PITON recognized 114 species. In a partial revision, however, BALAZUC and DESCARPENTRIES [1964] remark that practically none of the names given by PITON can be retained. They note that, with the exception of specimens enclosed in amber, rarely are insect remains sufficiently well preserved to allow generic determinations or specific descriptions. It follows, they add, that phylogenetic, paleo-faunistic or stratigraphic syntheses based on such fragile evidence should be received with scepticism. Their concluding suggestion, to the effect that Menat should be considered late Oligocene in age, bears out this warning.

The difficulty in dating the Menat sediments is due to the complete isolation of the basin in which they occur and the impossibility of effecting stratigraphic correlations. Only by chemical or paleontologic means is datation possible. Entirely surrounded by gneissic rock, the basin forms an ellipse, 900 by 450 m; the lacustrine lignites and bituminous shales contained within it reach a maximum thickness of 100 m. DANGEARD [1934] has emphasized that the greatest part of the deposit is due to the accumulation of algal remains of a single species, *Amphora subovalis*. The siliceous element of the sediment is related to the presence of these numerous diatoms as well as to fresh-water sponge spicules.

PITON [1940, pp. 298–300] has painted an extremely vivid verbal picture of ancient life at Menat but, unfortunately, few of his determinations are trustworthy and the scene depicted could well be mostly imaginary. As an example, it is unlikely that the strand-dwelling *Nipa* palm lined the shores of the intermontane lake that was Menat. A review of the flora is needed. The birds, mammals, and fish give little precise environmental information. Of the latter, *Amia* and *Thaumaturus* [WEITZMAN, 1960] can be cited, but other forms need verification. Turtles and crocodiles are rare, as at Walbeck, but it seems doubtful that the cause would be the same at the two localities. Concerning the insect fauna, the only major part of the fauna recently reviewed, BALAZUC and DESCARPENTRIES [1964] note that insects often prove valuable only as indicators of facies and not as stratigraphic age determinants. The dominance of one group or another is usually a question of biotope rather than age. Even the fossil remains preserved in lacustrine sediments, like those of Menat, can give a misrepresentation of the probable reality. A number of groups are incapable of drowning and at the same time are able to escape during periods of drought, which explains an otherwise misleading absence. The above authors cite other local factors, such as chance destruction, burial and fossilization, and the influence of geography and altitude which altogether create vast differences in contemporaneous fossil insect faunas. Moreover, the principal change undergone by insects during the Tertiary has been one of a biogeographic nature; their evolution since the beginning of the Mesozoic has been relatively little. The conclusion reached as to the value of the insects in demonstrating the environmental conditions at Menat as well as the age of the deposit is decidedly pessimistic. However, in contrast to PITON's statement that most of the insects are related to tropical or subtropical species, BALAZUC and DESCARPENTRIES [1964] suggest that the climate of Menat may have been comparable to that of the Mediterranean at present, rather than tropical. The role played by the primates in such a situation, therefore, could be visualized within a well-known framework.

V. Mons

The Mons Basin, southwest of Brussels, contains middle Paleocene sediments that represent the beginning of Tertiary sedimentation in Belgium; a lacuna separates them from the deposits of the late Cretaceous Maestrichtian Stage. The type locality of the Montian Stage corresponds to a shrinking and local deepening of the Maestrichtian sedimentary basin. Stratigraphy in the area is rather complex[MARLIERE,1964], but there is a clear superposition of marine sediments by continental deposits (clays, lignites, and fresh-water limestones) which constitute the final phase of the Montian sequence.

Students of early fossil mammals have eagerly anticipated the finding of mammalian remains in Montian deposits. Correlation of the latter by this means would be an important step in determining their exact relationships, and mammals are badly needed to shed light on an epoch in European vertebrate history that has remained undocumented. Stimulating this long-held hope has been the recent discovery of a few isolated teeth in the type locality. A preliminary report has been made by GODFRIAUX and THALER [1972].

VI. Paleocene Climate

On a world-wide basis, the climate of the early Paleocene is usually considered [KRUTZSCH, 1967; MONTFORD, 1970] to have been cooler than that of the late Cretaceous. SITTLER and MILLOT [1964], however, show that locally, for example in southern France, there was no climatic change in passing from the Mesozoic to the early Tertiary; an alternance of humid and dry seasons with a mean annual temperature of about 20 °C is suggested by an analysis of clay minerals and a floral assemblage. KRUTZSCH [1967], referring particularly to Central Europe, remarks that the floras of the early and late Paleocene indicate less tropical conditions than do those of the middle Paleocene (and, in general, the Eocene). The opinion to the effect that the Paleocene as a whole was a cooler epoch than that following is supported by studies on the distribution of corals and bauxite soils, as well as from oxygen-isotope paleotemperature measurements. Leaf floras of the northern hemisphere give evidence of extensive Paleocene temperate forests. A typical Central European Paleocene assemblage might yield pollen of suspected Juglandaceae, Myricaceae, Myrtaceae and Haloragaceae affinities. In lower percentages would be found pollen representing Nyssaceae, Ulmaceae, Betulaceae, Sapotaceae, Tiliaceae and Aquifoliaceae. PENNY [1969] notes that the absence or rarity of winged conifer pollen seems to be a characteristic feature of these floras.

In the Franco-Anglo-Belgian basin of Western Europe, the Paleocene floras had essentially a warm temperate character. But, being situated at the southern limit of the distributional area of warm temperate floras, the influence of periodic fluctuations in climate was strongly felt. During warm periods, genera and families of a subtropical, or sometimes, a tropical character, invaded the region. ROCHE [1970] has recently commented on a pollen assemblage from the middle Paleocene locality at Mons, Belgium. The dominant forms belong to the Palmae, Taxociaceae-Cupressaceae, Juglandaceae, Schizeaceae, and the Cyatheaceae, which leads one to believe that the temperate elements were dominated by those of warm temperate or even subtropical affinities. Normapolles, representing primitive angiosperm types, dominate in the Montian pollen diagrams.

The classical (late middle) Paleocene locality of Gelinden (Belgium) gave

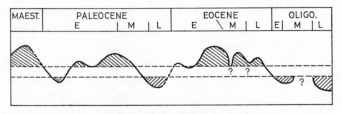

Fig. 2. Temperature maxima and minima during the late Cretaceous and Paleogene of Central Europe. [Modified from KRUTZSCH, 1967.]

its name to one of the Holarctic phytogeographic provinces as conceived by KRYSHTOFOVICH [1957]. Amentiflorae were abundant in this province, but evergreen plants were the dominant element, and the Lauraceae one of the most characteristic constituents. *Viburnum* abounded and *Sabal* was common. In recent years a palynologic study of the Gelinden flora has been undertaken by ROCHE [1969]. Younger Paleocene assemblages have long been known from Sézanne and Rilly in the Paris Basin. The dominant plants are the Lauraceae and the Juglandaceae. Although tropical or subtropical forms occur, the flora is essentially composed of temperate or warm temperate elements.

With recent microfloral studies now supplementing those made of the macroflora, the climate of the European Paleocene has become relatively well documented. Problems of local topography and correlation persist, but, in general, only details remain to be filled in.

VII. Dormaal

The locality of Dormaal, situated southeast of Brussels, was long known in the literature as Orsmael [TEILHARD DE CHARDIN, 1927]. References to its fossils began in 1884; excavations were carried out during the first quarter of the 20th century by Belgian scientists and, on a more intensive basis, from 1950 to 1965. The locality occupies a controversial position with respect to its age. Opinions based on geological data differ from those based on the nature of the mammalian fauna. Concerning the latter problem, RAT [1965] and VAN VALEN and SLOAN [1966] have suggested that the Dormaal fauna be considered earlier in age than the Sparnacian. Belgian workers [CASIER, 1967; QUINET, 1969; ROCHE, 1970] situate the locality in the late Landenian stage and also regard it as terminal Paleocene. Deposition took place between the end of the last phase of the marine early Landenian and the beginning of the Ypresian (Eocene) transgression, that is, during the regression of the Landenian Sea. The mammalian fauna, however, could easily be considered to be of early Eocene age.

An analysis of the Dormaal mammals reveals none that are elsewhere restricted to the Paleocene. A few genera are known from both the late Paleocene and the early Eocene, such as *Phenacodus, Oxyaena, Hyracotherium*, and *Plesiadapis*[1], but individuals of these genera are very rare in the Belgian fauna. Perhaps indicating the arrival of immigrants from North American and heralding the beginning of the Eocene is the presence in fairly large numbers of didelphids and rodents, both otherwise unknown in the late Paleocene of Europe. The primate *Teilhardina*[2], also frequent in the fauna, is elsewhere unknown in the Paleocene but occurs in the late Sparnacian and Cuisian of France.[3] Other genera characteristic of the Eocene, though rare in the fauna, include *Protodichobune, Miacis*, cf. *Palaeonictis*, and cf. *Prolimnocyon*. Furthermore, *Coryphodon* has been found at the contemporary locality of Erquelinnes.

The foregoing would be clearly indicative of Eocene age except for the presence of rather numerous arctocyonids, essentially absent in the European Sparnacian[4]. Also, the rarity of perissodactyls and artiodactyls (if not artificial, due to fluvial sorting) is unusual for an Eocene assemblage and could suggest that they had only begun to migrate into Western Europe.

Other vertebrates at Dormaal include selachians and many teleost and holostean fish. Of the latter, *Lepisosteus* and *Amia* form the principal elements. Turtle and crocodile remains occur, but usually reduced in size and rounded by transportation. Of Amphibia, two urodeles and several anurans have been cited [HECHT and HOFFSTETTER, 1962]. Seven genera of Lacertilia are mentioned, one amphisbaenid, and three types of snakes. The presence of bird bones has also been noted. Invertebrates are extremely rare in the deposit and those recovered appear to have been reworked. Blocks of fossilized wood are known, but no other references to the flora have been made.

1   The species *Plesiadapis orsmaelensis* Teilhard de Chardin, based on lower incisors, is not referable to that genus [GINGERICH, personal commun., 1972].
2   GINGERICH [personal commun., 1972] suggests that this genus might be derived from forms resembling the North American middle Paleocene *Palaechthon* and *Palenochtha*.
3   *Platychoerops* has just been identified in the Dormaal fauna; it occurs elsewhere in Sparnacian and Cuisian deposits.
4   A few undescribed arctocyonid teeth are known at Abbey Wood (England), a Sparnacian locality. Arctocyonids are present, but rare, in the early Eocene of North America.

The deposit is fluvial in origin and consists of a complex of gravel, sand and marly lenses. Most of the vertebrate remains were found in lignitic sands. The large river responsible for the series of channel sediments in which the locality occurs, flowed from the southeast toward a fairly distant sea in the north or northwest. No evidence of tidal influence was found. Geologically speaking, the deposit was probably made rapidly. The effect of fluvial transportation played an important part in the nature of the fossils recovered. Natural sorting, or gravimetric selection must be held responsible for the heterogeneity of the mammalian fauna (absent in the fish), for the near absence of fragile skull elements, and for the paucity of large bones or fragments. A few large specimens are known from Erquelinnes (Jeument), Leval, Vinalmont, and Orp-le-Grand, approximate age equivalents of Dormaal, but little collecting has been done at these sites where gravimetric effects might be less.

One may suppose that climactic conditions at Dormaal (during the opening period of the Eocene) were warmer than the climate prevailing during the time of the Cernay fauna, but still cooler than those shown by the London Clay forest that was to follow. Paleogeography indicates that the countryside around Dormaal was rather flat and that the locality, though situated on a coastal plain, was farther from the sea than was Cernay. It is difficult to evaluate with any precision the environment from the vertebrates. Like the Walbeck fauna, that of Dormaal curiously lacks multituberculates and shows an abundance of arctocyonids. The gregarious *Plesiadapis*, found at Walbeck, and in large numbers at Cernay, is also essentially lacking. The absence of these two groups is inexplicable, as their size is consistent with other recovered members of the fauna. *Pleuraspidotherium*, a presumably herd-type ungulate characteristic of the Cernay assemblage, being absent as well, would seem to indicate a denser forest at Dormaal. The principal primate inhabitant of this suspected forest was *Teilhardina*.

Much has been made of the primitive character of *Teilhardina*'s dental morphology. QUINET [1966a, b] envisaged the genus as occupying a place in the ancestral stock of the Old World anthropoids, and hence of man. Few other workers go so far. HÜRZELER [1948] has suggested that it was ancestral to the microchoerine tarsioids, and SIMONS [1961] has noted that it at least shows affinity to the tarsioid stock. In his classification of 1967, MCKENNA places the genus in the Omomyidae; SIMONS [1967] adds that members of this family 'could be ancestral to the ceboids, cercopithecoids, and hominoids', but cautions that this suggestion, based primarily on dental similarities, is difficult to confirm from the present state of our knowledge.

In conclusion, the age given by stratigraphic correlation equates the Dormaal locality with a period post-Cernaysian in time and pre-Sparnacian. The fauna, while not in contradiction, displays more Eocene (Sparnacian) influence than Cernaysian, but it could be possible to consider it also as representative of the final stage of the European Paleocene. If this be the case, the Eocene mammalian fauna appeared in Europe before the close of the Belgian Landenian stage. On the basis of molluscs, CURRY et al. [1969] have suggested that the terminal part of the last Paleocene stage in France (Thanetian) and in Belgium (Landenian) could be younger than the summit of the Thanet Beds in England. Graphically, they depict the Paleocene-Eocene boundary immediately after the time of the Cernay (Thanetian) deposits, and a little above the Thanet Beds, but within the duration of the late Landenian. This is corroborated by the mammalian evidence, if the beginning of the Eocene is taken to coincide with the appearance of the Sparnacian-Wasatchian faunas.

VIII. Sparnacian

In striking contrast to the paucity of vertebrate fossil localities in the European Paleocene, The Sparnacian early Eocene in the Franco-Anglo-Belgian Basin has furnished over 50 sites [RUSSELL, 1968]. Near the Mediterranean, in the south of France, a typical assemblage of this age has also been discovered [GINSBURG, et al. 1967]. Sparnacian localities inherited from the last century are often lost today under invading vegetation in abandoned lignite quarries, covered over by the relentless spread of urban development, or simply unknown because of inadequate geographic data. Most of them have yielded little material. The new localities found, however, coinciding with the initiation of intensive collecting techniques, have permitted a more complete recovery of the vertebrate fauna. The resulting increase in our knowledge of this epoch has been considerable. Detailed study of several groups and localities remains to be done, but the overall picture is well documented and the part to be filled in is relatively slight. The fauna, if not complete, is at least very representative.

One of the notable aspects of the Sparnacian fauna, in comparing it with the preceding Paleocene Cernaysian assemblage, is the rarity of forms that pass from one period into the next. Not only was the Sparnacian populated by new types of animals, but hardly any of the old ones persisted. Those that did, rapidly became extinct; none of them, in any case, figured in lineages leading to modern families.

In order to account for this difference, only the solution of a mass migration fulfills the necessary requirements. Not enough time elapsed between the Thanetian and the Sparnacian to invoke simple evolutionary change, and, in any case, the necessary ancestral forms lack in the Mont de Berru Cernaysian faunas. The Eocene mammals are so different from those that preceded them in the Paris Basin that one would have to search far back in the Tertiary (or even the Cretacous) for a common stock. The idea of migration has been cited since CUVIER in order to explain the profound changes observed in the succession of mammalian faunas. Breaches in hypothetical biogeographic boundaries similar in nature to Wallace's Line, or to the Mozambique Channel between Madagascar and Africa today, have often been held responsible for a rapid alteration in the fauna of a given region. Substantial evidence is now available indicating that the creation of the European Sparnacian fauna was due in very large part to the establishment of a complex of favorable ecologic factors at the latitude of a land connection between Europe and North America, and to the subsequent free faunal passage across it. A remarkable homogeneity of mammalian genera existed on this united landmass during the Sparnacian-Wasatchian[5]; over 50% of them are identical on the two continents, which is a higher proportion than that of any other time in the history of mammals. By contrast, in the late Paleocene, only about 20% of the genera are identical.

Whereas in most of Belgium and in the north of France the beginning of the Eocene is already marine (Argiles d'Ypres), the elevation of the Artois anticline initiated the development of a system of lagoons in the Paris Basin and at the same time ruptured the floral uniformity that had existed in the latter and its Belgian counterpart. The regressing Thanetian Sea left behind sand and limestone deposits over a large part of the Paris Basin. Of continental origin, the Argile Plastique was brought by fluvial action probably from the Massif Central and, to the North of Paris, the basin filled with lignitic clays and sands. Sparnacian life established itself on these sediments in a landscape made up of many lakes and brackish swamps or lagoons communicating with the fluctuating shores of a shallow sea. Estuaries or deltas, formed by rivers emptying into the basin, furnish today a rich concentration of fossil material.

This Sparnacian fauna contained few large animals. Only one identi-

5   SLOAN [1969] has suggested that this fauna is primarily of Central American origin.

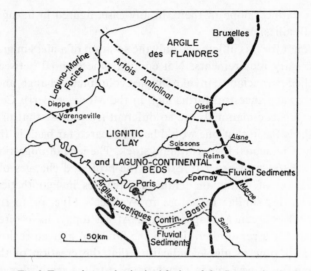

Fig. 3. Expansion and principal facies of the Sparnacian in the Paris Basin. [Modified from POMEROL, 1967.]

fiable element of the tapir-sized pantodont, *Coryphodon,* has been recovered during the last decade of intensive work. The European specimens of this form are for the most part the result of discoveries made in the preceding century during the exploitation of lignite quarries by hand labor. Our new collections consist almost entirely of small- or medium-sized animals. This can be explained by the techniques utilized in collecting, by the type of deposit in which they were found, and by the biotope which constituted the source of the fossil material. But, in any case, these primitive mammals were smaller than their descendants; in average size of individuals, a Sparnacian fauna could be expected to be made up of smaller forms than a fauna from the later Eocene, or Oligocene, for example. The presence of many species represented by a large number of small individuals would tend to indicate a dominantly forested environment. The scarcity of relatively large forms and of animals similar to those constituting supposed herds in the late Paleocene of the Mont de Berru supports this hypothesis.

The nature of the early Eocene vertebrate assemblage can be demonstrated thanks to the large collections now existing. Numerous fish teeth and bones occur in the Sparnacian sediments; depending upon the character of the locality they are predominantly marine [CASIER, 1966] or fresh water. Of particular interest in the latter category are the Characidae, a family including the living piranhas of South America, which are now known for the first time both in Europe and in the northern hemisphere [CAPPETTA *et al.* 1972]. Birds are generally rare although, in exception, the London Clay has produced nearly 30 genera. This avifauna is very distinctive and mostly composed of taxonomically new or extinct families; no modern genera are represented [CRACRAFT, personal commun., 1972]. In France, the remains of a giant form, similar to the Cernaysian *Gastornis*, have been found [LOUIS, 1969]. Turtles, represented by three genera, thrived and their fragmented bones occur in great numbers. The same two genera of crocodiles continue to frequent the region, passing with little change into the Eocene, but the *Simoedosaurus*, dominant in the Cernaysian faunas, has essentially disappeared. Lizards of several types (Anguidae, Lacertidae, Gekkonidae) occur in abundance. Though few in numbers, amphisbaenids are found in nearly all the collections. Snakes appear for the first time in the European Tertiary and are dominated by the Boidae, one form of which was aquatic. Amphibians, rather oddly, are very rare; urodeles are known from only one locality [for the lower vertebrates: RAGE, personal commun., 1972].

The mammalian component of the Sparnacian assemblage is represented principally by isolated teeth. Complete dentitions occur infrequently and associated skeletal material is known only from old accounts by quarry workers. Despite the inadequacy of the material, the fauna is of particular interest in that it contains the first European members of several living orders; modern ungulates and primates, and rodents. These newcomers constitute a dominant proportion of the Sparnacian mammals.

Sparnacian rodents occupy a primitive position with respect to all the living groups. Already diversified and numerous, they are found at all the localities and alone account for 25% of the mammalian remains. By their abundance and the rapidity of their evolution rodents constitute an element of prime importance for stratigraphic correlation. Their appearance at the opening of the European Eocene announces the availability of an extremely useful tool, in addition to the interest inherent in following the development of a versatile and successful group.

In contrast, two species of didelphid marsupials, other new arrivals at this time, underwent remarkably little change during their existence in the European Tertiary, and, in fact, show little evolution from their Cretaceous ancestors.

Insectivores are abundant: adapisoricids are the most frequently found, but also occurring are palaeoryctids, mixodectids, apatemyids, nyctitheriids, pantolestids and a number of forms still unstudied. Several genera

of small hyopsodont condylarths are now also known in the microfauna.
The large *Phenacodus* is rare and, except for *Landenodon* at Dormaal, arcto-
cyonids are essentially absent; mesonchyds are extremely rare. The tillodont,
Esthonyx, also figures among the forms seldom found.

From their Cernaysian abundancy, multituberculates become increasingly rare during the early Eocene and are not known in Europe thereafter.

The presence of minute teeth in Sparnacian localities referable to the
Dermoptera adds an exotic element to the mammalian assemblage. Although
similar genera are known from the late Paleocene and early Eocene of North
America, no intermediary forms have been found, in Tertiary or Quaternary
sediments, leading to the species presently inhabiting the forests of South-
East Asia. The Sparnacian London Clay flora is considered to have closely
resembled the forest of this area, indicating, perhaps, a persistant preference
in dermopterans for this sort of biotope.

Bats, the first in the Old World, demonstrate a frequency and variety in
Sparnacian sediments that is found nowhere else in the fossil record of this
time. In one locality, Mutigny, three genera represent more than 20% of the
microfauna, a rather astonishing percentage since the deposit gives no hint
of conditions usually favored by concentrations of bats. The discovery of a
completely preserved individual in approximately contemporaneous beds in
North America shows the precocious development of full flying ability.
Chiroptera, therefore, must have had an autonomous history extending far
back into the Paleocene.

With few exceptions, the Sparnacian carnivorous genera (hyaeno-
dontids, oxyaenids, and miacids) are common to the North American
Wasatch fauna or very closely related to its members. In no other group,
with an equivalent number of genera, is such an intercontinental homogeneity
demonstrated. As none of these forms have antecedents in the Cernaysian
fauna they can in large part be considered as immigrants, probably mostly
from North America. Speculation on the psosible contribution from Asia
is as yet gratuitous. The Miacidae, already appearing early in the North
American Paleocene, form the ancestral stock for modern Carnivora. Their
appearance in Europe with the beginning of the Eocene (and present even
at Dormaal) is an indication of their progressive spread and rise to dominance.

Sparnacian ungulates (perissodactyls and artiodactyls) are characterized
by their modest size and relative infrequence, quite in contrast to their later
development. Artiodactyls, for example, represent only about 5% of the
specimens recovered. Both groups have only one prevalent genus in the

Sparnacian fauna; *Hyracotherium*, a near-ancestral form for perissodactyls, and *Protodichobune*, a primitive bunodont artiodactyl. Only a few other genera, sparcely represented, are present in each group. But the abrupt appearance of these orders in early Eocene sediments announces the arrival of a new element in the mammalian assemblage that, increasing in numbers toward the end of the early Eocene, prodigiously multiplies in generic diversity, thus preparing the base from which sprang the hordes of middle and late Cenozoic ungulates.

Prosimian primates are common in the late Paleocene-early Eocene localities of Europe and are often present in great numbers. In each of the Paris Basin sites so far exploited, there exists, in company with infrequent, small but highly distinctive forms, one genus, of larger size, which often constitutes a dominant element in the mammalian fauna. This genus is sometimes exceeded only by rodents in number of individuals. *Platychoerops* is the first Sparnacian dominant form, present in such Paris Basin localities as Pourcy and Mutigny, and succeeds the late Paleocene *Plesiadapis*. In the basal Sparnacian site of Dormaal neither *Platychoerops* nor *Plesiadapis* is known (except for one lower incisor referred to the latter), but at Meudon (Paris Basin; older than Pourcy and Mutigny) a few isolated teeth indicate a form intermediate in morphology between *Plesiadapis* and *Platychoerops* [GINGERICH, personal commun., 1972]. With the concluding of the Sparnacian, *Platychoerops* is overlapped and replaced in its turn by *Protoadapis*. Whereas, as we have seen, considerable importance has been attached to the Sparnaco-Cuisian omomyid *Teilhardina* because of its probable affinities with later tarsioids, the appearance of *Protoadapis* is equally, if not more significant. Adapids are regarded as probably ancestral to living lemurs, but as pointed out by SIMONS [1963], there is, in fact, nothing in their known anatomy that would exclude them from being considered as possible ancestors to some or all higher primates. The European notharctid species of *Pelycodus* share the fate of *Platychoerops*, disappearing in the Cuisian with the rise of *Protoadapis*. *Phenacolemur*, a highly specialized genus, persists throughout the early Eocene.

In summation, the Sparnacian was an age of more than usual consequence. While the presence of a richly varied and abundant fauna and flora permits the imagining of a detailed picture of life at this moment in time, two events emphasize particularly well the importance of the Sparnacian-Wasatchian Stage: (1) the establishment of an easily practicable land connection across the North Atlantic which exerted a strong influence over a broad spectrum of ecologic factors in Europe and North America; (2) the appearance of a new mammalian fauna which laid the foundation for the development of forms existing today and which signaled the end of the Cretaceous-Paleocene stage of mammalian evolution.

IX. Cuisian

The latter part of the European early Eocene, the Cuisian, is represented by fewer localities and much less material than the Sparnacian. Exploitation of a group of quarries immediately to the south of Epernay, in the fluvio-estuarine Sables à Unios et Térédines, has furnished most of the Cuisian vertebrate fauna. The microfauna of these beds, however, has remained very poorly known, not only because the excavations made during the preceding 75 years tended to overlook small specimens, but because they are in fact quite rare; natural sorting during transportation and deposition of the sediment is responsible. In recent years, an attempt has been made to recover this microfauna, and the discovery of a new locality in beds of equivalent age at Sézanne-Broyes [LOUIS, 1970] has done much to fill the lacuna. Another locality in the south of France has also contributed remains of small vertebrates [CAPPETTA et al., 1968], and the Cuisian is represented in basal deposits of the Geiseltal (Germany) lignites. The Monte Bolca locality in Italy, whose reknown fish fauna is currently being revised under the direction of BLOT [1969], has recently been shown to be Cuisian [BLONDEAU, 1971]. More doubtfully referred to this age [HARTENBERGER, 1973] is the fauna of the lignitic basin of Messel (Germany).

Despite increasing knowledge of its microfauna, the Cuisian can be characterized by the presence of large animals contrasting sharply with Sparnacian mammalian faunas. The rare Coryphodon of the latter has disappeared, but the equally large Lophiodon is found rather frequently. Also present are other perissodactyls, Hyrachyus, Lophiaspis, and the especially common Propachynolophus; artiodactyls become more numerous in individuals but do not yet diversify. The abundance of these forms demonstrates probably a difference in biotope as well as a difference in generic dominance, although by far the major part of the Cuisian fauna is descended from elements in the Sparnacian.

Of particular interest is the presence of two undescribed primates, known only by a few specimens, from the late Cuisian Paris Basin locality of Grauves.

It has been suggested that the number of relatively large herbivores could indicate a drier, more savanna-like environment in the Paris Basin, but confirming evidence from other disciplines is slight.

As was noted for the mammals, there was, among reptiles, little change in generic composition in passing from the Sparnacian to the Cuisian. The numerous crocodiles, belonging to the genera *Diplocynodon* and *Asiatosuchus*, were joined by a newcomer, *Pristichampsus*. Turtles remained frequent and the smaller saurians were represented by the same Sparnacian forms. Birds are essentially unknown.

The Cuisian, then, represents in general a time of stabilization of the Sparnacian fauna. Its mammalian assemblage constitutes in large part an extension and a remainder of the great faunistic upheaval at the beginning of the Eocene caused by the influx of new elements from foreign regions, an invasion which brought about the nearly total elimination and replacement of the Paleocene fauna. At the same time, certain elements prepared the scene for the following revolution to take place with the opening of the middle Eocene, when the European mammalian fauna was again renewed.

X. Early Eocene Climate

Early Eocene times became, with a progressive rise in temperature, warmer than the last stages of the Paleocene. Localities containing floras of a more tropical nature are numerous, probably due in part to an increase in areas of favorable habitat as a result of late Cretaceous-Paleocene marine regression. The lowlands bounded by the Tethys Sea constituted highways for plant dispersal. The *Nipa* palm, important as an indicator of coastal conditions, spread along the shores of the continental rifts and in the early Eocene became worldwide in distribution. The less mobile vegetation of conifers and deciduous trees, derived from the Cretaceous floras of Laurasia, became the Arcto-Tertiary flora and continued to occupy the higher, drier lands.

In general, our knowledge of the Sparnacian flora has been obtained from the fruits and seeds of the London Clay and the leaves of the Paris Basin. The microfloras are known from both and provide a useful means of comparison, but few genera are known by both micro- and macroremains. A clear distinction exists between the Paleocene and Eocene microfloras. The Normapolles group, whose greatest expansion took place in the late Cretaceous and which remained well represented in the Paleocene, became of secondary importance in the early Eocene. The percentage of extinct types of vegetation, however, was still high. The flora of the Paris Basin Sparnacian possesses an essentially tropical or subtropical character but includes some temperate taxa. Certain elements termed tropical (for example, palm pollen) can be found in remarkable quantities, a phenomenon which persisted in the Paris Basin to the end of the Lutetian.

The most frequent Sparnacian plant association in the Paris Basin is that of a swampy forest, with either Myricaceae or *Coryphaeus* dominant [GRUAS-CAVAGNETTO, 1968]. Pollen indicating a drier, or a higher environment, are present in few numbers *(Quercus)*, or are rare *(Pinus, Podocarpus)*. This is in corroboration of the mode of sedimentation and of the paleogeography known for the region at that time. The relative paucity of some of these pollen suggests that the plants were situated rather distantly, or if they were near, that their pollen was masked in the pollinic spectra by the rich vegetation of the coastal plain. Conifers generally are not frequent in Eocene Paris Basin sediments, and since they produce great quantities of pollen especially well adapted to flight one might suppose that the originating trees were far from the deposit.

The London Clay flora, studied in detail by CHANDLER [see 1964], is particularly varied and one of the best known of Western Europe. Vegetation of the London Clay type, or very similar, appeared over a large part of the world, from South-East Asia, through Burma, India, the Ukraine, Central and Western Europe, West Africa, Brazil, Texas, Oregon, to Alaska, representing a uniform multispecific plant community. This flora has been considered as typifying lowland, tropical rain forest conditions, similar to those occurring today in South-East Asia (Indo-Malaya). Despite the transportation of some distant upland forms, most of the plants were thought to have grown near the region of deposition, with lowland species dominating because of proximity to the sea. As the London Clay is marine, the elements of the flora (fruits, wood, leaves, etc.) represent accumulations of current borne vegetation deposited beyond the mouths of estuaries. In some instances, these were estimated to occur as much as 80–115 km from the shore. Pollen analysis of London Clay specimens shows a remarkable agreement, in terms of the distribution of living counterparts, to the rich flora of fruits and seeds. This sporo-pollinic association differs somewhat from that of French localities of similar age [GRUAS-CAVAGNETTO, 1970] and is considered as late Sparnacian.

A critical reappraisal of the tropical rain forest nature of the London Clay flora has recently been made by DALEY [1972]. He considers the proportion of temperate elements in the flora unusually high for their presence to be explained by transportation from more or less distant high country. Furthermore, the existence of mountainous regions of sufficient altitude to have permitted the development of a temperate mountain flora (and relatively close to the London Clay Sea) is not supported by geologic evidence. That these elements were relics from the cooler Paleocene is also rejected. He doubts, even, that the climatic prerequisites necessary for the establishment of a tropical rain forest were present in southern England, and suggests that high percipitation, with adequate ground water, sufficied to support a tropical flora, but that, nearby, on slightly higher ground, more temperate plants flourished with less available moisture and a lower relative humidity. Temperatures as high as those found today in a tropical rain forest climate may not have been necessary. Other factors, like the absence of frost, were more important. Tropical vegetation successfully invaded temperate areas and the otherwise puzzling mixture of fossil forms was the result.

The plant succession in the Dorog Basin of Hungary, described by KEDVES [1960], can be taken as representative of the early Eocene flora of Central Europe. Low in the sequence is found a great quantity of Taxodiaceae-Cupressaceae pollen, indicative of a humid forest. On the higher and drier part of the basin, palms dominated, accompanied by a scattering of ginkgos. These forests, rather distant from the edge of the basin, alternated with a zone of Myricaceae, which included also a few Cyrillaceae-Clethraceae. Hypothetically, a *Sequoia* forest existed in the higher regions. KEDVES warns, however, that in other parts of the basin the vegetation could have been different.

Widespread subsidence of the European continent in the middle Eocene resulted in the development of extensive lacustrine, river-swamp, and embayment habitats. Oscillations of the strand line and luxuriant plant growth produced intercalations of vegetational debris which contributed to one of the major coal forming periods of earth history and provided a record of vast forest, thicket, and swamp communities growing under near tropical conditions. In Central Europe, the following major associations can be distinguished: (1) swamp forests with *Taxodium* and *Nyssa;* (2) riverbank and grove habitats of *Sabal* and other palms; (3) shrub thickets of Myricaceae-Cyrillaceae species, Sapotaceae-Symplocaceae species, *Aralia, Ilex,* and giolypodiaceous ferns; (4) hardwood forests of fagaceous species of *Fraxinus, Engelhardtia, Tilia, Ailanthus, Pterocarya, Carya,* and *Cornus,* and (5) conifer forests of *Sequoia* (or its Eocene taxodiaceous equivalent), *Pinus, Picea,* together with *Rhus* (vines or shrubs) and Schizaeaceous ferns (e. g., *Lygodium*).

With a peak in temperature during the middle Eocene, the period of relative tropicality that had characterized the late Mesozoic and early Tertiary of Europe came to an end. The global girdle of tropical rain forest was, in consequence, modified, reduced, and interrupted. A considerable decrease in annual mean temperature took place at the end of the Eocene. Estimates vary, but a lowering of from 8 to 10 °C is suggested. Supporting evidence has been obtained from studies of clay minerals [STEINBERG, 1969], oxygen-isotope paleotemperature measurements [DEVEREUX, 1967], variations in species number of planktonic foraminifera [JENKINS, 1968], as well as from paleobotany [WOLFE, 1971]. The late Eocene annual mean temperature drop was followed by a rise in the Oligocene, but the latter was accompanied by notably dry, even arid conditions; never again during the Tertiary was a humid, tropical climate attained to the degree reached before the late Eocene. Moreover, this fluctuation produced one of the Tertiary's greatest temperature variations and was responsible for extensive and important evolutionary consequences.

XI. Paleogeography

The distribution of the faunas under discussion and their change through time can be accounted for to a large degree be paleogeography. Two principal dispersal routes were apparently utilized, one joining Europe to North America, the other crossing the Russian platform to Asia.

The dry-land dispersal route between Western Europe and North America was controlled by climate, latitude and contiguous crustal blocks along the edge of the Arctic Basin. Contact between Europe and North America seems to have taken place at least in the middle Paleocene; earlier European evidence is inadequate. Despite various cases of generic similarity, or identity, European-North American faunal dispersal during the Paleocene never approached the free flow of elements across the North Atlantic that so distinguished the Sparnacian stage. If exchange between the Old and New Worlds was less intense during the Paleocene, it was due in part to the position of the rotational pole and a resultant higher paleolatitude for the North Atlantic corridor, and in part to a climate that was generally less equable than that of the Eocene. The intense intermigration that took place between the two continents at the beginning of the Eocene was effected by means of a communication through Greenland, Scandinavia, and the present Barents shelf [SZALAY and MCKENNA, 1971]. But with the rupture, about midway in

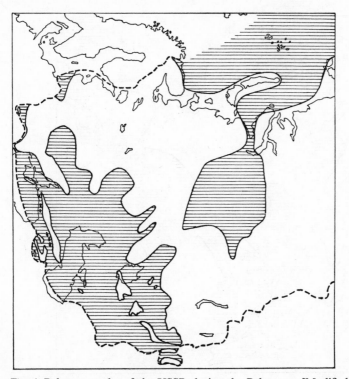

Fig. 4. Paleogeography of the USSR during the Paleocene. [Modified from VINO-GRADOV, 1967.]

the early Eocene, of the connection and the opening of the North Atlantic, Western Europe was left in a singularly isolated state. Possibilities for dispersal between the highly dissected parts of the continent were reduced and controlled by filters of fluctuating but considerable importance. These lasted until the retreat of the Tethys Sea at the time of the late Eocene Pyrenean orogeny.

For determining routes and times of intermigration between Asia and Western Europe, the paleogeographic history of Eastern Europe is of particular interest. The largest and in many respects the most important single element of Paleocene-Eocene topography in this area was the Russian platform. Although providing a vast territory for easy dispersal, it was limited on the east by the Ural Mountains and the Obik Sea, a marine arm from the Arctic Basin that tended to bisect the Asiatic landmass. Encroachment along the southern boundary of the platform by the Tethys Sea further reduced its potential as east-west passage.

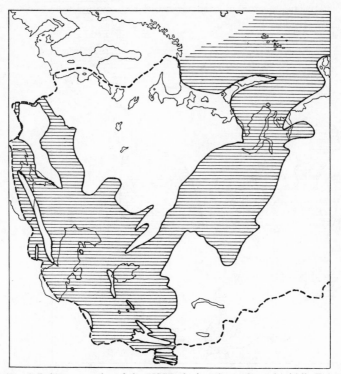

Fig. 5. Paleogeography of the USSR during the early and middle Eocene. [Modified from Vinogradov, 1967.]

Orogeny at the end of the Cretaceous caused a general regression so that during the Paleocene the Russian platform was relatively emerged. Despite this favorable geographic factor, Szalay and McKenna [1971] have shown by their study of the late Paleocene fauna from Gashato and the Nemegt Valley in Mongolia that no special similarity to Paleocene faunas of Europe existed. Nearly complete isolation is suggested. But contact with Western Europe appears to have been fairly extensive and a terrestrial route from Asia existed across the Turgai Straits; the latter, however, was subject to periodic flooding by the sea and constituted only an intermittent passage. Opinions seem to differ as to when the transgression of the Tertiary Tethys Sea began. Vinogradov [1967] shows a large part of the Russian platform, from the west side of the Black Sea, across the Caspian and Aral Seas to the present Chinese border, as being already submerged in the Paleocene. But Meszaros and Dudich [1966] state that the first transgression of the platform occurred in the post-Sparnacian early Eocene. Difficulties in correlation and varying

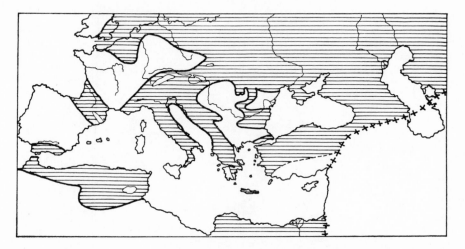

Fig. 6. Continents and areas of marine sedimentation during the middle Eocene. [Modified from Mészáros and Dudich, 1966.]

opinions of epoch boundaries are doubtlessly responsible for this divergence. Strong variability in marine litho- and biofacies indicate the formation of archipelagos and the presence of islands sprinkled on the surface of the shallow sea. But even reduced from its Mesozoic development and dismembered into partial basins, the Tethys Sea inevitably inhibited mammalian movement.

Northward advance of the sea continued in the middle Eocene, after a short period of uplift and erosion which caused a discordance between early and middle Eocene deposits [Kopek *et al.*, 1971]. Gidai [1971] notes that lignites are frequent in the early middle Eocene of Central Europe and adds that local emergence, reworking of early Eocene elements, and facies changes in marine sediments to fresh water or brackish conditions, indicate that middle Eocene time was characterized by considerable tectonic activity. The exact time of the definitive opening of the Turgai Straits and the formation of a broad epicontinental seaway extending from the Mediterranean to the

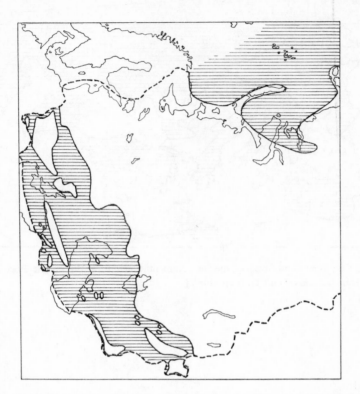

Fig. 7. Paleogeography of the USSR during the Oligocene. [Modified from VINO-GRADOV, 1967.]

Arctic is not clear. At least by the middle Eocene, if not considerably before, it was well established. In the late Eocene, after another oscillation which permitted scattered lignite deposits, the sea stabilized in expansion (with maximum transgression) and locally deepened. Then with orogeny at the end of the Eocene, the Russian platform emerged and the Eurasian landmasses were again broadly accessible to east-west mammalian movement.

XII. Conclusions

Evidence from the European Paleocene indicates that the elements responsible for mammalian faunal changes were apparently restricted to those originating in North America. Although the same relationship is suggested by the new assemblage that characterized the early Eocene, it is possible that arriving immigrants from Asia in the Sparnacian, particularly if few in number, were masked by the preponderance of forms in the process of active exchange across the North Atlantic. In any case, we have very little information on Asiatic species of this age for verifying places of origin. Mingling with the North American-European fauna they are indistinguishable or, at most, only suspected. At the end of the Sparnacian, however, with the North Atlantic connection broken, Asian forms became more evident: for example, *Lophiodon, Plesiarctomys*, and *Protoadapis*, the latter harbinging the adapid group of primates that played such an important role in Europe up into the early Oligocene. Arriving just before the middle Eocene sea dissected and isolated Western Europe were a variety of artiodactyls. The tarsioids appeared at the same time and were to dominate, with the adapids, the primate element. Most of the other new members of the middle Eocene mammalian fauna are derivable from forms already in Europe. A particular case was the rapid development of numerous perissodactyl genera (all derivable from hyracotheriine equids, according to RADINSKY [1969], or partially, according to FRANZEN [1968], and partially from phenacodontid condylarths). Another example is that of the gliravine rodents which evolved out of early Eocene microparamyines. More minor groups of unknown origin include the hipposideriid bats and the paroxyclaenid condylarths.

A summary of the history of European primate evolution is abbreviated by the absence of a fossil mammal record in this region during the early and most of the middle Paleocene, but by the end of the middle Paleocene plesiadapids, at least, were thriving. Illustrating the presence of a North American terrestrial passage, the primates of the Walbeck fauna include a species of *Plesiadapis* that closely resembles North American forms, and *Saxonella* represents a derivative from an ancestral stock common to the North American *Elphidotarsius-Carpodaptes-Carpolestes* lineage. The late Paleocene Cernay fauna equally contains genera identical or similar to North American contemporaries, although the ancestry of the *Plesiadapis* species is obscure and the relationships of *Berruvius* are uncertain. *Chiromyoides* could be derived from the Walbeck species of *Plesiadapis*, but GINGERICH [personal commun., 1973] reports a new species of the genus in the Paleocene of North America. In the late Paleocene plesiadapids attained their numerical maximum in Europe. With the basal Eocene, *Plesiadapis* became rare and was replaced by another plesiadapid, *Platychoerops*. *Teilhardina* appeared very early in the Eocene, disappeared from the regions known by fossil localities, then reappeared in the late Sparnacian and Cuisian. *Platychoerops* dominated in the

Sparnacian, but gradually died out in the Cuisian, signaling the end of the Plesiadapidae. Appearing timidly in the Sparnacian, *Protoadapis* flourished during the Cuisian and Lutetian. *Pelycodus* and *Phenacolemur*, immigrants from North America, are restricted to the early Eocene stages, and a rare form apparently referable to *Berruvius* carried over into the Eocene. Adapids and microchoerine tarsioids dominate in the middle and late Eocene, accompanied by a lone aberrant genus, *Alsaticopithecus*. As elements of the European fauna, primates declined in importance with the ending of the Eocene. The early Oligocene saw their disappearance from the continent, and an absence ensued until the Miocene appearance of hominoids.

XIII. References

AXELROD, D. I. and BAILEY, H. P.: Paleotemperature anylysis of Tertiary floras. Palaeogeogr. Palaeoclimat. Palaeoecol. *6:* 163–195 (1969).

BALAZUC, J. et DESCARPENTRIES, A.: Sur *Lampra gautieri* Bruyant et quelques autres Buprestidae fossiles des schistes de Menat (Puy-de-Dôme). Bull. Soc. entomol. France *69:* 47–108 (1964).

BERG, D. E.: Krokodile als Klimazeugen. Geol. Rdsch. *54:* 328–333 (1964).

BLONDEAU, A.: Esquisse phylogénétique des Nummulites. C. R. Acad. Sci. *272:* 377–380 (1971).

BLOT, J.: Les poissons fossiles du Monte Bolca classés jusqu'ici dans les familles des Carangidae-Menidae-Ephippidae-Scatophagidae. Mem. Mus. civ. Storia nat. Verona; Mem. out of normal series No. 2/*I* (1969).

CAPPETTA, H.; HARTENBERGER, J. L.; SIGÉ, B. et SUDRE, J.: Une faune de vertébrés de la zone de Cuis dans l'Eocène continental du Bas-Languedoc (gisement du Mas de Gimel, Grabels, Hérault). Bull. Bur. Rech. géol. min., s^e sér. *3:* 46–48 (1968).

CAPPETTA, H.; RUSSELL, D. E. et BRAILLON, J.: Sur la découverte de Characidae dans l'Eocène inférieur français (Pisces, Cypriniformes). Bull. Mus. nat. Hist. nat. *51:* 37–52 (1972).

CASIER, E.: La faune ichthyologique du London Clay. Mem. brit. Mus. nat. Hist., 2 vol. (1966).

CASIER, E.: Le Landénien de Dormaal (Brabant) et sa faune ichthyologique. Mém. Inst. roy. Sci. nat. belge *156:* (1967).

CHANDLER, M. E. J.: The lower Tertiary floras of southern England. IV. A summary and survey of findings in the light of recent botanical observations. Mem. brit. Mus. nat. Hist. (1964).

CHARLES-DOMINIQUE, P. and MARTIN, R. D.: Evolution of lorises and lemurs. Nature, Lond. *227:* 257–260 (1970).

CURRY, D.; GULINCK, M. et POMEROL, C.: Le Paléocène et l'Eocène dans les bassins de Paris, de Belgique et d'Angleterre. Mém. Bur. Rech. géol. min. *69:* 361–369 (1969).

DALEY, B.: Some problems concerning the early Tertiary climate of southern Britain. Palaeogeogr. Palaeoclimat. Palaeoecol. *11:* 177–190 (1972).

DANGEARD, L.: Observations sur les schistes bitumineux de Menat et sur une roche à grains de pollen du Puy de Mur et de Gergovie. Bull. Soc. Hist. nat. Auvergne, No. 12 (1934).

DEVEREUX, I.: Oxygen isotope palaeotemperature measurements on New Zealand Tertiary fossils. New Zeald. J. Sci. *10:* 988–1011 (1967).

ESTES, R.; HECHT, M., and HOFFSTETTER, R.: Paleocene amphibians from Cernay, France. Amer. Mus. Novit. *2295* (1967).

FRANZEN, H. L.: Revision der Gattung *Palaeotherium*, Perissodactyla, Mammalia, 2 vol. (Albert-Ludwigs-Universität, Freiburg 1968).

GIDAI, L.: Les relations stratigraphiques de la région nord-est de la Transdanubie. Ann. Inst. geol. publ. hung. *54:* 361–369 (1971).

GINGERICH, P. D.: Dental function of the Paleocene primate *Plesiadapis* (in press).

GINSBURG, L.; MENNESSIER, G. et RUSSELL, D. E.: Sur l'âge éocène inférieur des sables bleutés du Haut-Var et sur ses conséquences. C. R. Soc. géol. France *7:* 272–274 (1967).

GODFRIAUX, I. and THALER, L.: Note sur la découverte de dents de mammifères dans le Montien continental du Hainaut (Belgique). Bull. Acad. roy. Belg., 5 sér. *58:* 536–541.

GRUAS-CAVAGNETTO, C.: Etude playnologique des divers gisements du Sparnacien du bassin de Paris. Mém. Soc. géol. France *47* (1968).

GRUAS-CAVAGNETTO, C.: Aperçu sur la microflore et le microplancton du Paléogène anglais. C. R. Soc. géol. France *1:* 19–21 (1970).

GUTH, C.: Un insectivore de Menat. Ann. Paléont. *48:* 3–10 (1962).

HARTENBERGER, J. L.: Les rongeurs de l'Eocène d'Europe: leur évolution dans leur cadre biogéographique. Bull. Mus. nat. Hist. nat. *132:* 49–70 (1973).

HECHT, M. et HOFFSTETTER, R.: Note préliminaire sur les amphibiens et les squamates du Landénien supérieur et du Tongrien de Belgique. Inst. roy. Sci. nat. belge *38* (1962).

HÜRZELER, J.: Zur Stammesgeschichte der Necrolemuriden. Schweiz. palaeont. Abh. *66:* 1–46 (1948).

JENKINS, D. G.: Variations in the number of species and subspecies of planktonic Foraminiferida as an indicator of New Zealand Cenozoic paleotemperatures. Palaeogeogr. Palaeoclimat. Palaeoecol. *5:* 309–313 (1968).

KEDVES, M.: Etudes palynologiques dans le bassin de Dorog. I. Pollen Spores *2:* 89–118 (1960).

KEDVES, M.: Quelques types de Sporomorphes du bassin lignitifère de Menat. Acta. biol. *13:* 11–23 (1967).

KOPEK, G.; DUDICH, E., jr. et KECSKEMÉTI, T.: L'Eocène de la montagne du Bakony. Ann. Inst. geol. publ. hung. *54:* 201–232 (1971).

KRUTZSCH, W.: Der Florenwechsel im Alttertiär Mitteleuropas auf Grund von sporenpaläontologischen Untersuchungen. Abh. zentr. geol. Inst. *10:* 17–37 (1967).

KRYSHTOFOVICH, A. N.: Paleobotanika (Leningrad 1957).

LOUIS, P.: Note sur un nouveau gisement situé à Condé-en-Brie (Aisne) et renfermant des restes de mammifères de l'Eocène inférieur. Ann. Univ. ARERS *4:* 108–118 (1966).

LOUIS, P.: Les environs de Reims à l'Eocène inférieur. Soc. Et. sci. nat. Reims *136:* 3–11 (1969).

Louis, P.: Note préliminaire sur un gisement de mammifères de l'Eocène inférieur situé route de Broyes à Sézanne (Marne). Ann. Univ. ARERS *8:* 48–62 (1970).

Marlière, R.: Le Montien de Mons: état de la question. Mém. Bur. Rech. géol. min. *28:* 875–884 (1964).

McKenna, M. C.: Classification, range, and deployment of the prosimian primates. Colloques int. Cent. nat. Rech. sci. *163:* 603–610 (1967).

Mészáros, N. et Dudich, E., jr.: Esquisse comparative de la parallélisation stratigraphique et de l'évolution paléographique de l'Éocène de l'Europe centrale et sudorientale. Acta geol. hung. *10:* 203–231 (1966).

Montford, H. M.: The terrestrial environment during the upper Cretaceous and Tertiary times. Proc. geol. Ass. *81:* 181–204 (1970).

Penny, J. S.: Late Cretaceous and early Tertiary palynology; in Tschudy and Scott Aspects of palynology, pp. 331–376 (1969).

Piton, L. E.: Paléontologie du gisement éocène de Menat (Puy-de-Dôme) (flore et faune). Mém. Soc. Hist. nat. Auvergne *1:* 103 (1940).

Pomerol, C.: Esquisse paléogéographique du Bassin de Paris à l'ère tertiaire et aux temps quaternaires. Rev. Géogr. phys. Géol. dyn. *9:* 55–86 (1967).

Quinet, G. E.: *Teilhardina belgica*, ancêtre des Anthropoïdea de l'Ancien Monde. Bull. Inst. roy. Sci. nat. belge *42:* 14 (1966a).

Quinet, G. E.: Les mammifères du Landénien continental belge. Etude de la morphologie dentaire comparée des 'carnivores' de Dormaal. Mém. Inst. roy. Sci. nat. *158:* 64 (1966b).

Quinet, G. E.: Apport de l'étude de la faune mammalienne de Dormaal à la stratigraphie générale du Paléocène supérieur européen et à la théorie synthétique de la molaire mammalienne. Mém. Inst. roy. Sci. nat. *162:* 188 (1969).

Radinsky, L.: The early evolution of the Perissodactyla. Evolution *23:* 308–328 (1969).

Rat, P.: La succession stratigraphique des mammifères dans l'Eocène du bassin de Paris. Bull. Soc. géol. France *7:* 248–256.

Roche, E.: Etude playnologique de sédiments du Montien continental et du Landénien supérieur en Hainaut. Bull. Soc. belge, Géol. Paléont. Hydrol. *78:* 131–145 (1969).

Roche, E.: Flores du Paléocène et de l'Eocène inférieur des bassins sédimentaires anglais, belge et parisien. Intérêts climatique et phytogéographique, pp. 109–134 (Ass. nat Prof. Biol. Belg., 1970).

Russell, D. E.: Les Mammifères paléocènes d'Europe. Mém. Mus. nat. Hist. Paris *13* (1964).

Russell, D. E.: Sur *Menatotherium* et l'âge paléocène du gisement de Menat (Puy-de-Dome). Colloques int. Cent. nat. Rech. Sci. *163:* 483–490 (1967).

Russell, D. E.: Succession, en Europe, des faunes mammaliennes du début du Tertiaire. Mém. Bur. Rech. géol. min. *58:* 291–296 (1968).

Simons, E. L.: Notes on Eocene tarsioids and a revision of some Necrolemurinae. Bull. brit. Mus. nat. Hist. Geol. *5:* 45–69 (1961).

Simons, E. L.: A critical reappraisal of Tertiary primates; in Buettner-Janusch, pp. 65–129 (1963).

Simons, E. L.: Fossil primates and the evolution of some primate locomotor systems. Amer. J. phys. Anthrop. *26:* 241–254 (1967).

SITTLER, C. et MILLOT, C.: Les climats du Paléogène français reconstitués par les argiles néoformées et les microflores. Geol. Rdsch. *54:* 333–343 (1964).

SLOAN, R. E.: Cretaceous and Paleocene terrestrial communities of western North America. Proc. N. Amer. Paleont. Conv., pp. 427–453 (1969).

STEINBERG, M.: Sédimentologie des formations continentales de l'Eocène. Mém. Bur. Rech. géol. min. *69:* 353–357 (1969).

SZALAY, F. S. and MCKENNA, M. C.: Beginning of the age of mammals in Asia: the late Paleocene Gashato fauna, Mongolia. Bull. amer. Mus. nat. Hist. *144:* 269–319 (1971).

TEILHARD DE CHARDIN, P.: Les mammifères de l'Eocène inférieur français. Ann. Paléont. *11:* 9–116 (1922).

TEILHARD DE CHARDIN, P.: Les mammifères de l'Eocène inférieur de la Belgique. Mém. Inst. roy. Hist. nat. belge. *36:* 33 (1927).

VALEN, L. VAN and SLOAN, R. E.: The extinction of the multituberculates. Syst. Zool. *15:* 261–278 (1966).

VINOGRADOV, A. P.: Atlas of the lithological-paleogeographical maps of the USSR, vol. 4 (Ministry of Geology, Acad. Sci. USSR, Moscow 1967).

WEIGELT, J.: Die Aufdeckung der bisher ältesten tertiären Säugetierfauna Deutschlands. Nova Acta Leopoldina *7:* 1–13 (1939).

WEITZMAN, S. H.: The systematic position of Piton's presumed characid fishes from the Eocene of central France. Stanford ichthyol. Bull. *7:* 114–123 (1960).

WOLFE, J. A.: Tertiary climatic fluctuations and methods of analysis of Tertiary floras. Palaeogeogr. Palaeoclimat. Palaeoecol. *9:* 27–57 (1971).

Author's address: Dr. DONALD E. RUSSELL, Muséum National d'Histoire Naturelle, Institut de Paléontologie, 8, rue de Buffon, F-*75005 Paris* (France)

In SZALAY: Approaches to Primate Paleobiology
Contrib. Primat., vol. 5, pp. 62–103 (Karger, Basel 1975)

Palaeoenvironments in the East African Miocene

PETER ANDREWS and JUDITH A. H. VAN COUVERING

National Museum, Nairobi, and University of Colorado Museum, Boulder, Colo.

Contents

I. Introduction

In this paper, we will attempt to describe the Early Miocene environments of the East African primates. We use the word 'environment' in the sense of Webster: 'the aggregate of all the external conditions and influences affecting the life and development of an organism.' Specifically, for our purposes, we will be considering physiography and geography, climate, vegetation, and faunal associations. From this we will attempt to deduce on theoretical grounds, backed up where possible by geological and palaeontological evidence, what the situation in the Early Miocene might have been. The distributions of fossil African pongids will be briefly reviewed against this background.

Other workers such as MacINNES [1953], WHITWORTH [1953, 1958], CHESTERS [1957], BISHOP [1958, 1963, 1968], VERDCOURT [1963], WALKER [1969] and GENTRY, [1970] have discussed palaeoenvironments of the East African Miocene. Many of these authors have stressed the existence of a variety of environments, especially at Rusinga. LEAKEY, MacINNES and WHITWORTH have emphasized the dominance of savanna, or even arid [WHITWORTH, 1958] habitats while BISHOP, VERDCOURT and WALKER have pointed out that there is fossil evidence which demands some forested environment. CHESTERS, apparently in an attempt to rationalize the presence of fossil forest trees with the idea that the dominant habitat was savanna, suggested that the countryside was open with forest present along the water courses. It should be pointed out that BISHOP [see especially 1963, 1968] has done much to stimulate work along these lines in East Africa by his work at Napak and his discussions of fossil versus living environments.

We would like to specify that we use the terms Miocene and Pliocene to denote those periods of time from 22.5 to 5 Ma[1] and from 5 to 1.8 Ma as suggested by VAN COUVERING and MILLER [1971] and BERGGREN [1972]. In this calibration, Middle Miocene starts at 16 Ma and Late Miocene at 10.5 Ma. This is not what has ordinarily been meant by these terms, but we feel that our usage lies closer to the original meaning and also to the growing consensus among nonvertebrate palaeontologists [cf. BERGGREN, 1969, 1972; VAN COUVERING and MILLER, 1971, for detailed arguments].

II. Present Geography

Today the African continent, although physically joined to Eurasia through the Near East, is ecologically isolated south of the Sahara desert. Africa is characterized by extremes in altitude: from the low-lying basins of the Congo, Chad and Niger and the Sahara, which comprise 'Low Africa' to the extensive highlands of East Africa, where thousands of square kilometers lie at altitudes from 2,500 to 3,000 m. This and most of the southeastern part of the continent, which lies at an altitude greater than 1,000 m, comprise 'High Africa' [KING, 1967]. Eastern Africa is broken by great rift

1   Ma = Latin for million years.

valleys which extend 5,000 km from the northern margin of the continent at the Red Sea to the southern end of Lake Malawi and the Zambesi Lowlands.

Drainage is primarily to the sea, but there are large regions of the Sahara Desert, the Chad, Kalahari and Okovago basins, and the Eastern Rift Valley which have interior drainage. Short coastal rivers are important in draining the continental margins, but a large part of the interior of the continent is drained by a few large rivers. Major rivers draining to the Atlantic are: (1) the Niger, which has captured the former interior drainage of the Niger basin and has thus incorporated much of western Africa in its watershed; (2) the Congo, which drains almost all of Central Africa west of the Western Rift, including Lake Tanganyika, and (3) the Orange River, which claims most of southern Africa as its watershed. Feeding into the Indian Ocean are the Limpopo River, which drains the eastern part of southern Africa, and the Zambesi, which drains much of south-central Africa, including Lake Malawi. The Nile drains east-central Africa through a pre-Pliocene channel across the Sahara [Hsü, 1972].

Equatorial Africa, the area with which we are primarily concerned, is dominated in the west by the low-lying Congo basin. To the east, the country is broken by rifting. Between the Western and Eastern Rift Valleys is a wide shallow sag partly occupied by Lake Victoria. The Eastern Rift extends northeastward into Ethiopia and the Great Afar Depression where it meets the Red Sea and Gulf of Aden rift systems. Southwards, it extends into Tanzania where it changes into a block-faulted region.

A variety of mountains and highlands also break up the surface of eastern Africa. The Pleistocene fault block of the Mountains of the Moon (Ruwenzori) rises to a height of over 4,000 m between Lakes Albert and Edward. To the south, the Pleistocene and Recent Bufumbira (or Virunga) volcanic highland separates Lakes Edward and Kivu. Ancient volcanoes, active in the Miocene, stand in various stages of dissection along the Kenya Uganda border, while more recent volcanoes (Kenya, Kilimanjaro, Ngoro-ngoro) rise along the Eastern Rift forming some of East Africa's most spectacular scenery. The most important and extensive highlands, however, are those uplifted by Plio-Pleistocene rift movements. In addition, extensive areas are underlain by Late Miocene plateau volcanics. These highlands rise to over 4,000 m in Ethiopia and extend through Kenya southward into Tanzania.

As the landscape is diverse so also are the drainage, climate and vegetation. West Central Africa is drained by the Congo and its tributaries, which reach as far as 12° S. To the northeast, its upper reaches interfinger with the westernmost tributaries of the White Nile northwest of Lake Albert. The Nile is the main river draining east Central Africa: the Blue Nile drains the Ethiopian highlands and the White Nile receives water from Lakes Victoria, Kyoga, Edward, George and Albert and their tributary rivers. The Eastern Rift contains a series of interior basins, many with saline, alkaline playas. To the east of the Eastern Rift small, but ancient rivers flow to the Indian Ocean. We will explore the history of this drainage below.

III. Palaeogeography

The Tertiary palaeogeography of Africa has been reviewed in depth by MOREAU [1952, 1966] in an attempt to explain the distribution of recent birds, and also by COOKE [1958, 1968] and DE HEINZLEIN [1963]. We agree with most of what these authors have proposed, but will to a small extent bring their views up to date. Africa has essentially had its present shape since the end of the Mesozoic [KING, 1967]. However, several major events have taken place since that time: (1) in the Early Cenozoic, its connection with the northern hemisphere was broken [VAN COUVERING, 1972; WALKER, 1972]; (2) in the Oligocene, the sea withdrew from the Somali coastlands [MACFADYEN, 1952]; (3) in the Early Miocene, Arabia was separated from the main continental mass at the Red Sea Rift [ABDEL-GAWAD, 1969]; (4) between 18 and 14 Ma, the African-Arabian plate rejoined the northern, firstly in the region of the Zagros and Caucasus mountains [DEWEY and BIRD, 1970] and shortly afterwards in the Gibraltar region [BERGGREN and PHILLIPS, 1971], and (5) about 6 Ma, the Mediterranean became an isolated basin [HSÜ, 1972]. These events are important to our discussion in two ways. First of all, the geographical relationship of Africa with the rest of the world helped to determine the nature of its flora and fauna; and secondly, the changes in geographical relationships were probably responsible for changes in oceanic circulation which may have affected the climate.

Africa, in the Early Tertiary, was a continent of low swell and shallow basin such as 'low Africa' is today. The divides were low and many of the basins had interior drainage [DE HENZLEIN, 1963, fig. 7]. Near some of the divides, surfaces of higher elevation formed during earlier erosion cycles were preserved as plateaus, but nowhere was the relief very great. These erosion remnants and a few basement hills probably produced what little variety there was to the landscape. In the Oligocene, this began to change and by Pleistocene times the topography was, to a great extent, as it is now.

During the Early Tertiary, the African land surface in the equatorial belt rose gently from the Atlantic coast in the west to a continental watershed along the line of today's Kenya Rift Valley and down again over the much shorter distance to the Indian Ocean on the east [BISHOP and TRENDALL, 1967]. The divide was very low, much like the Congo-Zambesi divide today. An extensive low-relief erosion surface was formed (Buganda Peneplain and Kitale Surface) which is now characterized by a deep, and in places lateritic, residual soil. To the east, remnants of an earlier erosion surface stood at an elevation of at least 1,500 m along the watershed, preserved today as the Cherangani and southern Chemorongi Hills north of Lake Baringo.

In the Oligocene, small basement domes began to rise above the surrounding countryside along a NNE-SSW line in what is today eastern Uganda, due to the emplacement of alkaline volcanic rocks beneath them [KING and CHAPMAN, 1972]. To the west, the Kitale surface was slowly being destroyed by the headward erosion of westward flowing streams graded to a new level (Kyoga Surface). The limits of this headward erosion can be seen today in an ancient escarpment exposed by erosion of the volcanic cover between Moroto and Elgon [BISHOP and TRENDALL, 1967].

In the Early Miocene, large basement domes, later to become great volcanoes, had risen up to 1,000 m above the surrounding countryside along the line of the small Oligocene domes. The volcanoes are preserved today in various stages of dissection at Yelele, Moroto, Napak, Kadam, Elgon, Tinderet and Kisingiri. Basins of sedimentation, bounded by small fault escarpments, were formed by early rifting in the Albertine Rift and the northern part of the Kenya Rift (e.g. Kabuga, Karugamania, and Lower Nyamavi Beds [HOOIJER, 1970] of the Albertine Rift and 'Turkana Grits' of the Kenya Rift). In the Kenya Rift, this early faulting was followed by vulcanism (represented by so-called Tvb_1 strata) which has continued in different forms up to the present day [KING and CHAPMAN, 1972; BAKER and WOHLENBERG, 1971].

On the ancient 'western slope' of the East African watershed, the drainage was not disrupted to any great extent by these Late Oligocene and Early Miocene changes. Early rifting in the Kavirondo Valley of Western Kenya enhanced the earlier trend locally and a major east-west valley was formed with the Tinderet dome at its head (fig. 1). Songhor hill and the Nyando escarpment rose along the north side of this valley, and from the southwest, the growing Kisingiri dome extended into the valley [SHACKLETON, 1951; BISHOP and TRENDALL, 1967].

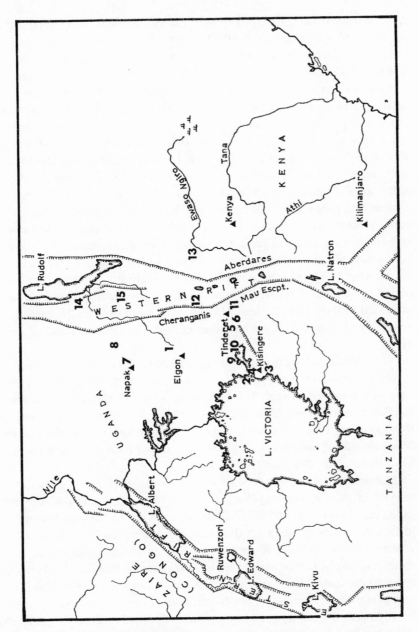

Fig. 1. Locality map. Fossil localities: 1 Bukwa, 2 Mfwangano Island, 3 Karungu,
4 Rusinga Island, 5 Songhor, 6 Koru, 7 Napak, 8 Moroto, 9 Ombo, 10 Maboko Island,
11 Fort Ternan, 12 Ngorora, 13 Kirimun, 14 Losidok and Moruarot, 15 Loperot.

The local palaeogeography of the area can be depicted in some detail. According to VAN COUVERING [1973], the Kisingiri Dome, which rose as an isolated mountain, some 1,000 m above its surroundings, was a center of explosive volcanic activity for some 2 Ma before it developed a lava and agglomerate stratovolcanic envelope. During this earlier eruptive activity, it ejected extremely alkaline particles of alnoitic lava, which were incorporated into the fossiliferous formations of Rusinga, Mfwangano, and Karungu. These beds accumulated in subsiding troughs north and south of the up-lifted dome, and were separated from it by escarpments along the Mfwangano and Kaniamwia fault lines. Inasmuch as the Kisingiri dome stood more or less on the flank of the Miocene Kavirondo Valley, the structurally depressed areas may have intercepted the Kavirondo drainage. The ground water in these depositional basins was probably highly alkaline and unsuitable for plant growth, as indicated by pervasive zeolitization of tuffaceous beds, clay-pellet dune deposits, lack of fossil soil horizons or root casts, and the presence of other sedimentary features indicative of barren ground. This evidence, which at first seems to point to an arid environment, seems better explained as the result of chemical breakdown of the unusually reactive and extremely alkaline ash falls where they collected in areas of impeded drainage.

The history of Tinderet is less well known. Zeolitized tuffaceous sediments like those on Rusinga have been collected at Songhor, and the existence of a marginal trough is indicated by a thick formation of tuffaceous conglomerates beneath Tinderet lavas from Muhoroni to Legetet Hill [SHACKLETON, 1951]. The fossil beds at Songhor appear, however, to have accumulated subaerially, probably on low slopes banked against the basement inselberg of Songhor Hill.

Little is as yet known of the effects on regional drainage of early tectonic activity in the Eastern and Western Rifts. In the Eastern Rift, sediments called the 'Turkana Grits' occur along the base of the Turkana Escarpment in isolated patches resting on the basement and covered by the first local volcanic eruptions (fig. 1). K-Ar dating and recent mapping suggest that these sediments may not all be of the same age [WALSH and DODSON, 1969; BISHOP, personal commun., 1973], and a detailed study of their relationships is only now being made. Although a north-south drainage route has been suggested [B. PATTERSON, personal commun.], three lines of evidence suggest that at least some of these deposits were part of an eastward-draining system. Firstly, there are Miocene marine sediments exposed along the Kenya coast [HAUGHTON, 1963], and it is possible that these occur in the subsurface as far west as 39° E latitude where the sub-Miocene and end-Tertiary erosion surfaces meet [SAGGERSON and BAKER, 1965]. Although covered today by Quaternary volcanics and alluvium, the area is still low-lying. Secondly, a fossil xiphiid whale, whose presence demands some fairly close access to the sea, has been found in the 'Turkana Grits' at Loperot [V. MAGLIO, personal commun.] which lies today near the southwest corner of Lake Rudolf (fig. 1) Thirdly, five fish tooth morphotypes, three of which have not been found yet in other sites, occur both at Loperot, in limestones within the first basalts, and at Kirimun (200 km south and 100 km east of Loperot, fig. 1). This suggests that these two areas may have had some drainage connections[1]. A river system draining east from Loperot and Kirimun to the Early Miocene Indian Ocean which met in a common floodplain or estuary could account for these facts.

Early Miocene faulting and sedimentation in the southern part of the Albertine Rift suggest that the drainage connection between East Africa and the Congo was locally interrupted at this time [GAUTIER, 1967]. There is some evidence, however, that the ancient trans-rift drainage pattern was maintained north and/or south of the down-faulted area. Fossil fish in Lower Miocene rocks at Bukwa and Rusinga are closely related to *Pelmatochromis*, a cichlid restricted today to Africa west of the Western Rift, and have not yet been found in sediments of the Eastern Rift. Furthermore, mollusc faunas from the Mohari Beds of the Albertine Rift and from Rusinga Island in the Kavirondo Rift are closely similar [GAUTIER, 1967]. These relationships suggest that in the Early Miocene there were, or had recently been, fresh-water connections between East Africa and the Congo Basin and that the ecological conditions in both areas may have been similar.

To the east of the watershed, gentle downwarping of the coastal region seems to have continued from the Early Miocene throughout most of the Neogene [SAGGERSON and BAKER, 1965], and the river systems draining the 'coastal slope' have probably remained unchanged since the early doming over the Eastern Rift. The occurrence of endemic species of the cichlid fish *Tilapia* in many of these streams [cf. TREWAVAS, 1966] adds further weight to this interpretation.

The landscape in East Africa changed fairly rapidly towards its present configuration after the Early Miocene. Several crucial events took place around 16 Ma: (1) Africa rejoined Eurasia and consequently a great faunal

1   It should be noted that the sediments at Loperot are fairly securely dated at ca. 17 m.y. [BAKER *et al.*, 1971] while those at Kirimun are only known to be older than ca. 12 m.y. [BAKER *et al.*, 1971], a date on the overlying basalt.

change took place (compare Rusinga and Fort Ternan faunas); (2) the first plateau volcanics were erupted, eventually culminating in local thicknesses of as much as 1,600 m [KING and CHAPMAN, 1972], and (3) interior basins were developed, containing saline, alkaline lakes within the Eastern Rift [J. MARTYN, written commun., 1969]. The stage was set for the final development of the landscape. Major rifting and sedimentation continued in both the Eastern and Western Rifts (although with a difference in type and periodicity [cf. BISHOP, 1971], and by the middle Pleistocene 1,700 m of uplift had taken place in some areas [BAKER and WOHLENBERG, 1971]. Coincident with the last major uplift and rifting was the development of massive central volcanoes and volcanic fields on the East shoulder and floor of the Eastern Rift Valleys. The final uplift of the rift shoulders, including the Ruwenzoris, severely impeded or reversed westward drainage, so that Lake Kyoga and, more importantly, Lake Victoria were formed in the middle Pleistocene [BISHOP, 1965].

IV. Present Climates

The rainfall density pattern on the African continent west of approximately 30° E longitude is distributed more or less symmetrically around the equator. It is shifted to the south during the November-April season and to the north from May to October. The heat budget remains more or less the same year-round in equatorial Africa, although in the west it is on the average 5 °C hotter in January than in July when temperatures average from 20 to 25 °C. East African temperatures do not have a wide average annual variation but vary a great deal from locality to locality due to the extremes of altitude and rainfall. Both North and South Africa are in the higher latitudes and experience the summer and winter temperature variations of semitropical and temperate climates.

Two basic meteorological facts should be kept in mind when considering rainfall today, or in the past. First of all, 'rainfall is heaviest where moisture bearing winds from the sea are forced up over the first mountain ranges or have vertical convection currents induced in them over the first strongly heated ground' [LAMB, 1961, p. 31]. Secondly, land in the leeward side or the rainshadow of mountain ranges is usually dry.

The moisture precipitated in equatorial West Africa and the Congo Basin is brought to it by warm, prevailing winds of the south Atlantic, enhanced by the Guinea Current. This, and the convergence of winds along the equator, leads to the long-standing low pressure systems of the equatorial trough resulting in the year-round distribution of rain and cloud, a positive water balance, and evergreen forest in the equatorial belt [GROVE, 1970]. The moisture-laden air of the equatorial belt penetrates inland to the western shoulder of the Western Rift Valley. These highlands capture the moisture remaining after the passage of the air mass over the Congo Basin. In East Africa, the northeast monsoon in the November-April season and the year-round southeast Trade Winds bring moisture off the Indian Ocean. This moisture precipitates predominately at the coast and over the highlands bordering the Eastern Rift. Thus, both of the rift valleys are in a permanent rainshadow and the inter-rift area is in a double rainshadow. This pattern is complicated by the permanent barometric depression over Lake Victoria which enhances year-round rainfall and supports evergreen forest around the northern margin of the Lake in what would probably otherwise be a drier region.

These general considerations will be taken into account when trying to estimate the Tertiary climate pattern of Africa. Present-day climates are exceedingly complex, however, and it is difficult to pinpoint any one feature as representative of any particular climatic, and thus vegetation, type. For instance, the annual temperature *range* is more important than the annual *mean*, but this is variable from place to place, especially in East Africa, and even the daily range often exceeds the difference between the means of the hottest and coldest months in equatorial Africa. It is these extremes that determine the vegetation more than the annual mean. Similarly the *distribution* of rainfall throughout the year affects vegetation most profoundly. Evaporation exceeds precipitation for most of the year throughout much of Africa today because the rain is concentrated in discrete seasons. This is true particularly in the rainshadow areas of East Africa. These complexities increase the difficulty of accurately describing past climates and determining their effect on vegetation.

V. Past Climates

A description of palaeoclimates, arrived at theoretically, can only be useful in the broadest sense. A true idea of past climates can only be arrived at by a synthesis of descriptive data from many sites [NAIRN, 1961]. Although this is not yet possible for the African Miocene, we will summarize what is so far known about actual sites (see table III). LAMB [1961] points out that 'the general circulation of the atmosphere and oceans is the mechanism of climate'. We will discuss past differences in this climatic mechanism with the help of recent work on Early Tertiary climates [FRAKES and KEMP, 1972].

If we examine FRAKES and KEMP's [1972] reconstruction of the Eocene and mid-Oligocene worlds, we see that in the Eocene Africa lay about 5° south of its present position and that the circulation between the northern and southern Atlantic was more restricted. This more southerly position might explain the occurrence in the Congo Basin of Early Cenozoic aeolian sands and siliceous duricrusts preserved in the extensive 'grés polymorphes' [cf. CAHEN, 1954]. On the other hand, FRAKES and KEMP [1972, fig. 1] concluded that the climate was wet in East Africa at this time due to the unobstructed circulation of warm waters from southeast Asia and ultimately from the Pacific. By the mid-Oligocene, this picture had changed: Africa was more or less in its present position relative to the equator and India was obstructing Pacific-derived circulation due to its position in the middle of the Indian Ocean. The Pacific current was very warm (34 °C), and, although diverted south of East Africa by the Indian continent, we suppose it would have flowed northwards through the 'Straits of Zinj' bringing rain to this region. In the season of the northeast monsoon, winds off the Tethys would probably have brought additional moisture to the area. On this reconstruction, therefore, the climate of eastern equatorial Africa would have been rather wetter in the Oligo-Miocene than at present, and the rain would have been moderately well-distributed throughout the year.

This pattern may have been interrupted locally by topographic changes. However, it has been shown that major uplift associated with rifting did not begin until Late Miocene, so that in the Early Miocene these interruptions would have been of much less significance than those of the present day. The physiographic position for the equatorial belt before the period of major uplift was one of low relief rising gently from the west coast to the continental divide more or less along the line of the present Eastern Rift, falling off from there to the east coast. This would result in an eastward extention of the high year-round rainfall zone that today characterizes the Congo Basin, to cover most of Uganda and western Kenya and Tanzania. Under the possible climatic conditions outlined above, the western equatorial rain forest, at present limited eastwards by the west rift highlands, may well have extended across the continent at least as far as the Miocene continental divide just west of the present Eastern Rift. The correlative effects of moister conditions coming off the Indian Ocean may have extended the high rainfall belt across the low continental divide and down to the east coast.

VI. Present Vegetation

As the density of rainfall is broadly distributed symmetrically about the equator to longitude 3° E, so also are the vegetation types. Equatorial lowland forest gives way northwards and southwards to moist woodland and savanna, thence to bushland and lastly to semidesert and desert belts. This pattern, like that of the rainfall, is interrupted at the wall of the Western Rift highlands. These broad categories of vegetation are shown in figure 2A, which attempts to depict in simplified terms the prehuman distribution of vegetation. Human actions which are considered to have caused important vegetational changes and which have been taken into consideration in this map are (1) forest clearance to produce forest savanna mosaics typical of the inter-rift area and the southern Congo; (2) woodland fires to produce savanna or grasslands; (3) woodland clearance to produce bushland; (4) savanna overgrazing to produce bushland, and (5) bushland fires to produce grassland or bushed grassland. These factors and others have so greatly altered the natural state of events that it is not possible to recover the whole picture. Figure 2 should, therefore, be considered as only an approximation.

A simplified and expanded version of GREENWAY's [1943] vegetation classification suitable for extension into the past will be fully outlined elsewhere [ANDREWS, 1973c] and is presented only briefly here in hopes that it will clarify use of terms for vegetational types. Five broad categories are specified: forest; woodland and savanna; bushland; semidesert; and edaphic grasslands. Each of these can be further subdivided into types of plant associations.

Forest is a continuous stand of trees, either evergreen or deciduous, with the crowns intermingling. The canopy is often complex and there is usually an herb layer on the forest floor; grasses are rare [GREENWAY, 1943]. Both the forest plant species and the habitats they form are distinctive and have little in common with those outside the forest. As a result, many of the animal species are also distinctive making it possible to identify a fauna with forest affinities from those living under nonforest conditions. Abundant year-round rainfall is required for forest growth in Africa, and as a result it flourishes in lowland conditions only in the equatorial zone. Edaphic forests occur locally outside the zone of year-round rainfall, and, although rare, are of zoogeographical interest because they act as faunal routes.

Woodland has a closed cover of trees, mostly deciduous. The crowns of the trees just touch, and there is usually a single canopy. Grasses and herbs

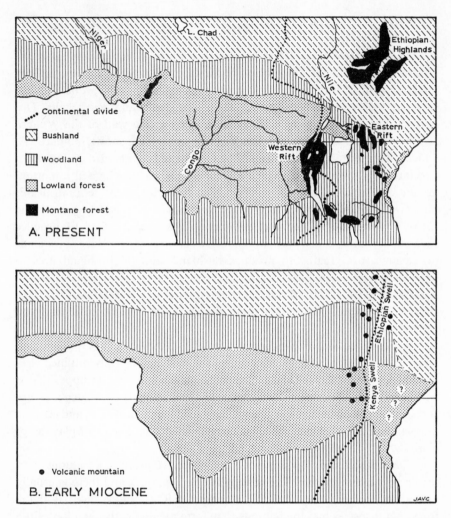

Fig. 2. Vegetation maps of equatorial Africa. *A* Present distribution pattern, much simplified and reconstructed to remove the effects of human influence as far as possible. *B* Probable Early Miocene vegetation distribution as discussed in the text.

constitute the dominant ground cover. *Savanna* is similar except that the trees are more widely spaced and shade less than 50% of the ground [GREENWAY, 1943]. It is widely held that savanna is derived from woodland as a result of overgrazing, fire and cultivation [AUBRÉVILLE, 1949; MOREAU, 1966; GLOVER, 1968; KINGDON, 1971; ANDREWS, 1973a].

Bushland has a thick cover of bushes, grass and small bushy trees. The ground has more than 50% cover of woody plants which are generally deciduous. Some large trees may be present, either in clumps or widely scattered [GREENWAY, 1943]. There are many bushland associations that appear to be of great antiquity, but this vegetational type can be seen today to be spreading at the expense of woodland and savanna, as a result of shifting cultivation, overgrazing, and overpopulation of elephants.

Semidesert consists of low bushes and stunted trees with a ground cover of succulent plants and a few scattered grasses. For most of the year, over 50% of the ground is devoid of plant cover [GREENWAY, 1943]. *Desert* itself has only ephemeral vegetation.

Grasslands do not form a part of the latitudinally zoned vegetation sequence in Africa but are edaphic in origin and local in distribution. The widespread areas of nonedaphic grassland in tropical Africa today are the result of burning [MICHELMORE, 1939; KINGDON, 1971].

The principal vegetation type of the low-lying Congo Basin is lowland evergreen forest and swamp forest. The forest-savanna mosaic of the southern Congo Basin is probably forest derived [AUBRÉVILLE, 1949]. This forest belt is broken along the line of the Western Rift; much of the forest in this area has now been cleared, but probably in the past the lowland forest merged directly into the montane forest of the west rift highlands. The inter-rift basin, in the rainshadow of the two areas of rift highlands, is covered by a number of associations of woodland and bushland [ANDREWS, 1973a], but the local climatic effects of Lake Victoria result in high rainfall, and therefore forest, on its northern shores. This forest, discontinuous now as a result of man's activities, was probably much more continuous in the past [LANGDALE-BROWN et al., 1964; KINGDON, 1971]. The Rift Valley floors are in even deeper rainshadow, and the woodland and bushland associations are of a correspondingly drier nature. There are, however, quite extensive tracts of edaphic woodland (e.g. on the shores of Lakes Naivasha and Nakuru) and sometimes even forest (e.g. the Lake Manyara forest) where ground water is available near the surface of these internal drainage basins.

The Rift Valley highlands have for the most part been covered by montane forest, although much of this has now been cleared for agriculture. Extensive glades of edaphic grassland in areas of impeded drainage have probably always been present, and above the tree line, which occurs at about 3,000 m, are zones of ericaceous and alpine vegetation. The isolated volcanic mountains have similar vegetation types.

East of the Eastern Rift, bushland and semidesert predominate. Along the east coast, there are the remains of what was probably at one time a continuous belt of lowland forest, in parts deciduous, and in parts evergreen [DALE, 1939; ANDREWS, GROVES and HORNE, in press].

VII. Past Vegetation

There is every reason to suppose that vegetation zones were symmetrically arranged around an equatorial belt of lowland evergreen forest in the past just as they are at present. Determining where these zones lay on the actual land surface and their latitudinal extent is complicated by two factors: (1) that the continents have moved relative to each other and the equator and (2) that changes in worldwide humidity or heat in the lower atmosphere have caused the equatorial zone to contract or expand with a resulting shift of the other vegetation zones. It is important to remember for our present discussion that these simple latitudinal zones extend today only up to the wall of the Western Rift.

Several vegetation types are thought to have had a long evolutionary history, relative to other types, by virtue of the endemism of the floral components and associated fauna. Three such types that together cover a large part of Africa are: (1) the equatorial forest of the Congo and Niger watershed regions [Tropical Rain Forest of RICHARDS, 1952]; (2) Miombo woodland [shown as 'moist woodland' on the AEFTAT vegetation map, KEAY, 1959], which covers a large block of land in East and Central Africa south of the equator, and (3) the bushland which covers large parts of Somalia and Kenya and southwest Africa [GREENWAY, 1943; MOREAU, 1952, 1966; VERDCOURT, 1969]. Vegetation types which seem to have a more recent, but still fairly ancient, origin are: (1) the montane forest of tropical Africa and (2) the coastal lowland evergreen forest along the shores of the Indian Ocean. We have little evidence which applies directly to the development of the three more ancient types of vegetation, but a moderately reasonable time of origin can be assigned to some of the more recent types.

The origin of African montane forests is related to the development of equatorial highlands. During the formation of the sub-Miocene surface, the only known relief in equatorial Africa was provided by the Kenya and Ethiopian swells, and an older surface which today is represented in the Cherangani and Chemerongi Hills. These hills rise about 3,200 m above sea level. The total uplift in the Cherangani area since the development of the sub-Miocene surface is estimated at a maximum of 1,700 m [BAKER and

WOHLENBERG, 1971]. When this is subtracted, the pre-uplift elevation of the remnant surface becomes 1,500 m. This is about the lowermost limit of montane forest development today [MOREAU, 1966], and it is, therefore, difficult to decide whether such features supported montane forest in the Early Cenozoic landscape.

The great stratovolcanoes which formed in the Early Miocene may have provided the earliest montane forest habitat on their upper slopes. Although the Kisingiri subvolcanic dome is known to have reached 1,000 m above the surrounding countryside [McCALL, 1958], it is not yet known from what altitude it rose or what elevation the later volcanic peak attained. Even less is known for Tinderet, but an indirect line of evidence is provided by gastropods which live today only in montane forest and which are closely related to species from the Miocene of the Tinderet sites: Songhor and Koru (see table II).

The East Coast lowland forest is a relic flora, and its antiquity can be estimated from the relative numbers of species and genera common to it and the western Congo Forest. This has been done by FADEN [in HAMILTON, 1973], who calculated that there are 13 plant species common to the east and west coast forests. This number is larger (up to 39) if the most easterly of the western forests (in Uganda and western Kenya) are included in the western group. He also lists two examples of species represented by distinct subspecies in the east and west, and six examples of genera with approximately equal numbers of species in the east and west. There are also genera with a predominantly western distribution having one or two species present in the east coast forests. This impressive number of disjunctions in plant distribution leads very plausibly to the conclusion that the East Coast forests were once continuous with and derived from the western forests. There must have been at least two periods of connection, one recent to account for the species in common and one more ancient to account for the shared genera.

A similar conclusion is reached by MOREAU [1966] on the basis of the low endemicity of the East Coast forest bird fauna. In addition, several small mammal species of predominantly western genera are also known along the east coast [DORST and DANDELOT, 1970]. On the other hand, the butterfly fauna of the eastern forests has a very high level of endemicity [CARCASSON, 1964] which suggests that the connections between the two regions may have been of a somewhat tenuous nature. However, butterfly generation time differs from that of the plants, birds, and mammals, and perhaps these groups are not strictly comparable.

The first appearance of grasses in the fossil record of Africa comes from the Eocene of Nigeria in which grass pollen occurs in relatively low densities [GERMERAAD *et al.*, 1968]. Grass remains are known from the Cretaceous in North America but are not abundant anywhere in the world until the Miocene. Although much of the grassland seen today in Africa is due to the influence of man, the record from the Mio-Pliocene of the Caribbean shows that they can occur in abundance naturally in the tropics [GERMERAAD *et al.*, 1968].

It has been seen to be probable that in the Early Miocene the absence of the Western Rift highlands and wetter conditions in the east would have promoted climatic conditions favoring lowland forest, at least to the west of the ancient pre-rift divide, and probably for short periods at least, to the east of it as well. All of the inter-rift Early Miocene fossil localities would, therefore, have been within the forest zone (fig. 1), which would probably have been mainly lowland in character, although, as the general elevation of this region in the Early Miocene is not known, this cannot be certain. Evidence has been presented earlier that suggests that at least two of the Miocene localities (Songhor and Koru) may have been at or near the montane vegetation limit.

An examination of the fossil flora makes it apparent that forest and woodland were widespread in East Africa during the Early Miocene. CHANEY [1933] described a fossil flora from the Early Miocene strata of Mt. Elgon which he thought was indicative of at least local woodland growth. The Bukwa flora, from basal layers of the Elgon volcanic sequence, is said to indicate forest conditions at or near the locality [HAMILTON, in WALKER, 1969].

The largest Early Miocene flora in East Africa is from Rusinga and Mfwangano Islands (see table 1) and consists in large part of seeds, fruit, and nuts, described by CHESTERS [1957]. The abundance of material attributed to climbing plants was considered by CHESTERS to indicate a gallery or riverine forest source. The data are summarized in table I with the probable habit of the most common taxa listed [CHESTERS, 1957; ANDREWS, 1973c]. From this it can be concluded that the proportion and relative abundance of climbers to forest tree remains is not sufficient to support this conclusion; the abundance of the mahogany *Entandrophragma* [CHESTERS, 1957; p. 32], suggests the presence of extensive tracts of evergreen forest; and the occurrence of *Celtis* makes it likely that the forests were, in part at least, lowland in character.

Table I. Rusinga and Mfwangano vegetation

Family	Genus and species	Habit
Anacardiaceae	*Antrocaryon*	forest trees of upper canopy
	Odina	bushland or woodland
Anonaceae	*Anonaspermum*	Family of tropical trees and shrubs
	5 spp.	
Apocynaceae	*Leakeyia*	trees with seeds adapted for water dispersal
Burseraceae	*Canarium?*	Upper canopy forest tree
Connaraceae	*Cnestis*	may be climbers
Cucurbitaceae	*Lagenaria*	may be climbers
	2 other species	
Euphorbiaceae	4 species	common, but of unknown affinities
Leguminosae	3 species, one close to *Pterocarpus*	large woodland/forest trees
Meliaceae	*Entandrophragma palaeocarpus*	a genus of upper canopy forest trees, close to *E. utile*, a large forest tree
Menispermaceae	*Cissampelos*	lianas very common in this family
	Stephania	
	Syntrisepalum	
	Triclisia	
Oleaceae	*Schreberoidea*	affinities w/*Schrebera* and montane, evergreen forest
Rhamnaceae	*Berchemia*	bushland, often riverine; only species in Africa today confined to East and South Africa
	Ziziphus	forest to bushland
Sapindaceae	*Sapindospermum*	may be climbers
	4 species	
Ulmaceae	*Celtis*	middle canopy forest trees in East African lowland forest today

VIII. Present Faunas

By far the most fundamental faunal difference is between forest and nonforest faunas. Out of a total of 362 species of East African mammals, no less than 92 live exclusively in forest [KINGDON, 1971]. This makes it possible to recognize a faunal assemblage as being representative of forest, but the converse is not necessarily true: absence of forest-specific mammals does not necessarily imply the absence of forest conditions. Moreover, the differences in faunal assemblage between nonforest vegetation types are much less significant and there is wide overlap between them.

A number of groups of animals occupying forest habitats and their usefulness as indicators of habitat will be commented on briefly. One of the most characteristically forest-living groups of mammals of the present day are the higher primates, the pongids and hylobatids in particular. Both families are represented in the African Miocene [LeGROS CLARK and LEA-KEY, 1951; ANDREWS, 1973b]. However, the Pongidae were represented by at least six species, some of them apparently very closely related, and this suggests that they were undergoing rapid adaptive radiation. Conclusions about their environmental significance are, therefore, probably not warranted, except insofar as to say that, being forest animals now, they must have been so for an unknown period of time in the past. Of the other primates relevant to this discussion, prosimians in tropical Africa are forest-living with the exception of one species of galagine which lives in woodland and bushland as well as in forest [KINGDON, 1971]. The lorisines today are exclusively forest-living [WALKER, personal commun.]. They are insectivorous, arboreal creatures, which have probably been relatively conservative in their ecological adaptations since the Miocene. This being the case, their presence can be taken as an indication of at least thick bushland and probably forest.

Insectivores as a group occupy a variety of habitats, but two genera are of particular interest to us here. These are the genus of elephant shrews, *Rhynchocyon*, and the family of golden moles, the Chrysochloridae. The former has a number of species, all of which occupy forest habitats mostly in lowland evergreen forest. They are conspicuous animals and rarely stray far from thick cover to which they run if attacked; they are diurnal, and make nests in the leaf litter for the night [RATHBUN, personal commun., 1972]. *Rhynchocyon*, therefore, would appear to have a long-standing adaptation to forest habitats, and the occurrence of any species of this genus would point to the presence of forest conditions.

The East African golden mole, *Chrysochloris stuhlmannii*, is even more habitat-specific, living in wet montane conditions above 2,000 m in soils rich in earthworms and insect larvae [DUNCAN and WRANGHAM, 1971]. Other species of *Chrysochloris*, and other chrysochlorids, occupy a variety of habitats, but all are subterranean and insectivorous, and require well-drained soils rich in microfauna. The presence of a chrysochlorid in the fossil record of East Africa, therefore, might indicate montane conditions, but conservatively it would probably mean no more than that soil conditions were suitable for a fossorial insectivorous mammal. These conditions could certainly include forest, and very dry soils would be ruled out because of lack of soil fauna for the insectivores to feed on, but a whole range of intermediate woodland and savanna soils would also be possible [WALKER, 1964].

Another group of insectivores found in Early Miocene faunas of East Africa, the tenrecs, are now limited entirely to Madagascar where they occupy a variety of habitats. They are less variable in habitat than true hedgehogs, however, and are more limited to woodland habitats, including forest. One fossil genus that is still extant is *Geogale*, a forest genus of western Madagascar, but *Tenrec* lives in drier forests and sandy bushland [WALKER, 1964]. In view of their survival only on Madagascar, they must be regarded as a relic form and their habitats would not necessarily be anything like those of the ancestral tenrecs.

Rodents occupy a great variety of habitats in Africa today. Thryonomyids are often found near water, but apart from this they occur in all vegetation types. As there was formerly a much greater variety of thryonomyid species, they would presumably have occupied many different kinds of habitat, and no direct parallel can be drawn between the one remaining genus and fossil taxa.

The flying 'squirrels', Anomaluridae, are more habitat-specific. The gliding habit of most of them presupposes the presence of trees, and in fact all the living species live today in large tracts of lowland forest, including the flightless species of *Zenkerella*, another genus represented in the Miocene of Kenya. There is no record of them occurring in narrow strips of riverine forest nor at great altitudes in montane forest. It may be that they are able to glide only in the sheltered windless depths of the forest and not in less sheltered spots outside or near the edge of the forest. Their occurrence, therefore, would indicate the presence of large tracts of lowland forest [WALKER, 1964].

Pedetids, represented today by just two species, live in nonforest conditions as befits their saltatorial and fossorial habits. Their prime requisites appear to be open conditions in which to jump, and light sandy soil in which to dig; the latter is particularly important, for, while their limb proportions are well adapted to jumping, they are not at all well adapted to digging, and it is probable that they would not be able to make their burrows in heavy soil [WALKER, 1964; MACINNES, 1957]. Since they do not feed underground as do the chrysochlorids, they are able to inhabit much more arid environments. There is no evidence that they have ever been more common or diverse than at present, and as highly specialized animals it is likely that they are conservative ecologically. Therefore, the appearance of pedetids in the Miocene could indicate dry sandy conditions. However, in this connection,

Table II. Ecology and zoogeography of African Miocene gastropods (extinct species are marked by asterisks)

Name of gastropod	Ecological indications	Distribution	Rusinga – Kaswanga 1971 level II–III	Rusinga – R3A (1947)	Mfwangano	Songhor	Other sites
Ampullaria ovata	large lakes rivers	widespread in Africa and Asia	×	×	×		Moruorot
Lanistes carinatus	rivers and swamps	East Africa only					Moruorot
Maizania 4 spp.	evergreen forests	mostly eastern Africa; the Miocene species are most closely allied to M. hildebrandti	×	×	×		Napak, Bukwa
Ligatella 2 spp.	bushland	southern and eastern Africa					Kirimun
*Cerastua miocenicus	widely distributed (genus)	Africa and Asia; the Miocene species is most closely allied to the living Somali species		×		×	
*Edouardia mfwanganensis	arboreal in lowland habitat (genus)	southern and eastern Africa; the Miocene species is most closely allied to living east coast species	×	×			
Homorus (Subulona)	wet evergreen forest	equatorial Africa	×	×	×	×	Napak, Bukwa
Burtoa nilotica	woodland	West Africa	×	×	×		Bukwa
Limicolaria 2 spp.	forest-bushland	widespread in Africa, absent from east coast		×	×		Napak, Bukwa
Bloyetia 1 sp.	arid bush	East Africa only	×	×		×	
Thapsia 1 sp.	widespread in forest	widespread in Africa		×		×	Bukwa
Trochonanina 5 spp.	evergreen forest and bush up to 2,500 m	widespread in tropical Africa	×	×	×	×	Napak

Table II. Continuation

Name of gastropod	Ecological indications	Distribution	Rusinga – Kaswanga 1971 level II–III	Rusinga – R3A (1947)	Mfwangano	Songhor	Other sites
Tayloria 2 spp.	drier evergreen forest and bush	mostly in eastern Africa, rare in West Africa		×		×	Bukwa
Gonaxis (Marconia) at least 9 spp.	wet evergreen forests, extensive in area	mostly east coast forests, some species also in west	×		×	×	Bukwa
Gulella 3 spp.	forest and thicket	world wide distribution	×		×	×	
Edentulina rusingensis	wet lowland evergreen forest (genus)	mostly eastern Africa, but the Miocene species is most closely allied to living west coast species		×	×		
Ptychotrema usiforme	lowland evergreen forest	West Africa	×		×		
Krapfiella angusta	montane evergreen forest	Kenya only				×	Koru
Primigulella	montane associations	East Africa only				×	
Saulea	widely distributed	West Africa only					Kirimun

Dr. ALAN WALKER has recently brought to our attention the existence of a large burrowing and jumping rodent that inhabits the forests of Western Madagascar. This is *Hypogeomys*, and it parallels the Miocene pedetid *Megapedetes* both in being better adapted for digging than the living *Pedetes* and also in retaining all five digits. Like both pedetids, it is well adapted for jumping. The fact that such similar forms can inhabit such different environments must cast some doubt on the environmental significance of the fossil form.

A more general aspect of rodent faunas is that forest habitats have a richer and more diverse one than do nonforest habitats: the number of species is greater, and there are usually several common species. In nonforest habitats, the number of species is usually less, and often one species is extremely common so that it constitutes nearly 80% of the total number of rodent individuals [ANDREWS, 1973c].

The larger mammals are by and large much less habitat-specific and are of limited usefulness in this discussion. There are of course certain bovid tribes and genera which are good indicators of habitat, but since bovids are rare and of primitive development in the Early Miocene they will not be commented on here. Tragulids (water chevrotains) are swamp or stream dwellers in dense evergreen forest and can be said to be semiaquatic [DORST and DANDELOT, 1970]. Their presence could indicate the presence both of water and forest, but their higher diversity in the Early Miocene precludes a firm conclusion as to their palaeoecological significance.

It is likely that the birds and reptiles would be useful indicators of habitat, but these have not yet been described for the Early Miocene of East Africa. Crocodiles and pleurodire turtles are exclusively water-living forms, and their occurrence, therefore, can be taken to indicate the nearby presence of rivers of lakes.

Gastropods are good environmental indicators. In large tracts of evergreen forest (lowland or montane) there is always a great variety of species of the Streptaxidae, a family that is restricted ecologically to forest conditions [VERDCOURT, personal commun., 1971]. Many additional examples are listed in table II. compiled from VERDCOURT [1963, 1964] and our recent work. The forest forms that are mainly East African in distribution include *Maizania* and *Gonaxis*; lowland forest forms include *Edouardia* and *Edentulina*; montane forest forms include *Krapfiella* and *Primigulella*; and nonforest forms include *Ligatella* and *Bloyetia* [VERDCOURT, 1972].

The lowland forests of the Congo and West Africa are the present centers of distribution of the forest faunas. Here occur the richest associations of prosimian primates, rhynchocyonine elephant shrews, anomalurid flying 'squirrels', rodents generally, and streptaxid gastropods. Many species are endemic to all or part of the area covered by the equatorial forest, but some cross the Western Rift into the Uganda lowland forests, and a few even penetrate to the lowland forests along the East African coast. The Uganda forests are impoverished relative to the western forests, as they do not have endemic forms to replace the species with restricted western distribution. This is even more true of the East Coast forests, where there are single species only of *Rhynchocyon* and *Anomalurus*, and none of the West African forest rodent species. There is possibly one endemic species of prosimian *(Galago zanzibaricus)*. The forest rodents of the western forests have been replaced on the east coast for the most part by woodland and bushland species [ANDREWS, GROVES, and HORNE, in press].

IX. Past Faunas

The Early Miocene faunas of East Africa are very distinctive. They have been compared in the past with the Burdigalian faunas of Eurasia, but recent work on absolute dating suggests that the East African faunas antedate the Eurasian ones by some millions of years [VAN COUVERING and MILLER, 1969; VAN COUVERING, 1972]. The strict time equivalent is with the Aquitanian Land Mammal Age of Europe, but there is little resemblance between the East African faunas and the European Aquitanian faunas. The explanation offered for this [VAN COUVERING, 1972] is that Africa south of the Sahara was isolated in the Oligocene and earliest Miocene and that a high proportion of endemic taxa had developed in the African fauna. Some time in the 18- to 14-Ma period, the Tethys, which up to this time had probably been the principal isolating factor, diminished in size and a land bridge was formed between Africa and Eurasia. This was the time both of the appearance of typical Burdigalian faunas in Eurasia [SAVAGE, 1967] and of an abrupt change in the East African faunas, from the Rusinga type to the Fort Ternan type.

The differences between the Early and Middle Miocene African faunas have been touched on briefly [BISHOP, 1967; LEAKEY, 1967] and more work is in progress now. The Early Miocene faunas of Rusinga Island, Songhor and Napak are dominated by deinotheres and rhinoceroses among the larger mammals, geniohyid hyracoids and tragulids among the medium-sized mammals, and thryonomyid and anomalurid rodents and ochotonids among the small mammals. Insectivores, especially macroscelidids, are also locally common. There is a notable proliferation of hominoid primates, seven species in all; and there are at least five species of prosimians. In the Middle Miocene faunas of Fort Ternan and Ngorora, there are, instead, procaviid hyracoids and very few tragulids. By far the greatest proportion of the fauna at Fort Ternan and Ngorora is composed of bovids and giraffids with strong Eurasian connections, and they may represent incursive elements [CHURCHER, 1970; GENTRY, 1970]. Small mammals and primates are rare. There are fewer rodents, and they belong to the Cricetidae, a family new in Africa at that time. A few specimens are known of anomalurid rodents and rhynchocyonine elephant shrews from Fort Ternan and hystricid rodents from Ngorora [PICKFORD, personal commun., 1973]. These differences are probably partly ecological and partly the result of the large-scale faunal exchange initiated during the time between the Early and Middle Miocene deposits.

Arising from the differences noted between the Early Miocene and later Miocene deposits it is not possible to trace the evolutionary history of many of the fossil groups mentioned in the previous section. The Rhynchocyoninae are first known from the Early Miocene of East Africa, and the fact that many of the characters of the subfamily are already present at that time is taken by BUTLER and HOPWOOD [1957] to indicate that the two subfamilies of the Macroscelididae had already differentiated. This being the case, the rhynchocyonines would appear to be a conservative monogeneric group, changing little through time.

The Miocene chrysochlorid, *Prochrysochloris miocaenicus*, is also the earliest member of the family known and throws little light on the evolution of the family. BUTLER and HOPWOOD [1957] emphasize the similarity of the Miocene and Recent species, and conclude that nearly all the distinctive features of the anterior part of the skull had already been acquired by the Early Miocene. From its short broad skull, like those of Recent species, it would appear that the Miocene species was already a burrowing form and, therefore, that its ecology would have been similar to that of Recent species.

Among the rodents, the Early Miocene of East Africa has the first record of the Anomaluridae and the Pedetidae, and little is known of their subsequent history. Some postcranial bones have been tentatively identified by us as anomalurid on the basis of their similarity to recent anomalurids, and this would indicate that some of the Miocene species had already evolved the flying habit characteristic of all but one of the recent species. The skull and tooth morphology is very similar to that of Recent species.

The Miocene pedetid, *Megapedetes pentadactylus*, is less specialized than the two recent species of the Pedetidae in that it retains a fully developed hallux and has a relatively shorter metatarsus [MACINNES, 1957]. It was already highly saltatorial, but evidently less so than the recent species. The relatively large front feet and massive clavicle suggest that the Miocene species was if anything more highly specialized for digging than the recent species [MACINNES, 1957].

The common rodents in the African Miocene sites belong to the endemic family Thryonomyidae, while those of Recent environments belong mainly to the Muridae. This difference makes it difficult to compare the structure and complexity of the Miocene rodent faunas with Recent ones, and any such approach must be highly tentative.

A. Miocene Faunas in Kenya

The Early Miocene sites in Africa are nearly all located in the inter-rift region of Eastern Africa. Therefore, most of the records available of Miocene faunal distribution for Africa are for East Africa, and it is apparently often assumed that this was an important center of distribution in the Miocene. This may have been the case, but the evidence we are presenting here leads rather to the conclusion that, with the probable eastwards extension of the western evergreen forests into this region, it would have been peripheral to the centers of distribution much further west. The forest-living species would be most common in the main body of the forest which then, as now, would have been in the area of the Congo basin. Although their ranges would have extended further east in the Miocene, assuming we are correct in saying the forests extended east, the peripheral nature of their distributions would imply that they were not entirely typical of the populations of other parts of equatorial Africa. Two consequences of this could have been that the eastern populations would be highly variable and that the fauna would have been impoverished to some degree [MAYR, 1963]. As peripheral isolates it would not be surprising to find active adaptive radiation in many groups, as was apparently the case in such groups as the Erinaceidae, Rhynchocyoninae, Pongidae, Galaginae, Rhinocerotidae, and Tragulidae.

There is no direct evidence from the Miocene faunas relating to this problem. Past collecting has tended to ignore minor stratigraphical differences between fossil horizons so that specimens coming from different horizons have been mixed. This has resulted, at Rusinga at least, in the faunas having a mixed ecological appearance: forest-dwelling, 'savanna'-dwelling, and lacustrine. No direct evidence on fossil environments has been available, either from sedimentary structures or from types of bone accumulation. Taxonomic treatment of the fossil species has put the stress on variability.

In order to try to throw light on some of these questions, we initiated a program of field excavations in 1971 in the Miocene deposits of Kenya. The first year we concentrated on Kaswanga point, Rusinga Island [see map in VAN COUVERING and MILLER, 1969], and a total of eight excavations were attempted. Some of these proved abortive, owing to the patchy distribution of the fossils, but others had more success. These were labeled KB, KF, KG, and KH. Overburden was removed by trowels, the sites marked out into meter squares, and the actual excavation done with dental picks and small paint brushes, taking the whole surface down parallel to the bedding plane. The aim was to locate bones *in situ* before they had been disturbed by the excavation; when a bone was located, it was carefully exposed on its top surface and its angle of dip and orientation were measured and recorded. For measuring the position of the bone, the meter squares proved cumbersome, and in the last two excavations (KF and KH) the position was measured with respect to two base lines at right angles to each other. Levels were initially divided at 6-cm intervals, and it was found that the fossils occurred at single levels at most sites. These levels did not necessarily correspond from one site to another. Having recovered bones in this manner they were catalogued and measured.

When the alignments of the bones were plotted on a rose diagram [cf. HILL and WALKER, 1972 for Bukwa], axes of preferred orientation appeared for all four sites. The direction of preferred orientation was similar at each site with one exception: at site KF, there appeared to be two fossil levels, one immediately on top of the other, the higher having a preferred orientation perpendicular to the lower one. The general direction of orientation was along a north-northeast to south-southwest line. This may indicate local water flow across the area along this line, and the orientation of crocodile teeth and the sedimentary structure of the deposits suggest that the direction of flow was towards the north-northeast, that is into the Kavirondo valley. With so few sites it is not possible to draw any wide-ranging conclusions on this evidence, and a great many more excavations should be made at Rusinga Island and other nearby Miocene localities.

Most of the good specimens from the 1971 collection came from the surface collection. In order to establish a record of where the specimens came from and from which level, a grid system was laid out over the whole of the Kaswanga Point deposits by JOHN VAN COUVERING. All the specimens found were tied in to this.

The following year (1972), one of us (P. A.) continued this work at Songhor. The Songhor deposits are much less extensive than the Rusinga ones, and the fossils appear to be concentrated in just two horizons, although they are scattered thinly through the rest of the deposits. Two excavations were put in, one in each of the fossiliferous horizons, but even in these the fossils were found to be scattered throughout the thickness of the bed and were not concentrated into discrete patches as were those excavated on Rusinga Island. The orientations of the bones do not show a preferred direction, and this, supported by the scattered distribution of the fossils, the fragmentary nature of the bones (broken after at least partial fossilization), the lack of association between skeletal elements, and the subaerial nature of the deposits, suggest that the deposits have been reworked after their initial deposition. When this was and when the fossils came to be associated with them, is not known at present, and it makes it very difficult to assess the possible palaeoenvironments in such a situation.

In our 1971 Rusinga collection, three distinct faunal assemblages were recovered from the four main excavations which were close together in both horizontal and vertical distance. These are shown in figure 3. The first consisted of a waterside community with abundant crocodiles and pleurodire turtles; *Rhynchocyon* was also present and may point to the nearby presence of forest. The second consisted of many small mammals nearly all related to forest-living forms of today; galagine prosimians, rhynchocyonine elephant shrews, anomalurid flying 'squirrels', and a multiplicity of rodents, all point to this being a forest fauna. The absence of large animals suggests that this accumulation of bones was selective, but it has been mentioned earlier that small animals tend to be more accurate habitat indicators, so that the absence of larger animals is probably not critical. The third assemblage shown in figure 3 is of uncertain status as there are no good habitat indicators in the fauna except possibly *Megapedetes*; their absence is interesting and adds weight to the presence of *Megapedetes* which might have occupied a similar habitat to that of present-day pedetids, namely open types of woodland and bushland (see pp. 73-75).

Gastropods were not associated with the mammalian faunas in any quantity in the Rusinga excavations. A good collection was made from another level, however, and as this consisted largely of streptaxid species it is probable that it represents a forest environment also. Past collections at Rusinga have yielded a great variety of gastropods, most of which have

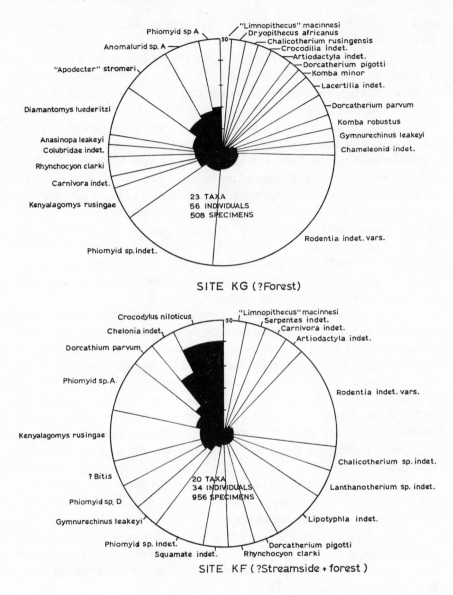

SITE KG (?Forest)

SITE KF (?Streamside + forest)

Fig. 3. Rusinga faunal associations. The constituent elements of the three associations so far recognized (?Forest, ?Streamside + forest, and ?Open), the results of four excavations in 1971, are shown. The divisions of the circles represent the relative abundance

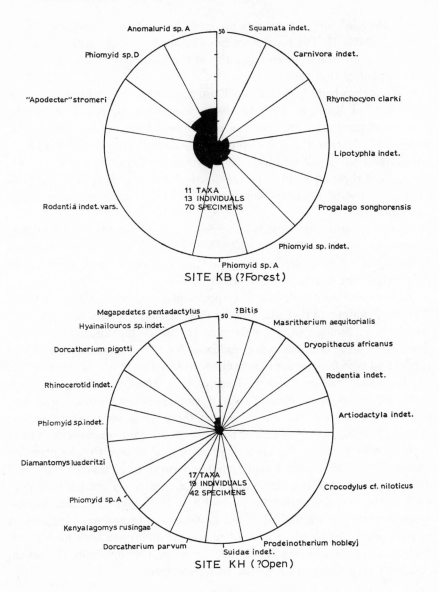

SITE KB (?Forest)

SITE KH (?Open)

of the taxa indicated and the radial bar represents the corrected number of specimens per individual [cf. SHOTWELL, 1955]. The total number of taxa, minimum individuals and specimens is given for each site.

closest affinities with present day lowland forest species [Verdcourt, 1963, 1972]: see table II. The fossil flora has already been commented on. Plant remains are not found in any quantity at the mammalian sites, so there is no direct correlation between them.

The faunas from Songhor and Koru appear to be more homogeneous than those of Rusinga. *Rhynchocyon* is less common, but anomalurids, tenrecs and prosimians are more common than at Rusinga. The number and variety of rodents is like that of present-day forest habitats. There is also a considerable variety of reptiles, which form a significant part of the fauna, but most of these have yet to be identified. *Prochrysochloris* occurs at both sites. Indicators of montane conditions are to be seen in the gastropod fauna (see table II).

Apart from the gastropods indicating montane conditions at Songhor, there are a number of faunal differences between Songhor and Rusinga indicating a major ecological (and perhaps temporal) difference between them. There are few large mammals from Songhor and few hyracoids, no ochotonids and no *Myohyrax*. The species of *Dorcatherium* that occurs at Songhor is unique to that site, and the three Rusinga species have not been found there. Similar species differences occur in the insectivore genera *Gymnurechinus* and *Rhynchocyon* and in at least four rodent genera. Some genera and even families are unique either to Songhor (e. g. Chrysochloridae, *Erythrozootes, Zenkerella*) or to Rusinga (e. g. *Myohyrax, Geogale, Crocidura*). The exact significance of these differences is not known, but it does imply one of several alternatives. Either there was a geographical barrier between Songhor and Rusinga, or there was a time difference between their deposition long enough for climatic or ecological change to take place, or finally, even if they were broadly contemporary, there was an ecological difference between them. These differences are relatively great if they were both sampling forest environments, and it is likely that some combination of the factors just mentioned would be necessary to account for them. A possible example would be a small ecological difference, such as between evergreen forests differing in altitude by some few hundreds of meters, coupled with a distribution barrier such as a river.

X. Miocene Primate Distribution

The relationships between the Miocene and Recent pongids are not known. The hypothesis put forward in a number of papers [LEGROS CLARK and LEAKEY, 1951; PILBEAM, 1969] that *D. africanus* was ancestral to the chimpanzee and *D. major* to the gorilla, is not substantiated by the greater variety of dryopithecine species now known to be present in the Early Miocene of East Africa [ANDREWS, 1973b]. There were apparently at least five species of *Dryopithecus* at that time, and there is no reason other than the nebulous one of size, for considering any of them to be directly ancestral to living forms.

The Miocene apes were in general smaller than those of the present day. The range of estimated sizes is from that of a small *Cercopithecus*, e.g. *C. aethiops*, to about the size of a female gorilla. The study of the limb bones so far completed suggests that the species of *Dryopithecus* had a generalized arboreal form of locomotion, more like quadrupedal monkeys, than the modern apes [LeGROS CLARK and LEAKEY, 1951; LeGROS CLARK, 1952; NAPIER and DAVIS, 1959]. At the same time, there is little evidence for the presence of monkeys in the Early Miocene environments. Monkeys are recorded from Napak [PILBEAM and WALKER, 1968] and they are common at Maboko, which is probably a late Early Miocene site, but the record from Rusinga is extremely doubtful [*contra* VON KOENIGSWALD, 1969]. It can be hypothesized, therefore, that the Miocene apes were in fact occupying many of the ecological niches now occupied by monkeys. This might be carried a stage further, and the Oligo-Miocene radiation of the genus *Dryopithecus* compared with the apparently recent radiation of the cercopithecine monkeys.

In the first case, that of the ecological parallelism between monkeys and Miocene apes, it appears that up to four species of dryopithecine can occur in the same deposits and apparently in the same palaeoenvironments, two or three of the species being very common and the others less so. In addition, *Limnopithecus legetet* and '*Limnopithecus*' *macinnesi* [see ANDREWS and PILBEAM, in press] also occur making a total of six species for at least two localities. This number of species, while unlike the distribution of modern apes, is very like that of forest monkey distributions, where there can be three to five species of *Cercopithecus*, one species of the more arboreal *Colobus*, sometimes a species of *Cercocebus*, and one species of the more terrestrial *Papio*, all in the same patch of forest. If the Miocene apes were occupying similar ecological niches of those to present-day monkeys, and were as ecologically successful as monkeys are in their forest habitats, there would be every reason to expect such a variety of species as apparently occurred in the Miocene, as was suggested by PILBEAM and WALKER [1968].

Table III. Miocene fossil environments

Locality	Sedimentary environment	Vegetation type
Karungu	lake, in part	not known
Rusinga island and	floor of Kavirondo Rift Valley	probably little vegetation
Mfwangano Island	(1) river flood plain	on the flood plain; large
	(2) narrow finger lakes	tracts of forest in the
	(3) depressions at edge of volcanic dome	immediate vicinity
Songhor	depressions between lower slopes of growing volcano and the basement Songhor Hill	evergreen forest, maybe with montane affinities
Koru	depressions at bottom of slopes of growing volcano and separate carbonatite centres	evergreen forest, maybe with montane affinities
Napak	depressions at bottom of slopes of volcanic dome rivers	probably little vegetation in the sedimentary basins, but lowland evergreen forest close by
Moroto	valleys in pre-Miocene topography	not known
Bukwa	lake side	evergreen forest
Ombo	valley in pre-Miocene topography	not known
Losidok Moruarot	swampy lakes	bushland or semidesert

This table has been compiled from our own data, from JOHN VAN COUVERING personal communications, and from OSWALD 1914, BISHOP 1963, 1968, WALKER 1969, and MADDEN 1972.

Table IV. Distribution of Early Miocene hominoids (numbers of individuals)

Site	L. legetet	D. africanus	D. nyanzae	D. major	D. gordoni	D. van-couveringi	'L'. macinnesi
Karungu	–	–	2	–	–	–	1
Rusinga Island	8	41	50	–	6	3	57
Mfwangano							
Island	–	5	2	–	1	1	5
Songhor	19	6	–	14	17	3	10
Koru	6	3	–	1	–	–	2
Napak	10	–	–	8	–	–	–
Moroto	–	–	–	1	–	–	–
Bukwa	1	–	–	–	–	–	–
Ombo	1	–	–	–	–	–	–
Losidok	–	1	–	2	–	–	–
Moruarot	–	1	–	1	–	–	–

In the second of the two comparisons, that of the radiation of the cercopithecine monkeys, there is considerable evidence that the Early Miocene ape populations represented part of an active Early Miocene radiation. At least two sets of species were so similar to each other to make their distinction merely one of size, namely *D. nyanzae/D. major*, and *D. gordoni/ D. vancouveringi*; the other two species of *Dryopithecus*, while more distinct, were still very similar to the others and to each other. The site differences between these three pairs suggests either temporal or geographical speciation, or perhaps a mixture of both.

These differences in distribution of closely allied pairs of species are very characteristic of the Early Miocene pongids. The differences encompass nearly all of the Early Miocene localities in such a way that the fossil pongids can be divided into two groups, a southern group and a northern group. The southern group comes from Karungu, Rusinga Island, and Mfwangano Island, where the common pongids are *D. africanus* and *D. nyanzae*, together with the hylobatid '*L*'. *macinnesi*. The northern group comes from Songhor, Koru, Ombo, Napak, Moroto, Bukwa, and Losidok: *D. major* is absent from the first two, and *L. legetet* from the third. In addition, *D. gordoni* is common at Songhor but rare elsewhere. These distributions are summarized in table IV.

It has been shown in the previous section that these site differences are a common pattern also in other orders of mammals, suggesting not only ecological differences between the two groups of sites but also the presence of a distributional barrier between them. Such a barrier today, separating forest faunas, would be such things as a river, mountain range, or a broad belt of more arid country. In this particular instance, there is nothing to suggest the presence of a mountain barrier between the sites, but evidence has been put forward to suggest the presence of a broad eastwest drainage channel, probably with little associated vegetation along the line of the Kavirondo rift valley. It is probable that a permanent river would have flowed through it, for with the greatly restricted catchment area of the present-day Kano Plain/Kavirondo gulf there are several relatively large rivers flowing into and through the valley. It is possible that the fossils preserved in the southern sites were derived in large part from the probably forested slopes of the Kisingiri dome, which was isolated from the northern sites by such a combination of river and inhospitable flood plain. The presence of a barrier such as this, isolating the two groups from each other, could account for the differences observed between them, particularly as regards the distribution of the pongid species and their apparently recent speciation.

A. Miocene Primate Distribution in Equatorial Africa

What is known of the palaeoenvironments of the Early Miocene lacalities is summarized in table III. This is correlated with the Miocene ape distributions in table IV, which shows the number of individuals known for each locality, calculated on the minimum number of individuals needed to account for the specimens available. The common primates in the Rusinga-Mfwangano deposits are '*Limnopithecus*' *macinnesi*[1], *Dryopithecus africanus*, and *D. nyanzae*, but these are not necessarily found together in the same deposits. The first of these species, '*L*'. *macinnesi*, was recovered from two excavations in 1971, associated in one case (site KF: see fig. 3) with the probable forest community of prosimians, rhynchocyonine elephant shrews and rodents. In view of the striking arboreal adaptations of the postcranial skeleton of this species [LEGROS CLARK and THOMAS, 1951], it is quite consistent with the evidence to regard it as a forest-living form.

D. africanus was found at site KH (fig. 3) in association with a number of large animals of no certain habitat affinity and few rodents.

Finally, at Rusinga, *D. nyanzae* has several times been recovered in past excavations associated with ecologically indeterminate large mammal associations of rhinos and hyracoids. One specimen of *D. nyanzae*, which was excavated by one of us (J.V.C.) in 1967, came from close to site KF with its waterside fauna (see fig. 3). The field catalogues of these excavations do not give sufficient detail to fully establish these associations, so they must also remain tentative until more data are available.

The common pongids at Songhor are *Limnopithecus legetet, Dryopithecus major*, and *D. gordoni* [ANDREWS, 1973b]. All three of these are found throughout the red-bed unit and there does not seem to be any unique association between any of them and a distinct fauna. Specimens of each have been recovered from excavations in 1972/73, together with remains of prosimians, anomalurids, chrysochlorids, and an abundance of rodents. It is possible that the red-bed sediments at Songhor have been reworked, in which case the faunal elements would be further removed from their source environments and thus associations of fauna more difficult to evaluate. It can only be stated that most of the small mammals from this site are almost certainly derived from forest environments and so it is likely that the associated pongids are also.

---

1 This taxon will shortly be removed from *Limnopithecus* and described as a new genus of the Hylobatidae [ANDREWS, PILBEAM and SIMONS, in press].

Little can be said of other Miocene sites as information is not available. Napak has both similar faunas and similar primate species to Songhor, although *D. gordoni* is not known from this site, and it is likely that the palaeoenvironment was similar. Koru is also similar to Songhor, but of the few pongid specimens recovered from this locality, *D. africanus* is the most common after *L. legetet* (see table IV). The numbers of specimens are too low for the significance of this to be evident. The only pongid from Bukwa is *L. legetet* and the gastropod, faunal, and floral evidence all point to the presence of forest at or near this site. In addition, it appears from recent work [HILL and WALKER, 1972] that the Bukwa fossils were deposited in a lake-shore environment.

In summary, there is good evidence associating '*L*'. *macinnesi* with predominantly forest-affiliated animals and for regarding it as a forest primate. There is less good evidence associating *L. legetet*, *D. major*, and *D. gordoni* with forest environments, mainly because of the uncertainties of Songhor, but the faunas there, and at Koru, Napak, and Bukwa, are all indicative of forest conditions, so it is likely that these species were also forest-living primates. As for *D. africanus* and *D. nyanzae*, it is possible that they lived outside the forest. There has been no record of their being associated with forest faunas except for the rare finds of *D. africanus* at Songhor and Koru; but they have, on the contrary, been found associated with faunas either of uncertain status or of possible nonforest affinities.

XI. Summary and Conclusions

(1) Some of the features of present-day geography of equatorial Africa are briefly described. The present situation is contrasted with that of the Early Miocene, a time when rifting and uplift in East Africa was still in its initial stages; the continental divide across the equatorial belt was probably just west of the line of the present Eastern Rift Valley, well to the east of its present position; and drainage was towards the Atlantic over most of Uganda, Western Kenya and Tanzania, although the pattern may have been broken locally by early faulting in the Western Rift Valley.

(2) Climatic patterns are discussed briefly, and it is concluded that Early Miocene climates in eastern equatorial Africa may have been wetter than at present. Correlated with the geographical changes just mentioned, it is concluded that the equatorial trough, at present limited eastwards by the west rift highlands and dry conditions in the east, could have extended across

Uganda and Western Kenya at least as far as the continental divide, with resulting high year-round rainfall in this region.

(3) Vegetation patterns are discussed in some detail as there are many misconceptions current among anthropologists concerning tropical African vegetation types. Lowland evergreen forest is the natural equatorial vegetation type where lowland conditions prevail; the forest belt is broken by the East African highlands with wet montane forests on the windward slopes and rainshadows downwind; and along the eastern slope between the highlands and the Indian Ocean, there is a bushland belt of apparently great antiquity, with a narrow strip of forest along the coast having some West African affinities. It is concluded that, in the absence of the East African highlands and the probable eastwards extension of the equatorial trough, the lowland forest belt may have extended eastwards as far as the continental divide and possibly to the East African coast. This conclusion is supported by the nature of the disjunct plant distributions and the fossil floras.

(4) Some small mammals and land gastropod taxa are mentioned briefly and their usefulness as indicators of environment assessed qualitatively. The presence of related taxa in the Miocene deposits is discussed and the following tentative conclusions are reached on environments represented by associations of these taxa: lowland forest at or near Rusinga and Mfwangano Islands, and Napak; lowland/montane forest at Songhor and Koru; waterside conditions, perhaps associated with forest, at Rusinga Island; and nonforest conditions of unknown type at Rusinga Island.

(5) Against this background, a number of conclusions on Miocene pongid distributions are reached: (a) no direct relationship between Miocene and Recent African pongids can be assumed; (b) the analogy is made between the Early Miocene dryopithecines and the Recent cercopithecines in terms of ecological adaptation and evolutionary radiation; (c) a number of fossil species have been found recently in direct association with faunas probably representing forest conditions; these are '*Limnopithecus*' *macinnesi*, *Limnopithecus legetet*, *Dryopithecus major*, *D. gordoni*, and maybe also *D. africanus*; (d) the East African Miocene sites represent a small corner of the presumed African equatorial forest and the pongids found there constitute only a portion of the Miocene pongid distributions, and (e) there is some evidence for a Miocene distribution barrier along the line of the Kavirondo rift valley, perhaps a large river, which could account for the differences in faunas observed between the Rusinga-Mfwangano-Karungu deposits to the south and Songhor-Koru-Napak-Moroto-Ombo deposits to the north.

XII. Acknowledgements

We should both like to thank the Wenner-Gren Foundation for Anthropological Research for financial support, one of us (P.A.) for year-round research support and the other (J.V.C.) for travel to Rusinga Island. Field work was supported by the Royal Society, London, and the Boise Fund, Oxford, in 1971, and by the Boise Fund again in 1972.

We would also like to thank all of those who worked with us on Rusinga Island, especially DEA ANDREWS for supervising the washing and screening and keeping body and soul together; GLENN CONROY for his hard work in the field; ERASTRO NDERE, MARTIN ODEL' and ALPHONSE OKELO for their consistent and reliable work in the field; and JOHN VAN COUVERING for supervising the surface collecting, acting as geological consultant, and draughting the figures. There were many others who also helped us in the field and whom we also thank.

In addition, we want to thank MARJORIE and DEA ANDREWS for typing the drafts of this paper, and JOHN VAN COUVERING and MARTIN PICKFORD for destroying the legibility of these drafts with editorial comments. Finally, we would like to thank ALAN WALKER for his many helpful comments.

XIII. References

ABDEL-GAWAD, M.: New evidence of transcurrent movements in Red Sea and petroleum indications. Amer. Ass. Petrol. Geol. Bull. 53: 1466–1499 (1969).

ANDREWS, P.: The vegetation of Rusinga Island. J. E. Afr. nat. Hist. Soc. 142: 1–8 (1973a).

ANDREWS, P.: New species of Dryopithecus from Kenya. Nature, Lond. (in press, 1973b).

ANDREWS, P.: The Miocene Apes (Pongidae, Hylobatidae) of East Africa; PhD thesis Cambridge (1973c).

ANDREWS, P., PILBEAM, D. R., and SIMONS, E. L.: Review of the fossil Hominoidea of Africa; in MAGLIO Mammalian evolution in Africa (in press).

ANDREWS, P.; GROVES, C. P., and HORNE, J. F. M.: Ecology of the Lower Tana River flood plain (Kenya). J. E. Afr. nat. Hist. Soc. (in press).

AUBRÉVILLE, A.: Climats des forêts et la sertification de l'Afrique. Soc. Edit. Géogr. marit. col., Paris (1949).

BAKER, B. H. and WOHLENBERG, J.: Structure and evolution of the Kenya Rift Valley. Nature, Lond. 229: 538–542 (1971).

BAKER, B. H.; WILLIAMS, L. A. J.; MILLER, J. A., and FITCH, F. J.: Sequence and geochronology of the Kenya rift volcanics. Tectonophysics 11: 191–215 (1971).

BERGGREN, W. A.: Cenozoic chronostratigraphy, planktonic foraminiferal zonation and the radiometric time scale. Nature, Lond. 224: 1072–1075 (1969).

BERGGREN, W. A.: A Cenozoic time scale – some implications for regional geology and palaeobiogeography. Lethaia 5: 195–215 (1972).

BERGGREN, W. A. and PHILLIPS, J. D.: Influence of continental drift on the distribution of the Tertiary benthonic foraminifera in the Caribbean and Mediterranean regions. Symp. Geology of Libya, Beirut, p. 263–299 (1971).

BISHOP, W. W.: Miocene mammalia from the Napak volcanics, Karamoja, Uganda. Nature, Lond. 182: 1480–1482 (1958).

BISHOP, W. W.: The later Tertiary and Pleistocene in Eastern Equatorial Africa; in HOWELL and BOULIERE African ecology and human evolution (Aldine, Chicago 1963).

BISHOP, W. W.: Quaternary geology and geomorphology in the Albertine Rift Valley, Uganda. Geol. Soc. Amer., suppl 84 (1965).

BISHOP, W. W.: The later Tertiary in East Africa: volcanics, sediments, and faunal inventory; in BISHOP and CLARK Background to evolution in Africa, pp. 31–56 (Chicago Univ. Press, Chicago 1967).

BISHOP, W. W.: The evolution of fossil environments in East Africa. Trans. Leicester lit. phil. Soc. *62:* 22–44 (1968).

BISHOP, W. W.: The Late Cenozoic history of East Africa in relation to hominoid evolution; in TUREKIAN Late Cenozoic glacial ages (Yale Univ. Press, New Haven, 1971).

BISHOP, W. W. and TRENDALL, A. F.: Erosion surfaces, tectonics and volcanic activity in Uganda. Quart. J. geol. Soc. Lond. *122:* 385–420 (1967).

BUTLER, P. M. and HOPWOOD, A. T.: Insectivora and Chiroptera from the Miocene rocks of Kenya Colony. Fossil mammals of Africa, No. 13. Brit. Mus. nat. Hist., Lond. (1957).

CAHEN, L.: Géologie du Congo Belge (Liège, 1954).

CARCASSON, R. H.: A preliminary survey of the zoo-geography of African butterflies. E. Afr. Wildl. J. *2:* 122–157 (1964).

CHANEY, R. W.: A Tertiary flora from Uganda. J. Geol. *41:* 702–709 (1933).

CHESTERS, K. I. M.: The Miocene flora of Rusinga Island, Lake Victoria, Kenya. Paläontographica *101 B:* 30–67 (1957).

CHURCHER, C. S.: Two new Upper Miocene giraffids from Fort Ternan, Kenya, East Africa; in LEAKEY and SAVAGE Fossil Vertebrates of Africa, vol. 2 (Academic Press, London 1970).

CLARK, W. E. LeGROS: Report on fossil hominoid material collected by British-Kenya Miocene Expedition. Proc. zool. Soc., Lond. *122:* 273–286 (1952).

CLARK, W. E. LeGROS and LEAKEY, L. S. B.: The Miocene Hominoidea of East Africa. Fossil mammals of Africa, No. 1. Brit. Mus. nat. Hist., Lond. (1951).

CLARK, W. E.; LeGROS, and THOMAS, D. P.: Associated jaws and limb bones of *Limnopithecus macinnesi*. Fossil mammals of Africa, No. 3. Brit. Mus. nat. Hist., Lond. (1951).

COOKE, H. B. S.: Observations relating to Quaternary environments in East and Southern Africa. A. L. du Toit Memorial Lecture No. 5. Trans. geol. Soc. S. Afr., No. 61 (1958).

COOKE, H. B. S.: Evolution of mammals on southern continents. II. The fossil mammal fauna of Africa. Quart. Rev. Biol. *43:* 234–264 (1968).

DALE, I. R.: The woody vegetation of the Coast Province of Kenya. Imp. Forest Inst., Pap. 18 (1939).

DAMME, D. VAN and GAUTIER, A.: Some fossil molluscs from Moruarot Hill (Turkana District, Kenya). J. Conch. *27:* 423–426 (1972).

DEWEY, J. F. and BIRD, J. M.: Mountain belts and the new global tectonics. J. geophys. Res. *75:* 2625–2647 (1970).

DORST, J. and DANDELOT, P.: A field guide to the larger mammals of Africa (Collins, London 1970).

DUNCAN, P. and WRANGHAM, R. W.: On the ecology and distribution of subterranean insectivores in Kenya. J. zool. Soc. Lond. *164:* 149–163 (1971).

FRAKES, L. A. and KEMP, E. M.: Influence of continental positions on Early Tertiary climates. Nature, Lond. *240:* 97–100 (1972).

GAUTIER, A.: New observations on the later Tertiary and Early Quaternary in the Western Rift: the stratigraphic and palaeontological evidence; in BISHOP and CLARK Background to evolution in Africa, p. 73–87 (Chicago Univ. Press, Chicago 1967).

GENTRY, A. W.: The Bovidae (Mammalia) of the Fort Ternan fossil fauna: in LEAKEY and SAVAGE Fossil vertebrates of Africa, Academic Press, London (1970).

GERMERAAD, J. H.; HOPPING, C. A., and MULLER, J.: Palynology of Tertiary sediments from tropical areas. Rev. Palaeobot. Palynol. *6:* 189–348 (1968).

GLOVER, P. E.: The role of fire and other influences on the savanna habitat. E. Afr. Wildl. J. *6:* 131–137 (1968).

GREENWAY, P. J.: Second draft report on vegetation classification. East African Pasture Res. Conf., Nairobi (1943).

GROVE, A. T.: African south of the Sahara (Oxford Univ. Press, London 1970).

HAMILTON, A. C.: History of the vegetation of East Africa; in LIND and MORRISON The vegetation of East Africa, (Longmans, London 1973).

HAMILTON, W. R.: The Lower Miocene ruminants of Gebel Zelten, Libya. Bul. Brit. Mus. nat. Hist., Geol. *21:* 75–150 (in press, 1973).

HAUGHTON, S. H.: Stratigraphic history of Africa south of the Sahara. (Oliver & Boyd, Edinburgh 1963).

HEINZLEIN, J. DE: A tentative palaeogeographic map of Neogene Africa; in HOWELL and BOURLIERE African ecology and human evolution (Chicago Univ. Press, Chicago 1963).

HILL, A. and WALKER, A.: Procedures in vertebrate taphonomy: notes on a Uganda Miocene fossil locality. J. geol. Soc. Lond. *128:* 399–406 (1972).

HOOIJER, D. A.: Miocene mammalia of Congo, a correction. Ann. Mus. roy. Afr. centr. Sci. géol. *67:* 163–167 (1970).

HSÜ, K. J.: When the Mediterranean dried up. Sci. Amer. *227 6:* 26–36 (1972).

KEAY, R. W. J.: Vegetation map of Africa (Oxford Univ. Press, London 1959).

KENT, P. E.: The Miocene beds of Kavirondo, Kenya. Quart. J. geol. Soc. Lond. *100:* 85–118 (1944).

KING, B. D. and CHAPMAN, G. R.: Volcanism of the Kenya Rift Valley. Phil. Trans. roy. Soc. Lond. (A) *271:* 185–208 (1972).

KING, L. C.: Morphology of the earth; 2nd ed. (Hafner, New York 1967).

KINGDON, J.: East African mammals: an atlas of evolution in Africa, vol. 1 (Academic Press, London 1971).

KOENIGSWALD, G. H. R. VON: Miocene Cercopithecoidea and Oreopithecoidea from the Miocene of East Africa; in LEAKEY Fossil vertebrates of Africa, vol. 1, pp. 39–52 (Academic Press, London 1969).

LAMB, H. R.: Fundamentals of climate; in NAIRN Descriptive palaeoclimatology (Interscience, New York 1961).

LANGDALE-BROWN, I.; OSMASTON, H. A., and WILSON, J. G.: The vegetation of Uganda (Government Printer, Entebbe 1964).

LEAKEY, L. S. B.: Notes on the mammalian faunas from the Miocene and Pleistocene of

East Africa; in: BISHOP and CLARK Background to evolution in Africa, (Univ. of Chicago Press, Chicago 1967).

MACFADYEN, W. A.: A note in the geology of the Daban area and the localities of the described nautiloids; in HAAS and MILLER Eocene nautiloids of British Somaliland. Amer. Mus. nat. Hist. Bull. *99 5:* 317–354 (1952).

MACINNES, D. G.: Miocene and Pleistocene Lagomorpha of East Africa. Fossil Mammals of Africa, No. 6. Brit. Mus. nat. Hist. (1953).

MAC INNES, D. G.: A new MIOCENE rodent from East Africa. Fossil mammals of Africa, No. 12. Brit. Mus. nat. Hist. (1957).

MADDEN, C. T.: Miocene mammals, stratigraphy and environment of Moruarot Hill, Kenya. Paleobios, No. 14 (1972).

MAYR, E.: Animal species and evolution (Univ. Press, Harvard 1963).

MCCALL, G. J. H.: Geology of the Gwasi area. Geol. Surv. Kenya, Rep. 45 (1958).

MICHELMORE, A. P. G.: Observations on tropical African grasslands. J. Ecol. *27:* 282–312 (1939).

MOREAU, R. E.: Africa since the Mesozoic: with particular reference to certain biological problems. Proc. zool. Soc., Lond. *121:* 869–913 (1952).

MOREAU, R. E.: The bird faunas of Africa and its islands (Academic Press, London 1966).

NAIRN, A. E. M.: The scope of palaeoclimatology; in NAIRN Descriptive palaeoclimatology (Interscience, New York 1961).

NAPIER, J. R. and DAVIS, P. R.: The fore-limb skeleton and associated remains of *Proconsul africanus*. Fossil mammals of Africa, No. 16. Brit. Mus. nat. Hist. (1959).

OSWALD, F.: The Miocene beds of the Victoria Nyanza and the geology of the country between the lake and the Kisii Highlands. Quart. J. geol. Soc. Lond. *70:* 128–162 (1914).

PILBEAM, D. R.: Tertiary Pongidae of East Africa: evolutionary relationships and taxonomy. Bull. Peabody Mus. nat. Hist., No. 31 (1969).

PILBEAM, D. R. and WALKER, A.: Fossil monkeys from the Miocene of Napak, Northeast Uganda. Nature, Lond. *220:* 657–660 (1968).

RICHARDS, P. W.: The tropical rain forest. An ecological study (Univ. Press, Cambridge 1952).

SAGGERSON, E. P. and BAKER. B. H.: Post-Jurassic erosion surfaces in eastern Kenya and their deformation in relation to rift structure. Quart. J. geol. Soc. Lond. *121:* 51–72 (1965).

SAVAGE, R. J. G.: Early Miocene mammal faunas of the Tethyan region; in ADAMS and AGER: Aspects of Tethyan biogeography (Systematics Association, 1967).

SHACKLETON, R. M.: A contribution to the geology of the Kavirondo rift valley. Quart. J. geol. Soc. Lond. *106:* 345–383 (1951).

SHOTWELL, J. A.: An approach to the paleoecology of mammals. Ecology *36:* 327–337 (1955).

TREWAVAS, E.: A preliminary review of fishes of the genus *Tilapia* in the eastward-flowing rivers of Africa, with proposals of two new specific names. Rev. zool. bot. afr. *74:* 394–424 (1966).

VAN COUVERING, J. A.: Radiometric calibration of the European Neogene; in BISHOP and MILLER Calibration of Hominoid Evolution (Scottish Adademic Press, Edinburgh 1972).

Van Couvering, J. A.: Evolution of the Kisingiri complex: evidence from Rusinga Island; in LeBas The Cainozoic volcanic province of Homa Bay, S. Nyanza, Kenya. Mem. geol. Soc. Lond. (in press, 1973).

Van Couvering, J. A. and Miller, J. A.: Miocene stratigraphy and age determinations, Rusinga Island, Kenya. Nature, Lond. 221: 628–632 (1969).

Van Couvering, J. A. and Miller, J. A.: Late Miocene marine and nonmarine time scale in Europe. Nature, Lond. 230: 559–563 (1971).

Verdcourt, B.: The Miocene non-marine mollusca of Rusinga Island, Lake Victoria and other localities in Kenya. Paläontographica 121 A: 1–37 (1963).

Verdcourt, B.: A new species of Krapfiella from the Miocene of Kenya. Arch. Mollusc. 92: 231–235 (1964).

Verdcourt, B.: The arid corridor between the north-east and south-west areas of Africa; in van Zinderen Bakker Palaeoecology of Africa, vol. 4 (Balkema, Cape Town 1969).

Verdcourt, B.: The zoogeography of the non-marine mollusca of East Africa. J. Conch. 27: 291–348 (1972).

Walker, A. C.: Lower Miocene fossils from Mount Elgon, Uganda. Nature, Lond. 223: 591–593 (1969).

Walker, A. C.: The dissemination and segregation of early primates in relation to continental drift; in Bishop and Miller Calibration of hominoid evolution (Scottish Academic Press, Edinburgh 1972).

Walker, E. P.: Mammals of the world (Johns Hopkins Univ. Press, Baltimore 1964).

Walsh, J. and Dodson, R. G.: Geology of northern Turkana. Kenya Geol. Surv. Rep., No. 82 (1969).

Whitworth, T.: A contribution to the geology of Rusinga Island, Kenya. Quart. J. geol. Soc. Lond. 109: 75–92 (1953).

Whitworth, T.: Miocene ruminants of East Africa. Brit. Mus. nat. Hist. Fossil Mammals of Africa, No. 15 (1958).

Authors' addresses: Dr. Peter Andrews, 4 Carpenters Close, Stratton, DT2 9SR (England) and Dr. Judith A. H. Van Couvering, University of Colorado Museum, Boulder, Colo. (USA)

II. Taxonomy and Phylogeny

In SZALAY: Approaches to Primate Paleobiology
Contrib. Primat., vol. 5, pp. 106–135 (Karger, Basel 1975)

Using Polar Coordinates to Measure Variability in Samples of *Phenacolemur:* A Method of Approach

PAUL RAMAEKERS

Royal Ontario Museum, Toronto, Ont.

Contents

I. Variability in a Sample

The definition of the limits of morphological variability in a population or sample is a necessity for our current system of classification and related fields such as ecology, biogeography and evolutionary studies. In 1961 [p.183], SIMPSON once again defined the role that the sample plays in our interpretation of taxa. Rather than summarizing the same points in an equal amount of space a lengthy passage is quoted here.

'It has already been emphasized often enough that taxa are inherently variable and that attention to their variability is essential in their description and necessary in their practical definition. That naturally demands taking into account all available specimens and involves the principle that no one specimen referred to the taxon is, for these purposes, any more important or any more typical than any other. Some specimens are of course more nearly average than others as regards particular characters in the sense of being nearer the mean, although this is rarely true of all characters of one organism.

The mere fact that a valid average is recognized means that *all* specimens have been taken into account and none especially weighted.

That procedure does not mean that description of single specimens has been abandoned, ... One practical way to discuss variation is to describe in detail one specimen, which need not and often should not be the nomenclatural type, and to note differences in other specimens in the course of that description. In accompanying statistical characterizations, however, all specimens are of course simultaneously and equally involved.

In identification, comparisons of individual specimens are of course made routinely, and not necessarily with the nomenclatural type, which is often far from the average in pertinent diagnostic characters. Final decision as to conspecific status depends, however, not on nearness to any one specimen, type or other, but on falling within or outside ranges of variation inferred for the whole taxon. Throughout both definition and identification, the specimens are considered as a sample from which characteristics of populations are inferred. If only one specimen is known, it is necessarily both the type and the sole basis of concrete description, but population inferences drawn from it, and not the specimen itself, remain the basis for comparison and identification. The usable sample will of course increase as more specimens are collected and will be different for different classifiers with access to different collections. The proper basis for definition and comparison thus changes and, as a rule, improves as time goes on. That basis is not a fixed type that can be designated once and for all in the original description.'

The frequency with which this book is quoted and the general acceptance of most of its ideas might lead one to believe that it is now a standard practice to quantify the morphology of samples and use this data to infer the ranges of variation of the morphology of the whole taxon in order to ascertain membership in a particular group. A glance at the recent literature concerning early Tertiary mammals shows that the standards set out in *Principles of Animal Taxonomy* are rarely applied in practice, at least in this field. This is doubly unfortunate in this case because of the similarity between members of different orders in the early phases of the Eutherian radiation. One might expect to see more than the usual attention paid to variability in this situation.

Faunal papers especially tend to fall short of the standards set by the theoretical works in taxonomy and classification. It may be useful to examine the reasons.

Authors intending to write faunal papers are in a dilemma. On the one hand, there is the need to notify the scientific community in as great detail as possible of the existence of a new fauna or the extension of a previously known one. On the other hand, with the present proliferation of information, to complete such a task merely adequately demands great breadth of knowledge and presupposes a familiarity with a quantity of literature that is all but impossible to maintain, even if no extraneous pressures such as curating or lecturing existed. To adequately describe a population and make the proper statistical inferences before assigning specimens to conspecific or any other status is virtually never done in such papers. To do so would require not only the quantification of one's own collection, but also the quantification of samples of the species to which one's own specimens are compared. Adequate samples of the latter are generally not found in a single museum, and to my knowledge none have been described quantitatively in more than bivariate fashion, measured with varying degrees of comparability, as work by DELSON [1971] indicates.

Faced with this predicament, workers have either abandoned writing faunal papers or have published work which is essentially typological in nature, despite their convictions and occasional statements to the contrary.

It may be noted that despite the theorists' statements about accompanying the description and diagnosis of specimens with statistical characterizations involving all specimens, the only ones used routinely are analyses of length and width considered univariately or as ratios. Many of the characters listed in diagnoses are not treated statistically and hence no inferences are made as to their distribution in the population. Instead we see a frequent use of such terms as 'relatively longer', 'broader anteriorly than in …', etc., with the words in the comparative degree generally undefined.

It is virtually never explained how the specimens were measured, what their orientations were during comparison, or for that matter, if it was possible at all to orient them with respect to some criteria that other workers can duplicate.

II. Criteria for Adequately Characterizing Populations

Judging from SIMPSON's statement quoted above, a good morphological definition of a sample should describe in detail a specimen close to or at the average in pertinent diagnostic characters and provide statistical measures of the variation in these characters in which all specimens are equally involved. In view of the state of affairs in current literature, it should perhaps be noted that all characters must be considered simultaneously, as multiple univariate comparisons do not necessarily result in the same conclusions as a multivariate comparison using the identical measurements. Care should be taken to make the description complete enough not only to differentiate the population being characterized from previously described taxa, but also to distinguish it from any other possible shape. It should not be necessary to redescribe a population every time a new but related form is found.

Making the description complete enough must involve quantifying the vague comparative terms encountered so frequently. The problem here lies in the difficulty of measuring the characters referred to. These generally are curvatures of lophs or edges, features whose extremities are often ill-defined. They frequently run or fade into other features imperceptibly. It is difficult to describe their exact size and location because, as normally considered, there are no fixed, invariable points on a tooth or bone surface.

The increasing availability and in the early Tertiary often the sole availability of completely disarticulated material such as isolated teeth should be reflected in any methodology. Advantage should be taken of the fact that disarticulated specimens are more accessible to measuring devices because they are not obscured in any way by adjacent material.

The limitations imposed by small sample size are probably the most crippling to proper assessment of paleontological samples. Generally speaking, there must be at least one more specimen than there are variables under consideration, merely to establish confidence limits. These will be very large in such cases, and useful only to indicate that a larger sample is needed to make reliable statements about the population. This may be a negative conclusion, but nevertheless it is an important one to make if the literature is to become less cluttered by unnecessary synonyms.

Other considerations listed below are of a purely practical nature and the importance of these is seldom acknowledged, but in fact they may be the main reason why there is such a gap between theory and practice in vertebrate paleontology.

A method of describing a sample must possess the following characteristics to be of wide use in early Tertiary paleomammalogy:

(1) The mean configuration of the features of the sample's specimens must be established.

(2) Confidence regions within which a specified percentage of the sample's specimens will fall must be established around these means.

(3) The potential to use all of the specimens' features must be present. One should not be restricted to a particular subset of features such as meristic ones or measurements between fixed points.

(4) The method must be suitable for use with small samples.

(5) It should be usable on collections of isolated or disarticulated material.

(6) The end product must be concise and easy to grasp requiring a minimum of specialized knowledge.

(7) The method should be practicable in terms of time, cost, and equipment needs.

(8) As with any scientific procedure, the results should be duplicable.

III. Method

Equipment needs. The basic problem in quantifying the morphology of a specimen is mapping its topography in such a way that equivalent points on different specimens can be established and compared. Ideally such a method should involve the three-dimensional shape of the object. This is certainly possible at the present by working from stereophotos and using established photogrammetric methods. These methods are laborious and require equipment not generally available to most vertebrate paleontology departments.

A compromise can be made by considering only one view at a time and using the resulting line drawing, a type of representation recommended by MAYR [1969, p. 282]. This would reduce the equipment needs to a suitable microscope with a camera lucida attachment or a camera, and still allow full quantification in one plane.

The great amount of data to be generated and processed does require the use of a computer and plotter, both of which are becoming widely availably. The use of a digitizer would speed up parts of the process of collecting data, especially if hooked up to a magnetic tape or a keypunch. More important, its use would greatly reduce the number of errors made and consequently reduce the time, cost and difficulty of finding and correcting them.

Orienting the specimen. It is obvious that to compare two specimens precisely they must be in the same relative orientation or at least their different orientation must be known. This problem applies to all nonmeristic descriptive work and its lack of solution is one source of error common to all existing work of this kind. It is equally present in quantitative work as in the more usual efforts based on visual comparisons, although in the latter case it is impossible to check how much or how little error is introduced. The amount of error introduced this way probably varies from worker to worker as well as in time with the same individual.

If a three-dimensional map of a tooth surface were available, a non arbitrary reference plane could be set up from a regression surface of the heigth values of the specimen using the unworn, undamaged areas bounded by specified features such as cusps or lophs. However, as mentioned before, for lack of time, equipment and finances, our investigation is restricted to the two-dimensional case.

Any arbitrary orienting algorithm can be used with the understanding that the details of the resultant patterns of variation in the sample are artifacts of our orienting algorithm and are comparable only to similarly oriented samples. Within this restriction, the resultant pattern of variation uniquely characterizes the population measured. Whatever the algorithm, it should be applicable to all members of the sample and it should be repeatable. Reasons for choosing a particular orienting algorithm stem from our prior knowledge of the biology of the forms represented in the particular study.

After a specimen is oriented properly, it is either photographed or drawn accurately in this position. Using a camera lucida with a binocular microscope has the advantage (or disadvantage) that only those features we wish to measure the variability of need be drawn, and that these can generally be recognized more clearly through a microscope than from a photograph.

After drawing the desired features on tracing paper, a reference point is decided on. The same kind of considerations affect the choice of such a point as are involved in the previous procedure. To facilitate work it should lie centrally within the drawing and should be picked so as to avoid or minimize the number of reentrant angles.

Next an algorithm for orienting the drawing relative to the reference point and the measuring grid must be devised.

Once the drawing is placed over a polar coordinate grid with its center coinciding with the reference point and properly oriented the location of any point on the specimen can be measured. For linear features such as lophs, edges, circumferences, etc., the polar coordinates of points on these features at suitably small intervals are measured. Taking them in a set sequence will facilitate data processing. If measurements on successive specimens are taken in the same order, then this process will result in a data matrix of the same order for each speciment in which the nth measurements represent the same point on successive specimens and thus are directly comparable.

From this point on, the type and amount of data processing depend solely on the desire of the researcher.

Univariate or multivariate confidence limits of various types to the variability of points such as cusp locations or linear features can be calculated. Areas of various parts of the specimens, such as cusps, trigonids, etc., can be calculated and compared to other areas of the same specimen or of others. Samples or specimens can be sorted on the basis of one or more or any number of criteria.

Note that the method of measuring is open-ended. If more detail is required, more points are measured. Thus, it satisfies point three of our summary of requirements.

Test of orienting procedure. An advantage of having a standard algorithm for the orienting of specimens lies in the fact that we now can test how repeatable our results are or how much error is introduced by the technique. To do this, the same specimen is measured a number of times starting from the very beginning of the process. Superimposing the resultant plots shows at a glance the amount of variation introduced by the method as well as its location. By using different criteria for superimposing the plots, different results are obtained, giving an indication of the variance due to this part of the orienting procedure. By repeated experimentation in this fashion, one can arrive at an optimum method for orienting specimens (for a particular purpose), another advantage of this approach over traditional methods.

Data reduction. The result of the previous process will be a large number of measurements per specimen, describing much of the morphology. Most of the measurements, however, will be highly correlated. This means that in a typical early Eocene sample of 15 specimens we could use only 14 of these variables at the very most if any confidence regions were to be calculated.

Principal component analysis, developed by HOTELLING [1933] and others, can be used to replace the correlated variables by uncorrelated 'factors' or 'components' in such a way that significantly fewer factors can account for most of the variance expressed by the much greater number of correlated variables. Computer programs are available to do such ananalysis. A modification of the one provided by COOLEY and LOHNES [1971] was used here

Interpretation. Besides effecting a more parsimonious description of the data it is often possible in principal component analysis to identify the factors in their own right using output provided by the computer program. COOLEY and LOHNES [1971] and BLACK-ITH and REYMENT [1971] provide good explanations and, in the case of the latter, many illustrations from different fields. Identifying the factors facilitates interpreting and finding meaning in the factor scores of the specimens, as will be seen in the example using a sample of *Phenacolemur* teeth.

Confidence limits and tests of significance of difference can now be made using the factor scores in multivariate analysis of variance and HOTELLING's T^2 programs such as can be found in COOLEY and LOHNES [1971] and DAVIES [1971]. Scattergrams with confidence regions can be plotted of choosen factors. These plots are more useful than the traditionally used plots of length versus width because they account for much more of the variance in the sample.

From the original data a program, to be described in a publication of the Royal Ontario Museum, calculated the centroid of each sample data matrix and then uses this to provide a plot of the 'average' specimen with univariate confidence limits at each point. These provide an indication of the variability of different points relative to each other as a coefficient of variation would. The drawing of the mean values gives an accurate picture of the size and shape of the specimen that is the mean for the sample. If the sample was unimodal (ie. if the data were multivariately normally distributed), then this would correspond with the most common size and shape in the sample. It would be these dimensions that should serve as a basis of comparison between this sample and others. Figure 1 is an example.

To check whether the sample is unimodal and to provide an easily comprehensible way of displaying the total variability in the sample, the raw data are plotted with all specimens superimposed over their reference centers and in the same orientation.

Frequently, size is not an important consideration at some stage in the analysis of a particular sample. To aid in the understanding of the variability in shape, in such instances, the entire data matrix is replotted, but with each specimen transformed to the same size. Examples of each of these plots can be found in the figures with this article, which were taken directly from the output of a Calcomp plotter.

A length-width scattergram is also made, primarily because this sort of plot is frequently encountered in the literature.

Assumptions. For the statistical tests to be valid, the characters measured in the samples tested must have a multivariate normal distribution and have equal dispersions. Tests for the latter exist and can be found in COOLEY and LOHNES' [1971] MANOVA program.

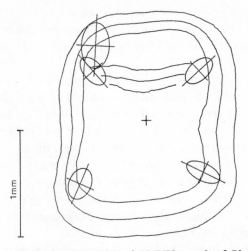

Fig. 1. Small-sized M_2 of AMNH sample of *Phenacolemur*. Table of 95-percent bivariate confidence areas for the cusp locations and 95-percent univariate confidence limits for points on the circumference and on the crest between protoconid and metaconid. The middle line of each group of three indicates the mean for the sample (n = 14) for that particular point, while the other two lines bordering this one mark the 95-percent confidence limits of the mean for these points. Endpoints of axes of ellipses mark the 99-percent ellipse. The data matrix for this sample is plotted in figure 7c. The millimeter scale shows 1 mm. The same scale applies to all plots shown in other figures.

No adequate way of testing the former exists at present. In dealing with teeth we have the good fortune that the crowns, once erupted, have no further growth, and that post-eruption changes, such as produced by wear, are detectable. This makes it more likely that our data are normally distributed. Collecting samples from a single stratigraphic horizon, preferably from a single quarry, increases our chance of sampling a single population and with it the probability that our data will have the desired distribution. If samples contain mixtures of populations, the sample variances are less likely to be homogeneous, increasing the difficulty of treating them statistically.

IV. *Systematics of Samples of* Phenacolemur

Samples. To demonstrate the use of polar coordinates in analyzing morphology, two samples of *Phenacolemur* (Paromomyidae, Primates) are used. The first, from the Royal Ontario Museum (ROM) and the National Museum of Canada (NMC) collections, is from the early Eocene Willwood

Formation of Northwestern Wyoming. The second sample is from the Hiawatha member of the Wasatch Formation of Northwestern Colorado and was collected by the staff of the American Museum of Natural History (AMNH).

The AMNH sample, consisting entirely of isolated teeth, was from the same locality as the University of California sample described by McKenna [1960] who drew attention to its great variability but reserved further comment pending the discovery of more complete material. It was collected entirely from a 1-inch-thick stratigraphic interval from a single quarry. Because the first and second molars in *Phenacolemur* are very similar, it was impossible to separate them visually with any degree of confidence.

The ROM and NMC material, collected over the years by surface prospecting are from a variety of stratigraphic horizons. They cannot always be precisely allocated to a specific level because of the method of collecting and the difficulty of tracing beds in the Willwood formation. The samples are too small to be considered separately and of necessity must be pooled to form one sample coming from a 300-foot-thick stratigraphic interval. This aggregate sample, all from the lower Gray Bull fauna as used by Simpson [1955], is composed of a number of isolated teeth as well as some jaw fragments in which it is possible to identify each molar with certainty. Our problem then is to sort out from this very variable group the first from the second molars, as well as to determine whether or not more than one species, or at any rate, distinct populations of *Phenacolemur* are present. Because of space limitations, the premolars and the upper molars are not treated here. Their analysis duplicated that of the other molars in every respect and led to same conclusions. A detailed treatment would add to the length, but not to the substance of this paper.

Orienting procedure. For *Phenacolemur* lower first and second molars the method of orienting specimens consisted of the following steps.

(1) The tooth was approximately aligned under the cross hairs of the microscope with its anteroposterior axis coinciding with the vertical cross hair.

(2) The tooth was rotated around its lingual-buccal axis until the distance between its posterior edge visible in crown view and the most posterior part of the crest joining the hypoconid and the entoconid was $1/15$ of the total length of the tooth, i. e. in figure 2A distance x is $1/15$ of the total length.

This step in my opinion was the least satisfactory part of the orienting algorithm for these particular teeth. It made the orientation around one of

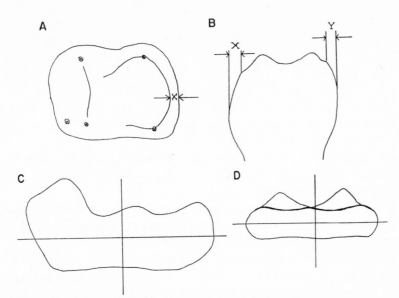

Fig. 2. A Orientation of M_{1-2}. Crown view of tooth. Specimen is rotated around its lingual-buccal axis until distance x, the distance between the posterior end of the lophid joining hypoconid and entoconid and the posterior margin of the tooth is $1/_{15}$ of the total length of the tooth. *B* Orientation of M_{1-3}. Lingual-buccal section through trigonid. Specimen is rotated around its anteroposterior axis until, in a crown view, the amount of the *sides* of the tooth showing lateral to the protoconid and metaconid is equal. That is, distance x equals distance y. *C* Orientation of M_3. Side view of tooth through eyepiece of microscope via a mirror on stage. Tooth is rotated around its lingual-buccal axis until the vertical cross hair bisects the area of the side of the talonid. *D* Orientation of M^3. Posterior view of tooth through eyepiece of microscope via a mirror on stage. Tooth is rotated around its lingual-buccal and anteroposterior axes until neither top nor bottom surface of the tooth is visible and the vertical cross hair bisects the area of the posterior side of the talon visible.

the axes of the tooth dependent on a rather minor feature that was difficult to adjust through the microscope eyepiece, even if it was supplied with a grid. In the sample of 65 teeth used, the ratio of 'x' to the length of the teeth after they were finally drawn proved to vary from $1/_{10}$ to $1/_{21}$. The advantage of the method is that the researcher can set his own standards of precision, simply by redrawing those specimens which were not within the specifications he set himself. In this case, in order to find out how much the subsequent analysis would be affected by defects of this sort, the drawings were left as they were.

(3) In a crown view, the tooth was rotated around its anteroposterior axis until the areas of the *side* of the tooth visible laterally to the protoconid and the metaconid were equal. This would mean rotating the tooth until distance 'x' equalled distance 'y' in figure 2B.

(4) At this point the tooth was drawn.

(5) The intersection of the protoconid-entoconid and the metaconid-hypoconid lines was designated as the reference point to be superimposed on the origin of the polar coordinate grid.

(6) The bisector of the metaconid-reference point-protoconid angle was taken as the anteroposterior axis of the tooth, and made to coincide with the 90-degree axis on the polar coordinate grid.

Different orienting procedures proved necessary for the third molars. The lower third molars were oriented as the M_{1-2} except for step (2), for which the following was substituted:

In a lateral view, at 90° rotation from the crown view, (obtained with the use of a mirror mounted on the stage at exactly 45° from horizontal, and parallel to the vertical cross hair), the tooth was rotated around its lingual-buccal axis until the vertical cross hair bisected the area of the talonid as shown in figure 2C.

For the third upper molars, a rather different scheme was used:

(1) In a crown view, the tooth was approximately aligned with the bi-sector of the paracone-protocone-metacone angle parallel to the vertical cross hair.

(2) In a posterior view, the tooth was rotated around both the lingual-buccal and the anteroposterior axes until the vertical cross hair bisected the posterior side of the talon when the area of the talon visible in this view was minimized as in figure 2D, i.e. with neither the crown surface nor the underside of the tooth visible. As this tooth is roughly triangular in shape with one apex of the triangle at the extreme posterior end of the tooth, this single step will orient the specimen in a plane defined by the anteroposterior and the lingual-buccal axes.

(3) The tooth was drawn.

(4) The bisector of the paracone-protocone-metacone axis was made to coincide with the 90-degree axis of the polar coordinate grid.

(5) On this bisector, the midpoint of the line segment bounded by the protocone and the intersection of the bisector and the line joining the paracone and metacone was designated as the reference point and made to coincide with the origin of the polar coordinate grid.

Other orienting procedures are possible and perhaps preferable. This one does satisfy the basic requirement of repeatability and in *Phenacolemur*, where tooth wear is seldom extreme, the vast majority of specimens can be used. In other genera where a wide range of wear is evident, teeth with the same degree of wear may have to be used, or else, a procedure not using cusp extremities should be used.

No single procedure was found that could deal effectively with all types of teeth.

Measuring. To facilitate comparisons, all left teeth were transformed into right teeth by reversing the tracing paper before measuring. The distance from the origin to the edge of the tooth was then read off in 5-degree intervals and recorded. For the lower teeth, the location of the crest between proto-conid and metaconid was recorded in similar fashion as well as the coordinates of the main cusps. For M^3, the anteroexternal cingulum was also recorded.

Relative to the origin, the trigonid crest begins and terminates in varying places in different teeth, with the result that in the data matrix the ends of the crest are not represented by the entire sample for some values of the angular measure. For this reason, the confidence belt around those values is larger. In the principal component analysis, missing data, recorded as zero, may have great effect and care should be taken to eliminate those elements in the data matrix that do not have their full complement of measurements.

Data reduction. First and second lower molars. The data were then put through the computer program which calculated and plotted mean values, univariate confidence limits, etc., as outlined before. The AMNH sample was processed as one unit at this stage, while the ROM sample was split into three: two containing either M_1 or M_2 only and the third containing the isolated teeth which could not be identified with certainty. The graphical output for the AMNH sample is shown in figures 3–5. It suggests strongly that this sample is trimodal. The equiprobability ellipses in figure 5 are based on the (here evidently wrong) assumption that the sample is taken randomly from a single population with a bivariate normal distribution. They define, from inner ellipse to outer (1) the 95-percent confidence area for the bivariate population mean, (2) the 95-percent confidence region containing the area wherein a single specimen being compared to the sample would fall, and (3) the 99-percent confidence limits for this kind of ellipse. To avoid clutter, this last ellipse is not drawn, but its location is defined by the ends of the axes. The bivariate mean is located at the intersection of the axes.

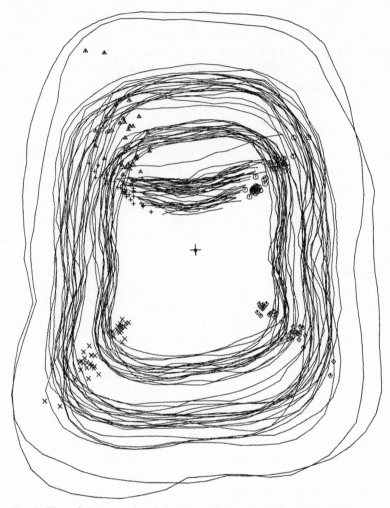

Fig. 3. Plot of data matrix of the M$_{1\text{-}2}$ of the AMNH sample of *Phenacolemur*, showing the reference point (+), circumference, crest joining anterior cusps and the location of the apices of the five main cusps of each tooth. Note the threefold size division.

Next the data of all samples together plus that of three type specimens of species and subspecies of *Phenacolemur* were run through the principal component analysis program. Since adjacent measurements on a crest or tooth edge are extremely highly correlated, only every other one was used. This resulted in a matrix of 53 variables for each tooth. If the only matter of importance was the treatment of this particular sample many fewer points

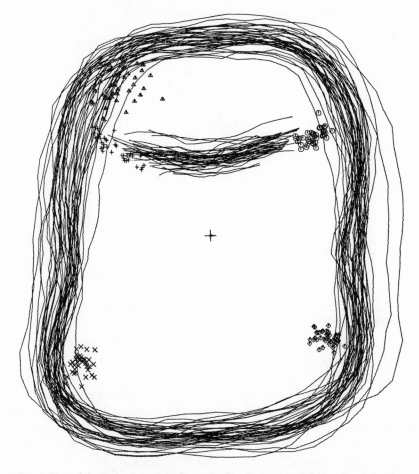

Fig. 4. Plot of the same data as in figure 3, after the measurements of each tooth are standardized so that all teeth are drawn with the same length, but keep their original proportions. Note the bimodality of some of the cusp locations and of some areas of the circumference.

on the crest and the tooth edge would have sufficed; however, it may be useful to keep the results to test future finds against them. Data points describing the outline or crests should be close enough together, so that small cingular cusps or bulges are not missed, as they might be if measurements were taken further apart than 10°.

The program was run so as to extract the first ten principal factors,

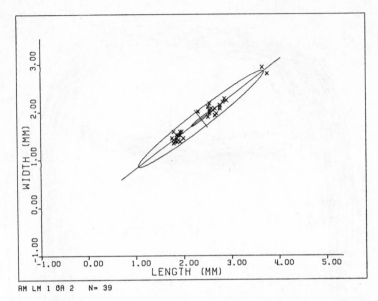

AM LM 1 OR 2   N= 39

Fig. 5. Length/width scattergram of the AMNH M_{1-2} sample (n = 39). The meaning of the ellipses is explained in the text.

which accounted for 98.6% of the total variance in the data matrix. As the sample was only moderately large, following COOLEY and LOHNES [1971], only those factors with eigenvalues of over 1.0 were retained for further analysis. These four factors accounted for 94.2% of the variance in the data matrix. With the M_3, again the first four factors were found to have eigenvalues larger than 1.0, and these were kept for further analysis. They contained 94.2% of the variance. With the upper third molars, three factors were retained accounting for 9.13% of the variance.

Interpretation. The main point of this article is to illustrate a way of quantifying the morphology of a sample and not how to analyze the resulting data matrix. Hence, no complete analysis of the factor scores is undertaken here. For the same reason, only a fraction of the computer output is presented here. What is presented is meant to indicate the type of output available for each sample and subsample. It is on the basis of the aggregate of these outputs that the conclusions are based.

Lower first and second molars. Interpretation of the factors proved rather simple, both from the factor pattern and from other ways of displaying the data. (The factor pattern is a matrix providing the correlations of each vari-

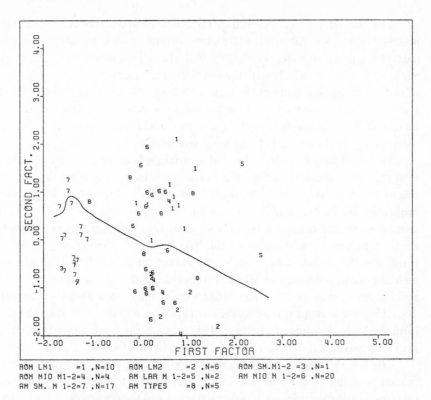

Fig. 6. Scattergram of the first and second factors of all M_1 and/or M_2. Line divides M_1 (above) from M_2 (below) as separated by the discriminant function. American Museum types (AM types), designated by the symbol 8 on the plot are, from smallest to largest on the axis of the first factor: M_1 of *P. jepseni simpsoni*, M_1 of *P. citatus*, M_2 of *P. citatus*, M_1 of *P. praecox*, and M_2 of *P. praecox*. ROM and AMNH (AM) samples are sub-divided into small (SM), middle (MID), and large (LAR) sized groups.

able with each of the factors.) All variables had a high correlation with the first factor, and, as all had the same sign as the corresponding variable in the data matrix, this factor can be interpreted as an overall size factor. RAO [1964] gives the mathematical justification for this reasoning. A comparison of a plot of the first factor with the length/width scattergram (fig. 5, 6) bears out this interpretation if all entries on the plots are identified. Output from the principal component analysis program indicates that this size factor accounts for 83.0% of the variance in the total sample. Qualitatively the same could be learned from a comparison of the standardized and unstandardized plots of the data matrix (fig. 3, 4).

The second factor, accounting for 5.6% of the variance in the data matrix, is shown to be relatively highly correlated with the anterointernal part of the tooth outline, very highly with the degree of anteriorness of the paraconid, and very highly with the location of the metaconid. A comparison of the superimposed plots of the teeth definitely known to be M_1 with those known to be M_2 shows that these are the areas where the first and second molars differ. Factor two then is a factor describing those parts of the morphology that distinguish first from second molars.

The third factor, with 3.5% of the variance, is correlated strongly only with the measurements of the crest between protoconid and metaconid. As this is the highest part of the tooth and centrally located, it is the part proportionally most heavily influenced by the orientation of the tooth around its buccal-lingual axis. This suggests that this factor may predominantly reflect the inadequacies of this aspect of the orientation procedure. If this was the case, however, one would expect the correlation of this factor with the length coordinates of the protoconid and metaconid to be of the same order or greater than the correlations of this factor with the variables describing the crest positions, as the cusp apices are in line with the crest and form the highest points of the teeth. This is not so, indicating that the orienting error was not the predominant influence in this factor.

The fourth factor, with 2.6% of the variance of the data matrix is correlated to some extent with the buccal side of the teeth and the posterior margins. As our sample size is only moderate, and the correlations low, interpretation may not be warranted. Factor rotation might aid in identifying this factor.

The factor scores of all teeth whose relative position in the molar row was certain were analyzed in a discriminant function program for two groups to obtain discriminant coefficients that would enable us to subdivide the rest of the specimens into first or second molars. This program, here slightly modified, is described in DAVIES [1971]. The discriminant coefficients were 3.074, 12.945, —1.358 and —4.556. Note that the heavy weighting of the second factor supports our interpretation that this factor represents a shape factor distinguishing first from second lower molars. MAHALANOBIS' $D/2$ for the two groups was 2.645 indicating that if the variables of each group were distributed normally and their variances were homogeneous (assumptions underlying the whole test) then about 99.6% of the rest of our sample would be classified correctly using these discriminant coefficients [DAVIES, 1971, p. 291]. Hotelling's T^2 test gave an F value of 29.18 with 4 and 16 degrees of freedom showing that the two samples are significantly different

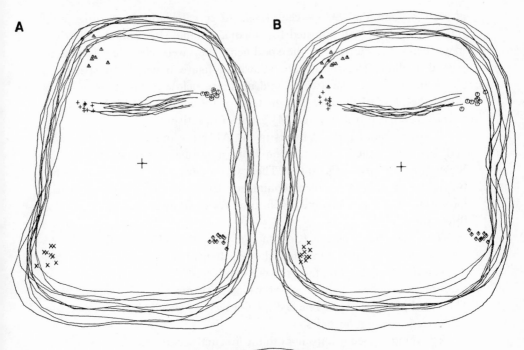

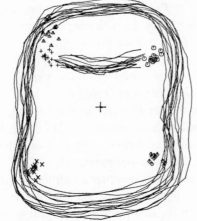

Fig. 7. Plot of data matrix. *A* AMNH M_1 middle-sized group, unstandardized. *B* AMNH M_2 middle-sized group, unstandardized. *C* AMNH M_2 small-sized group, unstandardized.

at well beyond the 99.9-percent level. After calculating a cutting point [Davies, 1971, p. 290] adjusted for unequal sample size, and the discriminant scores for the rest of the specimens, they were classified as first or second molars. The results are shown in figure 6 where all specimens above the dividing line are M_1 and all those below it M_2. Figure 7A, B shows the plots of the raw data matrix for the middle-sized AMNH specimens allocated to the M_1 and M_2 groups respectively.

The lengths of the specimens in the AMNH samples of M_1, M_2, M_3, and M^3 were tested for deviation from a normal distribution with the Kolmogorov-Smirnov (K-S) test. There was a deviation from normality between the 80- and 95-percent confidence levels for the various samples, indicating that there is some justification for dividing the samples into the three size groups.

The Manova program [Cooley and Lohnes, 1971, p. 230], run when the sizes of the samples permitted, showed that when the factors were tested one at a time only the first factors were significantly different, indicating that size, and not shape, was the basis on which the samples were in fact divided. This conclusion should be reached qualitatively as well by inspecting figures 7B, C, 3, 4, and similar pairs.

The middle-sized groups in both the first and second molar samples contained both the types of *Phenacolemur praecox* and *Phenacolemur citatus* as defined by Matthew [1915]. In reviewing the species, Simpson [1955] decided, on the basis of a t-test on the lengths, that there was a significant difference between Granger's samples from the lower Gray Bull and Sand Coulee levels, i.e. *P. praecox*, and those from the Upper Gray Bull level, i.e. *P. citatus*. Although Simpson [1955] listed it, he chose not to include Jepsen's sample from the Sand Coulee level in his test, thus going against his later advice that population inferences should be based equally on all available specimens. Inclusion of Jepsen's sample would have altered the result of the t-test as one of the three specimens, as Simpson recognized, clearly falls into the *P. citatus* group and the other straddles the difference between the two. More important, however, it would alter or remove the logical basis for making the t-test in the first place as now there is no evidence that small size is simply correlated with stratigraphic level. If we still wish to posit that there are two species of *Phenacolemur*, differing only in size, in the Willwood formation, it must be acknowledged that they were contemporaneous at least in the Lower Gray Bull. If this is so, then a t-test showing that the larger specimens differ in size from the smaller ones proves little. To prove the existence of two contemporary sympatric species or subspecies, differing only

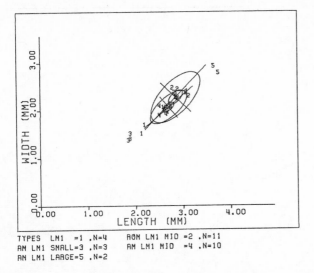

TYPES LM1 =1 ,N=4   ROM LM1 MID =2 ,N=11
RM LM1 SMALL=3 ,N=3   RM LM1 MID =4 ,N=10
RM LM1 LARGE=5 ,N=2

Fig. 8. Length/width scattergram of *Phenacolemur* M_1. Types, designated on the plot by the symbol 1, from smallest to largest are of: *P. jepseni simpsoni*, *P. j. jepseni*, *P. citatus* and *P. praecox*. The smaller set of equiprobability ellipses refers to the AMNH middle-sized sample, the larger set to the ROM sample. The sets of ellipses have the same meaning as those in figure 5, explained in the text.

in size, it must be shown at least that the sample containing both the posited forms does not have a normal size distribution. A K-S test for SIMPSON's data shows that there is no evidence for rejecting the null hypothesis that they are normally distributed. The test value is far below the critical value for the 80-percent confidence level. The ROM sample, entirely from the Lower Gray Bull falls all but one specimen between the two types, thus reducing even more the probability that there were two species present. This article, however, is not the proper place for a taxonomic revision.

The AMNH middle-sized samples are slightly smaller than the ROM middle-sized ones, as is shown by figures 6, 8 and 9. The difference is significant between the 90- and 95-percent confidence levels in the case of the M_2 and between the 97.5- and the 99-percent level in the case of the first lower molars. Statistical significance is not the same as biological meaningfulness, however, and considering the difference in locality and the probable time interval I do not think that it is enough to warrant the erection of separate taxa in this case. When the samples are pooled, a K-S test does not show any significant deviation from normality for the combined sample.

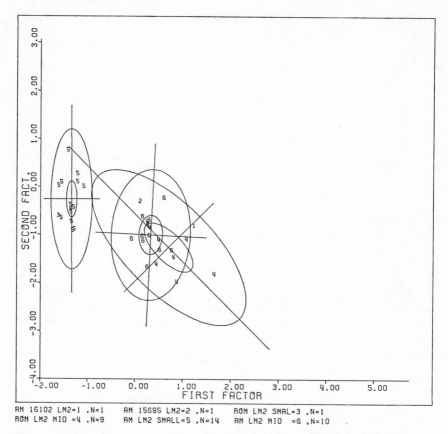

AM 16102 LM2=1 ,N=1   AM 15695 LM2=2 ,N=1   ROM LM2 SMAL=3 ,N=1
ROM LM2 MID =4 ,N=9   AM LM2 SMALL=5 ,N=14   AM LM2 MID =6 ,N=10

Fig. 9. Scattergram of the first and second factors of *Phenacolemur* M₂ as calculated by the Principal component analysis program. The largest set of ellipses refers to the ROM middle-sized sample; the medium-sized set to the AMNH middle-sized sample and the smallest set to the AMNH small-sized sample.

The types of *P. praecox* (AMNH 16102) and *P. citatus* (AMNH 15695) were compared with the middle-sized samples after pooling the ROM and AMNH specimens. Using Hotelling's T² test on the first four factors, the only significant difference found was between the M₂ of *P. praecox* (AMNH 16102) and the M₂ sample. Considering that it was the second largest tooth out of a sample of 19 this is not surprising. Figures 6, 8 and 9 summarize part of this data. They also indicate the confidence areas if only the first two factors are considered and the separate ROM and AMNH samples are used as opposed to the pooled samples. Although other sub-

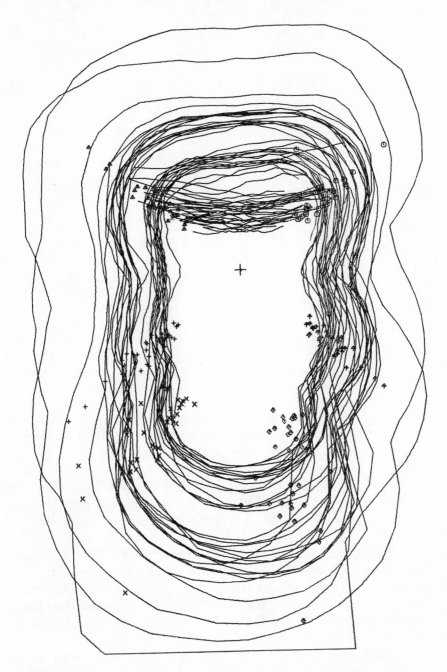

Fig. 10. Plot of data matrix of AMNH *Phenacolemur* M₃ sample. Note the three-fold size division. Due to size limitations of the plotter, the posterior end of the largest specimen in shown truncated.

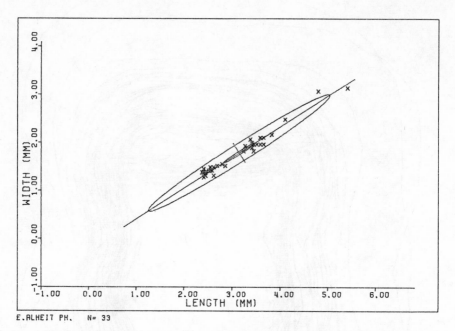

E.ALHEIT PH.   N= 33

Fig. 11. Length/width scattergram of AMNH *Phenacolemur* M₃ sample. Note the threefold size division. The ellipses are drawn as in figure 8.

samples are plotted as indicated, the ellipses are calculated only for the AMNH small- and middle-sized samples and for the ROM midlle-sized sample. Where two ellipses overlap, in these particular figures, the one referring to the ROM sample is always the largest.

When all the evidence is considered, there seems to be no compelling reason to assume that the middle-sized samples represent more than one population. As is always the case, further collecting could alter this conclusion.

Third molars. With the M_3, the first four factors were found to have eigenvalues larger than 1.0 and were retained fort further analysis. These factors contained 84.2, 4.4, 3.6 and 2.0% of the variance in the data matrix respectively for a total of 94.2%. With the upper third molars, three factors accounting for 71.5, 16.2 and 3.6% of the variance were retained for a total of 91.3%.

In both cases, the first factor proved to be a size factor. With the M_3, the remaining factors were so small as to demonstrate only how uniform the shape of the specimens were throughout the sample.

The M^3 sample was quite small (n = 14) and showed the division along

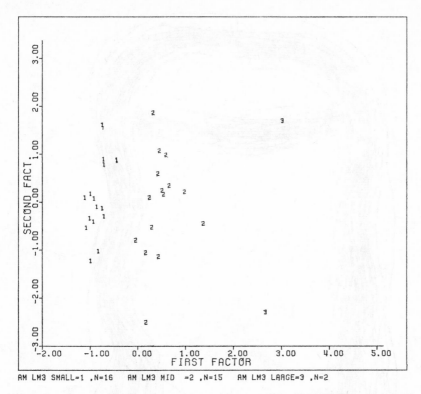

Fig. 12. Scattergram of the first and second factors of the AMNH *Phenacolemur* M$_3$ sample.

size lines much less clearly. If such a division were to be made, two specimens would go into the small and large groups each with the remaining ten forming a medium-sized group.

The second factor here was correlated strongly and negatively with the measurements of the anterior edge of the teeth and less so, but positively, with those of the posterior edge. It is thus an indication of the location of the cusps along the anteroposterior axis relative to the edges of the specimen.

With the M$_3$, there was a clear division into three size groups as shown by figures 10–12, with little variability in shape as figure 13 shows. The MANOVA test again led to the same conclusion.

Use of Hotelling's T^2 test showed that the AMNH middle-sized sample was not significantly different from the ROM sample, i.e. *P. praecox*. The discriminant coefficients showed that of what little difference there was, size was the most important factor by far.

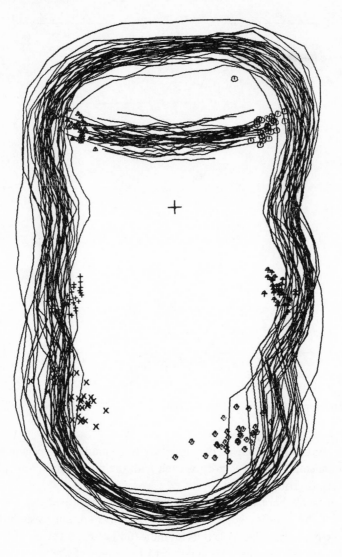

Fig. 13. Plot of data matrix of AMNH *Phenacolemur* M$_3$ sample, standardized to a uniform size.

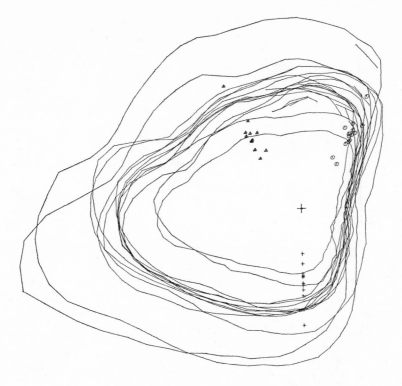

Fig. 14. Plot of data matrix of AMNH *Phenacolemur* M³ sample, unstandardized.

With the upper third molars, the division into three size groups was much less clear, with the widths showing the division more clearly than the lengths. Figures 14–17 show the relevant plots. Again, the AMNH middle-sized group was inseparable from the Willwood *P. praecox* in the ROM sample.

Conclusion. The isolated teeth of the AMNH sample consistently show a threefold division on the basis of size only. The middle-sized group is of the same shape as *Phenacolemur praecox* from the Lower Gray Bull level of the Willwood formation and is slightly smaller, but not enought to warrant erection of a separate species or subspecies. The small-sized subgroups in the AMNH and ROM samples definitely represent an undescribed form. (It is at present being described at the Yale Peabody Museum from better material collected by their staff from the Willwood Formation.) The large-

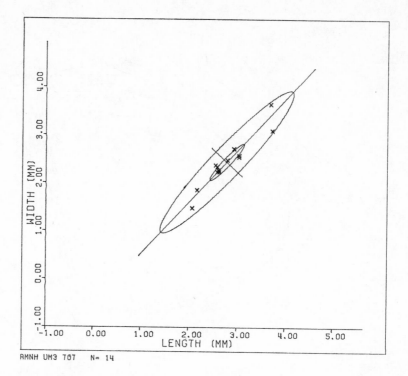

AMNH UM3 707   N= 14

Fig. 15. Length/width scattergram of AMNH *Phenacolemur* M³ sample. Ellipses drawn as in figure 5.

sized group in the AMNH sample in all probability represents another un-described species, not found in the Willwood formation thus far. However, as there is very little of this material, description is best postponed until more and hopefully better specimens are obtained.

V. Summary

Polar coordinates can be used to quantify shapes on which homologous points cannot be readily identified such as lophs, edges, circumferences, outlines of areas, etc., This allows one to describe quantitatively characters, often diagnostic, that at present are not usually quantified and hence do not figure in the statistical analysis of variability that ought to accompany morphological descriptions in taxonomic work.

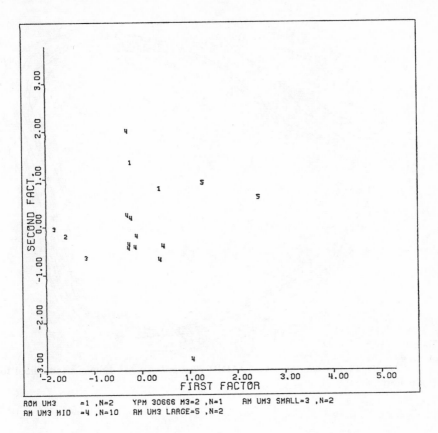

Fig. 16. Scattergram of first and second factors of *Phenacolemur* M³ sample.

Computer-driven plotters, using this data, provide an efficient means of graphing it, with great accuracy, in a variety of ways that can be used as an aid in qualitatively assessing variability in a sample or as a basis for further quantitative work.

Principal component analysis of the data is essential to reduce the great quantity of measurements to a more manageable amount. It is also a valuable aid in interpreting the information.

In the example using the *Phenacolemur* sample, it was shown with the data obtained from the use of polar coordinates in measuring that:

(1) Discriminant functions can be used to distinguish between teeth that are very similar in shape and size, such as the *Phenacolemur* M_1 and M_2, and with which it is difficult to make such a distinction visually in an objective fashion.

(2) It can be shown quantitatively that size is the only valid distinguishing criterion in splitting the AMNH *Phenacolemur* sample into three species and that the variability in shape is not significant.

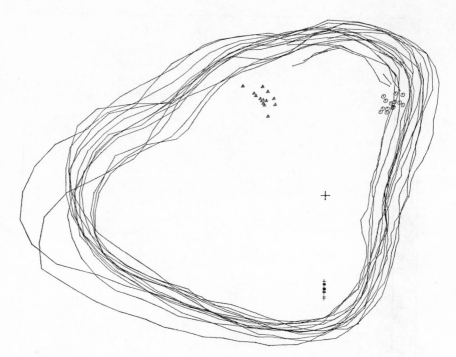

Fig. 17. Plot of data matrix of AMNH *Phenacolemur* M³ sample, standardized to a uniform size.

VI. *Acknowledgements*

I wish to thank Dr. DALE A. RUSSELL of the National Museum of Canada, Dr. MALCOLM MCKENNA of the American Museum of Natural History, and Dr. ELWYN L. SIMONS of Yale University for making the collections at these institutions available to me for study. Dr. FRED SZALAY of Hunter College, Dr. J. E. PALOHEIMO and Dr. F. G. SMITH of the University of Toronto have contributed much valuable advide. Dr. A. G. EDMUND and Dr. C. S. CHURCHER have critically read the manuscript of this paper and offered useful suggestions for its improvement. Dr. L. S. RUSSELL and Dr. A. G. EDMUND, under whose direction this project was done, have contributed much in time, advice and encouragement. Workspace and collections were provided by the Royal Ontario Museum through Dr. A. G. EDMUND. The author was supported while working on this project by funds from Province of Ontario Grants, from National Research Council of Canada grants to Dr. L. S. RUSSELL and by funds from the Royal Ontario Museum through Dr. A. G. EDMUND.

VII. References

BLACKITH, R. E. and REYMENT, R. A.: Multivariate morphometrics (Academic Press, London 1971).

COOLEY, W. W. and LOHNES, P. R.: Multivariate data analysis (Wiley, New York 1971).

DAVIES, R. G.: Computer programming in quantitative biology (Academic Press, London 1971).

DELSON, E.: Fossil mammals of the early Wasatchian Powder River local fauna, Eocene of Northeast Wyoming. Bull. amer. Mus. nat. Hist. *146:* 305–364 (1971).

HOTELLING, H.: Analysis of a complex of statistical variables into principal components. J. educ. Psychol. *24:* 417–441, 498–520 (1933).

MATTHEW, W. D.: A revision of the lower Eocene Wasatch and Wind River faunas. IV. Entylonychia, primates, insectivora (part). Bull. amer. Mus. nat. Hist. *34:* 429–483 (1915).

MAYR, E.: Principles of systematic zoology (McGraw-Hill, New York 1969).

McKENNA, M. C.: Fossil mammalia from the early Wasatchian Four Mile fauna, Eocene of Northwest Colorado. Univ. Calif. Publ. geol. Sci. *37:* 1–130 (1960).

RAO, C. R.: The use and interpretation of principal components analysis in applied research. Sankhyā *26:* 329–358 (1964).

SIMPSON, G. G.: The Phenacolemuridae, new family of early Primates. Bull. amer. Mus. nat. Hist. *105:* 411–442 (1955).

SIMPSON, G. G.: Principles of animal taxonomy (Columbia Univ. Press, New York 1961).

Authors' address: Dr. PAUL RAMAEKERS, Department of Vertebrate Palaeontology, Royal Ontario Museum, *Toronto, Ont.* (Canada)

In SZALAY: Approaches to Primate Paleobiology
Contrib. Primat., vol. 5, pp. 136–166 (Karger, Basel 1975)

Phylogenetic Relationships of *Plesiadapis* – Postcranial Evidence

FREDERICK S. SZALAY, IAN TATTERSALL and RICHARD LEE DECKER

Hunter College, CUNY, and The American Museum of Natural History, New York, N.Y.

Contents

I. Introduction

Our aim in this paper is neither to describe completely the postcranial morphology of *Plesiadapis* nor to present a detailed functional analysis. We aim, rather, to make a preliminary assessment of the phylogenetic affinities of *Plesiadapis* in light of the relevant postcranial remains. We attempt to show the relative recency of common ancestry among *Plesiadapis* (and other paromomyiforms), adapids and other higher primates, and further, between the former and such relatively primitive eutherians as the condylarths, carnivorans, hyaenodontans, erinaceotan and palaeoryctoid insectivorans.

Our analyses suggest that the postcranial morphology of *Plesiadapis* shares a host of derived character states with adapids, as has been amply demonstrated for its cranium. This compels us to conclude that *Plesiadapis*, along with the other groups of paromomyiform primates, shares a more recent common ancestor with 'undoubted' primates than with any other known group of the Eutheria. In addition, an (admittedly preliminary) functional analysis of *Plesiadapis* seems to indicate an arboreal mode of life, as do the remains of the earliest lemuriform or the more derived tarsiiform primates.

In the relevant sections of his three-part monograph on the Tiffany fauna, Simpson [1935a, b] has dealt with the history of study and discovery of *Plesiadapis* up to that time. It seems unnecessary to reproduce that history here. However, since the publication of Simpson's paper in which he described what was then the most complete known material of a Paleocene primate (the jaws, teeth, and postcranial specimens of *P. gidleyi* from the Mason Pocket of the Tiffany beds, south-western Colorado), several major new discoveries of *Plesiadapis* material have been made.

Foremost among the excavations leading to the latter are those carried out in 1958/59 by Dr. Donald E. Russell, of the Musée National d'Histoire Naturelle, Paris. Some three-quarters of a century after they had originally been worked, Russell returned to the classic Thanetian (late Paleocene) deposits near Cernay-les-Reims (France), where Victor Lemoine had recovered the type material of *Plesiadapis tricuspidens* Gervais 1877. From a new locality, the Mouras quarry, Mont de Berru, Russell recovered a crushed but complete skull of *P. tricuspidens*, numerous upper and lower jaws and isolated teeth, and a variety of postcranial bones, some of them associated [see D. E. Russell, pp. 28–61].

Full descriptions of the skull and dentition have already been provided by Russell [1964] and later comparisons of the ear region [Szalay, 1972] have unequivocally established that the bulla was formed by the petrosal, thus firmly aligning *Plesiadapis* with other primates. Later excavations in the same area by M. Pierre Louis, a paleontologist from the region, led to the recovery of further postcranial material of *Plesiadapis*, some of which is discussed here through the courtesy of M. Louis.

Subsequent work was carried out by Russell in 1960 at Lemoine's original quarry at Cernay. Finds of *Plesiadapis* from this site consisted largely of isolated teeth. According to P. D. Gingerich [personal commun.] two species of *Plesiadapis*, one large and one small, are represented at Cernay, the latter preponderating. The large species is, of course, *P. tricuspidens*, while the prior name for the smaller one is *P. remensis* Lemoine. The dental evidence indicates that *P. tricuspidens* is almost exclusively represented among the dental remains from Mont de Berru, although there is some indication that *P. remensis* may have been present. This is also borne out by the postcranial bones from the site which, with very few exceptions, are large in size and obviously attributable to *P. tricuspidens*.

In 1967, Russell published the results of a study of the type and only skeleton of '*Menatotherium insigne*' reported by Piton in 1940 from early late Paleocene coal deposits in southern France. Russell concluded that the specimen is in fact attributable to *Plesiadapis*. Unfortunately, the preservation of this individual, geologically somewhat older than the Cernay and Mont de Berru materials, is inadequate for detailed anatomical study. Approximately contemporaneous with '*Menatotherium*', however, is *Plesiadapis walbeckensis* from fissure-fill deposits at Walbeck (DRG), of which several specimens were recovered by Weigelt in 1939. Later work by Dr. G. Krumbiegel led to the recovery of further material in 1960, and through the courtesy of Dr. Matthes of Halle we have been able to include the postcranial bones of *P. walbeckensis* in this survey.

Although most recent authors who have studied *Plesiadapis* [e. g. Simpson, 1935a, b, 1940; LeGros Clark, 1959; Russell, 1964; Russell et al., 1967; McKenna, 1967; Szalay, 1968, 1972; Simons, 1972; Tattersall, 1970] have been unanimous in regarding the form as a primate, suggestions to the contrary have appeared in the literature of the past several years. Thus, for instance, Martin [1972, p. 302] believes that the allocation to Primates of the 'bizarre' Plesiadapidae is 'of questionable value'. But despite the fact that Martin has strongly advocated something closely akin to a purely cladistic approach to primate classification, his reasons for wishing to exclude Plesiadapidae from the order would appear to have more to do with notions of grade than of clade.

Following a somewhat different but not entirely unrelated line of reasoning, Cartmill [1972] has also argued for the exclusion of *Plesiadapis* from Primates. Cartmill believes that 'the ancestral primate adaptation involved nocturnal, visually-directed predation on insects in the terminal branches of the lower strata of tropical forests' [p. 121]; by extension, '...a monophyletic and adaptively meaningful order Primates may be delimited by taking the petrosal bulla, complete postorbital bar, and divergent hallux or pollex as ordinally diagnostic' [p. 121]. Such a definition would, of course, eliminate Plesiadapidae and other representatives of the Paleocene radiation from consideration as primates. Irrespective, however, of whether or not Cartmill's initial hypothesis is valid, if these latter forms can be shown to share nonconvergent derived character states with those Eocene and later forms which are universally agreed to be primates, i. e. that they can be regarded taxonomically as the sister group of the latter, then we may conclude that all belong in a monophyletic order Primates. This is so even if no direct ancestor-descendent relationships can be demonstrated between any *known* species belonging to the Paleocene and Eocene primate radiations.

The discovery of the new *Plesiadapis* material from France and Germany has clarified our concept of the genus to the extent that it is now possible to employ the entire skeleton, postcranial as well as cranial, in weighing the affinities of the form. In considering the implications of the various character states in the skeleton of *Plesiadapis* for primate phylogeny, we believe that we are able to provide a relatively firm basis for the delimitation of the order. We speak of the genus as a whole because the morphological differences between the various species, despite great disparities in size, are sufficiently slight to allow the taxon to be treated as a unit.

II. Assessment of Relationships

In pursuing the problem outlined above we have extensively compared the well preserved parts of the forelimb and hindlimb of *Plesiadapis* to those of adapids and to various early Tertiary tarsiiform specimens. In addition, we have also compared the *Plesiadapis* material to a relatively large number of relevant early Tertiary and living eutherians.

The most significant primitive nonprimate eutherian postcranial sample utilized was that collected by R. E. SLOAN in the Cretaceous Bug Creek Anthills of Montana, and presently under study by SZALAY, NELSON, and SLOAN. Among this material, there are two distinctive sets of humeri, ulnae, astragali and calcanea which probably belong to species of *Procerberus* and *Cimolestes*, palaeoryctoid Insectivora, and to *Protungulatum*, a condylarth. The late Cretaceous palaeoryctoid remains, at least in the tarsal elements, appear to be considerably more derived than those of the contemporaneous condylarth *Protungulatum*. An equally reliable assessment of the elbow joint cannot yet be made, but, as discussed below, in this feature both groups are apparently primitive compared to all known primates. Comparison with postcranial remains of erinaceotan Insectivora unfortunately cannot at present be carried out as no certain associations exist between early Tertiary dental taxa and postcranial remains from the same localities.

A. Vertebral Column

Vertebrae (fig. 1)

The best known vertebrae of *Plesiadapis* are still those of AMNH 17379 from the Mason Pocket described by Simpson [1935, pp. 10–12]. As they add very little to the assessment of affinities at this time, they are not redescribed or compared to those of other primates. The reason for our reluctance to study the vertebrae in detail at this point stems from our general ignorance of early eutherian vertebral morphology. Before we can assess whether features recognized by us are primitive eutherian or specifically primate or, of course, specifically plesiadapid, we must learn more of this region of the postcranium among early Tertiary Theria.

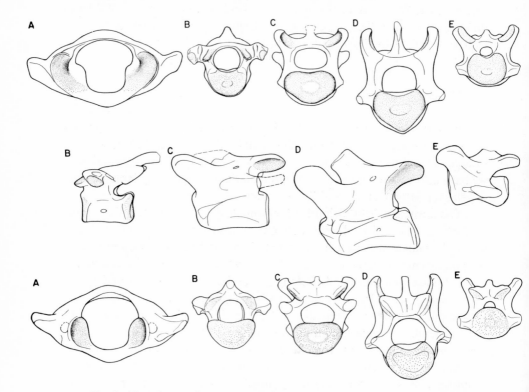

Fig. 1. Plesiadapis gidleyi, representative vertebrae, AMNH 17379, Mason Pocket, Tiffanian. A = Atlas; B = cervical; C = thoracic; D = lumbar; E = caudal vertebra, respectively. Proximal (above), lateral (middle), and distal (below) views.

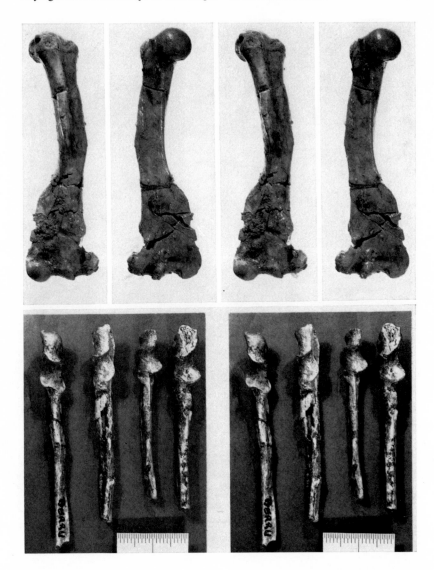

Fig. 2. Plesiadapis tricuspidens, Thanetian. Above: MNHN BR 3-L, right humerus, anterior (left) and posterior (right) views. Below: one right (on the left) and three left (on the right) ulnae, anterior view. Subdivisions on scale represent 0.5 mm.

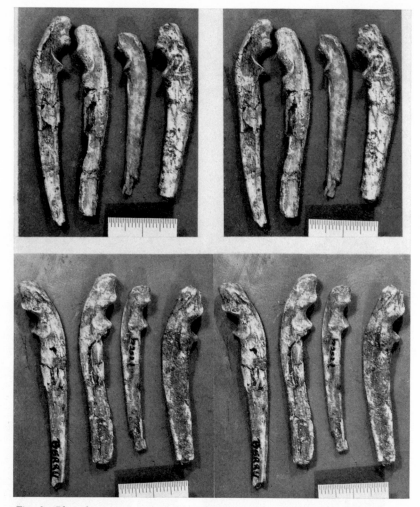

Fig. 3. Plesiadapis tricuspidens, Thanetian. Lateral (above) and medial (below) views of ulnae. Subdivisions on scale represent 0.5 mm.

B. Forelimb

Scapula

We found it impossible to assess conclusively whether the fragmentary scapula known in *P. gidleyi* [see SIMPSON, 1935, p. 13] is primitive eutherian or adapid-like in conformation.

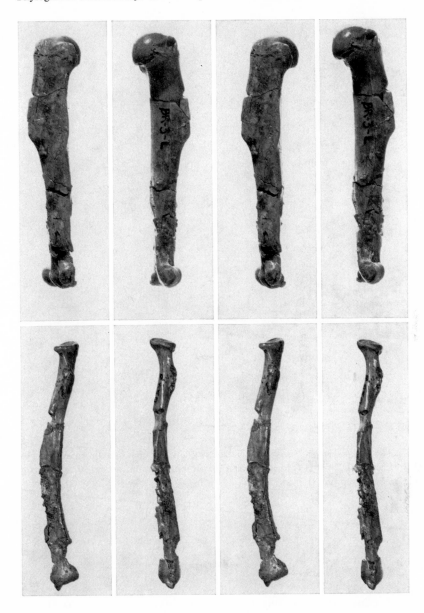

Fig. 4. Plesiadapis tricuspidens, Thanetian. Above: MNHN BR3-L, right humerus, lateral (left) and medial (right) views. Below: MNHN R 550, right radius, medial (left) and frontal (right) views.

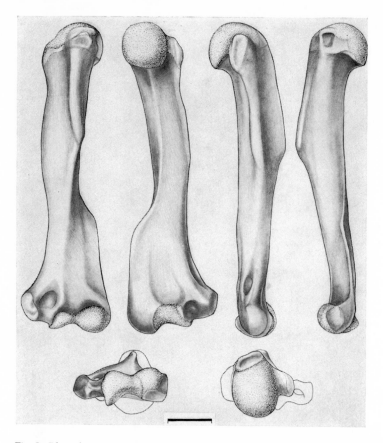

Fig. 5. Plesiadapis tricuspidens, Thanetian, left humerus based primarily on MNHN BR 3L and B4L. From left to right: anterior, posterior, medial, lateral, distal, and proximal views. Scale represents 1 cm.

Humerus (fig. 2, 4, 5)

The humerus is among the best represented bones of the skeleton in *P. tricuspidens* and *P. walbeckensis*.

The shaft is robust with a strong sigmoid curvature in lateral view although there is some variation among the species (e.g., this curvature is somewhat less marked in *P. tricuspidens* than in *P. gidleyi* or *P. walbecken-*

sis). The head is subhemispherical, facing mostly backwards. The delto-pectoral crest is thin, prominent, gradually narrows distally, and is possibly more robust in *P. tricuspidens* and *P. walbeckensis* than in *P. gidleyi*. The greater tuberosity is weak and pitted laterally for infraspinatus insertion, and the lesser tuberosity is strongly developed and pronounced medially and slightly superiorly. The bicipital groove is shallower in *P. tricuspidens* than in *P. gidleyi*. The teres tuberosity is oblong, rounded, and is pronounced in *P. gidleyi* but less well developed in the other species. The medial epicondyle faces directly medially and it is pronounced. The capitulum is spheroidal to subspheroidal and faces slightly superiorly; the supracapitular depression is deep. The trochlea is not broader than the capitulum and there is no distinct ridge separating the two. The former is also very narrow or con-stricted at its junction with the capitulum, but expands laterally. The lateral epicondyle is weakly developed. The distolateral crest ('brachioradialis flange') is large and flared slightly backwards. The olecranon fossa is present as a small shallow depression. A deep pit of small diameter is present on the posterior surface of the medial epicondyle adjacent to the trochlea; it is probably for the medial ligament. This feature is not evident in the humeri of *P. gidleyi*, but it is just possible that this is due to poor preservation of the bone. The entepicondylar foramen is large and opens mediosuperiorly.

As SIMPSON [1935, p. 13] noted, one of the major differences between the humeri of *Plesiadapis*, on the one hand, and lemuriforms, tarsiiforms, and many New World monkeys, on the other, is the absence in the former of a semicylindrical shape to the trochlea, together with a ridge demarcating the trochlea from the capitular area; all are similar in possessing a ball-shaped capitulum. Comparison of living and Cretaceous insectivorans, Tertiary carnivorans and extant squirrels indicates that the primitive euthe-rian capitulum is more spindle-shaped, as in the late Cretaceous palaeoryc-toids, condylarths, and early Tertiary hyaenodontans. Thus, in the latter the degree of its curvature in the coronal plane is less than in the parasagittal plane, whereas in *Plesiadapis*, as in other primates, the two are subequal. However, the form of the trochlea, although variable, is unlike the semi-cylindrical form of lemuriforms and many other primates. We believe that the spherical capitulum of *Plesiadapis* is a shared derived specialization with other primates whereas the trochlea represents either the primitive primate, or *sui generis Plesiadapis*, morphology. Incidentally, the conformation of the trochlea in *Plesiadapis* resembles that of arctocyonid condylarths rather than those of Cretaceous and Paleocene Insectivora.

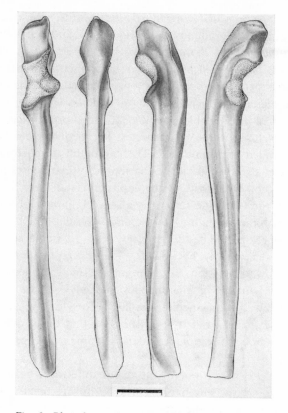

Fig. 6. Plesiadapis tricuspidens, Thanetian, left ulna, based primarily on MNHN R 564. From left to right: anterior, posterior, medial, and lateral views. Scale represents 1 cm.

Ulna (fig. 2, 3, 6)

The account below is based on MNHN CR 546, a right ulna which is relatively undamaged and complete except for its extreme distal end.

The shaft is robust, with distinct anteroposterior sigmoid curvature. In cross-section, it is lunate at its midpoint because of a strong excavation laterally in the area of origin of abductor pollicis longus. The olecranon process is large and moderately long; its posterior margin is strongly inclined forward, and its superior and posterolateral aspects are rugose for triceps insertion. Medially, the olecranon process is compressed by the proximal portion of the large groove accommodating the flexor carpi ulnaris. The trochlear notch is not particularly deep; inferiorly the notch slopes steeply down, and laterally it is not well differentiated from the radial facet.

As SIMPSON [1935, pp. 15–16] noted, the ulna, particularly the proximal half, strongly resembles the homologous region in *Notharctus*. In particular two characters show this resemblance: the relative shallowness of the trochlear notch and the slight forward curvature of the proximal end. However, distally, the shaft changes direction and curves posteriorly in *Plesiadapis*, a configuration exhibited by many insectivorans and other eutherians and probably primitive for the Eutheria. It is more continuously bowed in *Notharctus* and recent lemuriforms. The radial facet is very slightly shaped as is the homologous region of this facet in *Notharctus* and recent lemuriforms. Although the olecranon process is by no means small, it is distinctly relatively smaller than in paleoryctoid Insectivora and arctocyonid Condylarthra known to us, and at least approaches the lemuriform condition.

Radius (fig. 4)

The virtually complete MNHN CR 550 is the best preserved radius known in the genus; it is somewhat damaged in the distal portion of the shaft.

The robust shaft is bowed laterally (not anteriorly) and possesses a sharply-defined interossous border. The head is tilted to face somewhat anteriorly and laterally. The articular surface of the head is suboval in profile and is quite well excavated in a subspheroidal fashion corresponding to the capitular shape. The small eminence on the lateral margin of the capitular depression, articulating with the anterior aspect of the capitulum (and characteristic of the primitive eutherian conformation), is, as in lemuriforms and other primates, absent in *Plesiadapis*. The ulnar facet is incomplete in CR 550; in other specimens it is slightly oval, extensive, and quite sharply curved. Detail has been obliterated on the area of pronator quadratus insertion, but this was apparently quite extensive. The outline of the carpal articular surface is pear-shaped, slightly convex, and tilted to face slightly medially and anteriorly. The styloid process is small and rounded; distally, on the posterior aspect, there is a strongly-marked groove for the carpal extensor tendons. The neck, although short, is well defined and the bicipital tuberosity, as in other primates, is well developed.

The articular surface of the head, and indeed the head in general, is clearly more similar to lemuriforms than to other early Tertiary Eutheria; we believe this similarity represents a homologous derived condition.

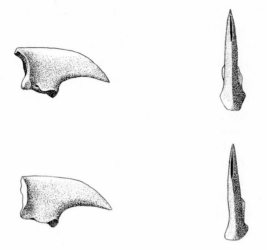

Fig. 7. Plesiadapis tricuspidens, Thanetian, two terminal phalanges showing an extreme bilaterally constricted condition, lateral (left) and dorsal (right) views.

Manus

Only a few elements of the carpus and cheiridia are known in *P. tricuspidens*. Moreover, the association of these elements is not at all certain due to occurrence of similarly sized arctocyonids in the same quarry. These bones will be analyzed, and this association discussed elsewhere.

Note should be made here of the basally robustly fissured and strongly curved claws (fig. 7) found at Cernay. As all condylarths known to us usually have very differently modified, hoof-like terminal phalanges, it is likely that these bones belonged to *P. tricuspidens*.

C. Hindlimb

Innominate (fig. 8, 9)

The pelvis is well known in *Plesiadapis gidleyi*, *P. tricuspidens*, and *P. walbeckensis*. The pelvis of the Mason Pocket *Plesiadapis gidleyi* is best represented by AMNH 17409, a nearly complete left innominate lacking only the symphysis and a small part of the cranial end of the ilium. There

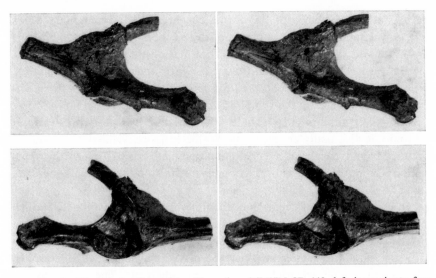

Fig. 8. *Plesiadapis tricuspidens*, Thanetian, MNHN CR 448, left innominate fragment, lateral (above) and medial (below) views.

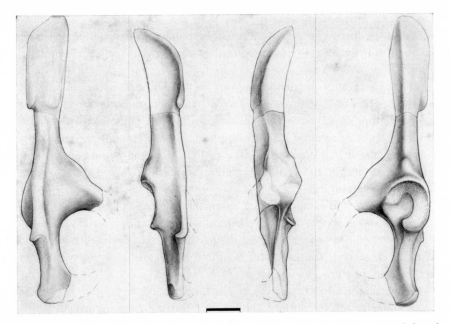

Fig. 9. *Plesiadapis tricuspidens*, Thanetian, left innominate, partly restored, based on MNHN CR 448, CR 409, and CR 413. From left to right: medial, dorsal, ventral, and lateral views. Broken lines indicate restoration. Scale represents 1 cm.

are two other specimens of this form, a right and left innominate which have only a small additional portion of the acetabulum not seen in AMNH 17409 and add a little more (3 mm) to the anterior of the ilium, respectively. There is also an additional ilial fragment which adds no new information. Representing Cernay *Plesiadapis tricuspidens* is MNHN CR 448, a left innominate also lacking the symphysis and the anterior of the ilium. There are also a left ischium (MNHN CR 406) and two ilial parts (MNHN CR 409 and CR 413, badly eroded). The Walbeck fissure yielded one excellent specimen from the right side. The ilium is complete except for a small cranial portion; the pubic portion of the bone is missing, and the ischial tuberosity is eroded.

In general, the conformation of the innominate is not unlike that which we infer to be the primitive eutherian condition. As in squirrels, but unlike known adapids or omomyids, the obturator foramen is elongated. In all three species of *Plesiadapis* in which the pelvis is available, the iliac spine, just craniad to the acetabulum, is a blunt tuberosity. The ilium has the primitive eutherian triangular cross-section. Three faces may be distinguished: an inner or medial face, a dorsolateral face, and a ventrolateral face. The dorsolateral face is somewhat broader than the ventrolateral one, more so cranially; the former is even more greatly expanded in undisputed primates, although this is also found, independently, in many other mammalian groups.

The abundant postcranial remains from the Walbeck fissure have among them the pelves of both Insectivora and Condylarthra, both of which show the condition displayed by the paromomyiforms. Thus, in this respect, plesiadapids are probably primitive eutherians. The tupaiid pelvis appears to be specialized in a direction similar to that of modern primates. Among living species of this family, the dorsolateral surface is enlarged at the expense of the ventrolateral one.

Plesiadapis shows at least one character of the acetabulum which sets this genus, and probably all other paromomyiforms, apart from the primitive eutherian condition displayed by condylarths, Paleocene Insectivora from Walbeck, and also, for example, *Sciurus carolinensis:* a relatively larger acetabular diameter.

Three major areas of the lunate facet, corresponding to the regions of the innominate (pubis, ilium, ischium), may often be distinguished. These areas may be connected to one another by a constricted area of the facet or separated by a nonarticular gap. These junctions often roughly approximate the lines of fusion in the innominate. In the shape and relative dimensions of the lunate facet the specimens of *Plesiadapis gidleyi* and *Plesiadapis tricuspidens* are essentially similar. The length of the ischial region, of nearly uniform diameter, is more than half that of the entire facet; it narrows very slightly and gradually from the posterior corner forward to the ilioischiac constriction. Thus, in this region, the breadth of the constriction is not much less than that of the posterior corner. The ilioischiac constriction is marked in one specimen of *Plesiadapis gidleyi*. The maximum breadth of the iliac region (at the level of the iliac spine) is somewhat greater than that of the posterior corner; the ratios of the former to the latter are 1.1 and 1.2 in two specimens of *Plesiadapis gidleyi*. As exhibited by the one almost complete facet in a specimen of *Plesiadapis gidleyi*, and partially in *Plesiadapis tricuspidens*, the iliac region narrows into the short pubic region of the facet without any obvious discontinuity.

The only major difference between the lunate facets of *Notharctus* and *Hemiacodon*, on the one hand, and *Plesiadapis* (based on two specimens of *P. gidleyi*), on the other, is an enlargement of the iliac region. The ratio of the diameter of this region to that of the posterior corner is approximately 1.50 and more (1.7) in *Notharctus* and 1.50 in *Hemiacodon*, as compared to the lower values for *Plesiadapis* given above. *Plesiadapis* exhibits more nearly primitive proportions in this respect. The ischial region tapers somewhat more craniad in *Hemiacodon*, giving a teardrop appearance to this area.

Femur (fig. 10, 11)

The femur is best known in *Plesiadapis tricuspidens;* that of *P. gidleyi*, although clearly very similar to the former, is still poorly known [see SIMPSON, 1935, pp. 18–20]. The French material is known from six very well preserved specimens (BR-15-L, BR-16-L. MNHN CR 408, CR 438, CR 444, and CR 450) and from an additional half dozen fragmentary bones. No femora are known from Walbeck. Although the French sample is moderately variable in size, the following account applies to all of the specimens.

The shaft is straight, robust and short relative to the large proximal moiety. Suboval in cross-section at its midpoint, it is, however, flattened on its posterior surface. The head is subspherical and faces medially and superiorly. The fovea for ligamentum teres is fairly distinct and lies slightly inferior and posterior to the center of the articular surface of the head. The neck is short and robust, but moderately well defined, and with the head

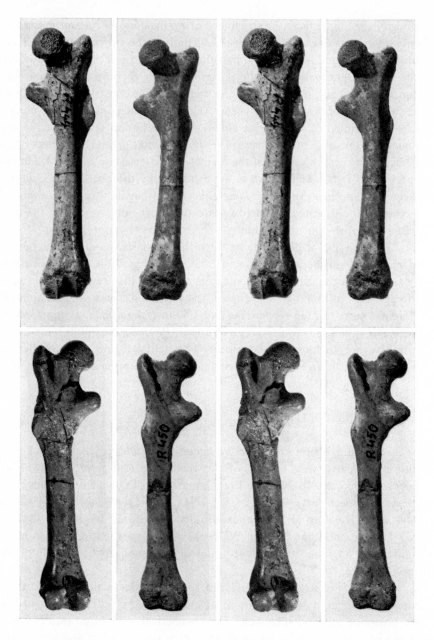

Fig. 10. Plesiadapis tricuspidens, Thanetian, MNHN R 444 (left) and R 450 (rigth), left femora. Anterior (above) and posterior (below) views.

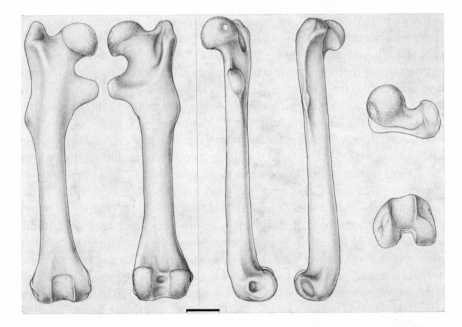

Fig. 11. Plesiadapis tricuspidens, Thanetian, right femur based on MNHN BR-15-L, BR-16-L, R-408, R438, R444 and R450. From left to right: anterior, posterior, medial, lateral, dorsal, and distal views. Scale represents 1 cm.

forms an approximate angle of 120° with the shaft (less than in *Notharctus* and *Hemiacodon*). The greater trochanter is robust and subequal in height to the head. As in *Notharctus*, a roughened area for muscular attachment on this trochanter is disposed on its lateral aspect and elongated in the direction of the shaft axis; in *Hemiacodon*, this area is tilted to face more superiorly and is elongated anteroposteriorly. The trochanteric fossa is long and quite deeply excavated, more so in *Plesiadapis* than in *Hemiacodon* and *Notharctus* (relatively, it is nearly one and a half times as long as in *Hemiacodon*). The groove between the head and the trochanter is deep in *Plesiadapis* (giving a constricted appearance to the neck), but is shallower in *Notharctus*, and shallower still in *Hemiacodon*. The lesser trochanter is large and projects medially to a distance subequal to that of the femoral head. It is smaller in *Notharctus* and *Hemiacodon* where it is also directed more posteriorly as well as medially. *P. tricuspidens*, like *Notharctus* and *Hemiacodon*, differs from *P. gidleyi* in the absence of an excavated depression on the anterior surface. The third trochanter is laterally projecting; it is long in its vertical dimension, but is also fairly rounded in profile. The inter-trochanteric crest is faintly distinguishable throughout its course from greater to lesser trochanter in *P. gidleyi*, less so in *P. tricuspidens*, and is certainly discontinuous in *Hemiacodon* and *Notharctus*. The third trochanter is immediately below the level of the lesser trochanter in *Plesiadapis* whereas it is at the level of the lesser trochanter in *Notharctus* and is slightly above the mean level of the lesser trochanter in *Hemiacodon*.

In distal view, the proportion of the femoral and contrast with those in *Hemiacodon* or *Notharctus*. These latter two genera exhibit relatively high distal ends (ratios of height to breadth: *Hemiacodon*, 1.29; *Notharctus*, two specimens, 1.25 and 1.21) compared to *Plesiadapis* (*P. gidleyi*, 1.01; *P. tricuspidens*, two specimens, 0.88 and 0.83). Much of this discrepancy in height is precondylar; the base of the patellar groove does not extend beyond the anterior level of the shaft as in *Notharctus* and *Hemiacodon*. The patellar groove is much shallower than in the other two. Its medial and lateral ridges are only weakly developed whereas *Notharctus* and *Hemiacodon* exhibit pronounced rounded lateral crests. This accounts, in part, for the greater lateral height in *Notharctus* and *Hemiacodon*, which contrasts with the opposite condition in *Plesiadapis*. Lorisines and galagines provide a similar contrast in most of the above characters, though the patellar groove is not as broad and shallow in *Plesiadapis* as in lorisids. The medial and lateral epicondylar surfaces in *Plesiadapis* are srongly pitted for the collateral ligaments of the knee.

Tibia and fibula

These bones are known in *P. tricuspidens* only from the associated MNHN CR 410, but both are too badly crushed to be of much value. The proximal and distal ends of the tibia are preserved in *P. gidleyi*, AMNH 17379, and have been described by Simpson [1935, pp. 20–21].

Tarsus (fig. 12–13)

Two tarsals of the foot of *Plesiadapis*, the astralagus and calcaneum, have been discussed in some detail by Szalay and Decker [1974]. Only these two elements of the tarsus can be allocated with certainty to *Plesiadapis*, although a cuboid, a mesocuneiform, and several entocuneiforms known

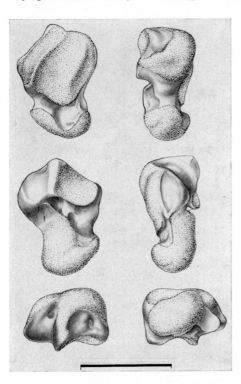

Fig. 12. Plesiadapis tricuspidens, Thanetian, right astragalus based on MNHN R-610. From left to right: dorsal, lateral, ventral, medial, proximal, and distal views. Scale represents 1 cm.

also probably belong to *P. tricuspidens*. The tarsals other than the astragalus and the calcaneum, because of their very rare associations with dental taxa, are of limited value for this study. They will be described in detail elsewhere, together with metatarsals, phalanges, and elements of the manus.

Comparisons of the astragali and calcanea of *Plesiadapis* to some eutherians and adapids are presented on tables I and II. It appears to us that, in spite of the fact that in many ways these bones show the primitive eutherian condition, there are a number of derived characters which may be considered as specializations shared with lemuriforms and other primates. These primitive primate (but derived eutherian) characters are:

(1) Pronounced groove for the flexor fibularis tendon on plantar side of calcaneum. (2) Astragalar head both broad medially as well as laterally and oriented slightly dorsoventrally, with an enlarged facet for the plantar

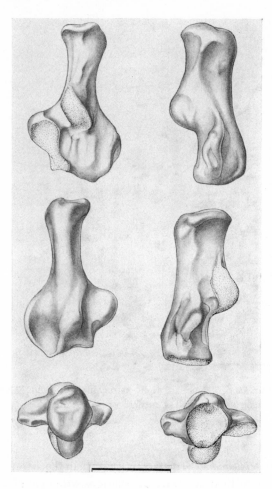

Fig. 13. Plesiadapis tricuspidens, Thanetian, left calcaneum, based on MNHN R-611. From left to right: dorsal, lateral, ventral, medial, proximal, and distal views. Scale represents 1 cm.

calcaneonavicular ligament bridging the gap between the sustentacular and navicular facets. (3) Tibial trochlea of astragalus longer than wide, the upper ankle articulation showing a trochlear radius disparately larger laterally than medially, and the axis of the tibial shaft forming an acute angle laterally with the transverse plane of its articulation with the trochlea. (4) Helical-shaped posterior astragalocalcaneal articulation.

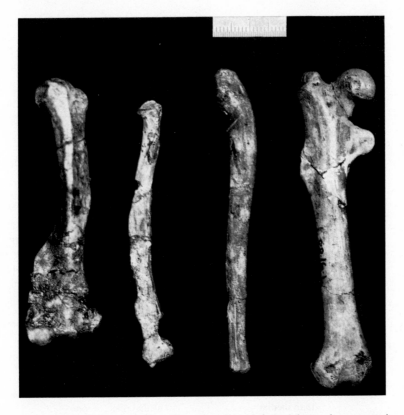

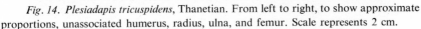

Fig. 14. Plesiadapis tricuspidens, Thanetian. From left to right, to show approximate proportions, unassociated humerus, radius, ulna, and femur. Scale represents 2 cm.

III. Notes on Mechanical Function

JENKINS [1971] has analyzed the movements of the forelimb and hindlimb during locomotion in an echidna, opossum, tree shrew, hamster, rat, ferret, hyrax, and a cat. What complicates the understanding of the morphology of the forelimb is that its role in the manipulation of food and other free objects may be as important as its role in locomotion. It is still possible, however, that the forelimb may facultatively fulfill the former roles with few, if any, special adaptations for this purpose on many mammals. It is remarkable how similar movements are in the relatively unspecialized therians studied by JENKINS. The differences were relatively minor (except in the hyrax and cat compared to the other forms among the therians) in the

Table I. Comparison of some morphological characters of the astragalus of some of the known latest Cretaceous eutherians, the late Paleocene *Plesiadapis*, and some Eocene adapids; the primitive eutherian condition is best represented by *Protungulatum*, whereas those of palaeoryctoids and primitive primates show taxon-specific derived morphology from a *Protungulatum*-like ancestry; adapids are in many ways representative of the derived condition of higher primate astragali.

Astragalus	Condylarth morphotype represented by *Protungulatum*	Palaeoryctoid morphotype represented by *Procerberus* and *Cimolestes*	Primate morphotype represented by *Plesiadapis*	Some adapids: *Notharctus*, *Leptadapis*, and *Adapis*
Tibial trochlea	very broad short trochlea on shallow body	broad short trochlea on shallow body	long, narrow trochlea on deep body in particular lateral aspect	long, narrow trochlea on deep body in *Notharctus;* not so narrow and deep on *Adapis*
	shallow trochlea	distinct groove to lateral side of trochlea	shallow trochlea	shallow to moderately grooved trochlea
	lateral trochlear crest less rounded than medial one	both trochlear crests sharp	lateral trochlear crest very sharp, medial one rounded	lateral trochlear crest somewhat sharper than medial one
	trochlea limited to body	trochlea limited to body	trochlea extending onto neck	trochlea limited to body in *Notharctus* (unusual in Lemuroidea), not so in adapines
	fibular shelf present	fibular shelf present	fibular shelf absent or small, lateral aspect steep	fibular shelf present (except in *Adapis*)
Trochlear length/ astragalar length	0.54	0.49	0.66	*Notharctus:* 0.65, *Leptadapis:* 0.74, *Adapis:* 0.71

Table I. Continuation.

Astragalus	Condylarth morphotype represented by *Protungulatum*	Palaeoryctoid morphotype represented by *Procerberus* and *Cimolestes*	Primate morphotype represented by *Plesiadapis*	Some adapids: *Notharctus*, *Leptadapis*, and *Adapis*
Flexor fibularis groove	distinct	distinct	distinct	shallow, more distinct in adapines
Astragalar sustentacular facet	facet separate from distal facets of astragalus	facet very close distomedially to spring ligament facet	facet disto-medially close to spring liga-ment facet; continuous in several spec-imens as a result of enlargement of the spring liga-ment facet	facet usually con-tinuous with spring ligament facet as a result of spring ligament facet enlargement
Calcaneal astragalar facet	slightly concave	slightly concave	strongly concave	strongly concave
Distal calcaneo-astragalar facet	convex	convex	convex	convex
Secondary fibular facet	absent	absent	absent	present
Trochlear shelf	absent	absent	absent	present

Table II. Comparison of some morphological characters of the calcaneum of some of the known latest Cretaceous eutherians, the late Paleocene *Plesiadapis*, and some Eocene adapids; the primitive eutherian condition is best represented by *Protungulatum*, whereas those of palaeoryctoids and primitive primates show taxon-specific derived morphology from a *Protungulatum*-like ancestry; adapids are in many ways representative of the derived condition of higher primate calcanea.

Calcaneum	Condylarth morphotype represented by *Protungulatum*	Palaeoryctoid morphotype represented by *Procerberus* and *Cimolestes*	Primate morphotype represented by *Plesiadapis*	Some adapids: *Notharctus*, *Leptadapis*, and *Adapis*
Fibular facet	present lateral to astragalo-calcaneal facet	missing	?missing	missing
Peroneal tubercle	very extensive, bordering on margin of cuboid facet	extensive, bordering on margin of cuboid facet	prominent, opposite the calcaneal sustentacular facet, and posterior to the margin of the cuboid facet	distinctly present, but reduced to a usually nondescript tubercle opposite the calcaneal sustentacular facet
Anterior plantar tubercle	small, narrow	small, narrow	large, broad, blunt	large, broad, depressed by fovea
Groove for flexor fibularis	indistinct	indistinct	pronounced	pronounced
Astragalo-calcaneal facet	not very arched (large radius) long axis forms relatively large angle (35°) with long axis of calcaneum	arched (relatively small radius) long axis forms relatively large angle (45°) with long axis of calcaneum	arched (relatively small radius) long axis forms very small (20–25°) angle with long axis of calcaneum	arched (relatively small radius) long axis approximately parallel to long axis of calcaneum
Cuboid facet	shallow to transversely	shallow; small pit medio-	deeper and more rounded;	semilunar and depressed ventro-

Table II. Continuation.

Calcaneum	Condylarth morphotype represented by *Protungulatum*	Palaeoryctoid morphotype represented by *Procerberus* and *Cimolestes*	Primate morphotype represented by *Plesiadapis*	Some adapids: *Notharctus*, *Leptadapis*, and *Adapis*
	elongate; transverse groove for calcaneocuboid ligament plantad and slightly distal to cuboid facet	ventral to cuboid facet for calcaneocuboid ligament	plantad at distal astragalocalcaneal facet the border is straight; small pit distomedial to the cuboid facet for calcaneocuboid ligament	medially for calcaneal projection of cuboid
Ratio of length of proximal part/ length of distal part (both measured along long axis)	3.4	2.4	2.1	*Notharctus:* 1.6, *Leptadapis:* 2.1. *Adapis:* 2.5

abduction-adduction arc of the humerus relative to a parasagittal plane, the elevation-depression excursion arc of the humerus relative to the horizontal, and the excursion arc of the radius and ulna relative to the horizontal when one compares the opossum, tree shrew, hamster, rat, and ferret. Perhaps, forelimb movements in early primates were not very distinct from those in these species under similar conditions. It should be noted that the mammals were not given challenging problems to test their full capabilities. Nonetheless, realizing that any attempt to understand fully the locomotor mechanics in early Tertiary mammals is premature, we make a few tentative suggestions in regard to the locomotor adaptations of *Plesiadapis*.

The elbow joint of *Plesiadapis* clearly did not permit full extension of the antebrachium. The olecranon fossa is too shallow and the shaft of the ulna is bent forward. However, the deeply excavated supracapitular fossa and the anterosuperior orientation of the capitulum indicate that considerable flexion was possible; thus, we may reasonably infer that the forelimb was habitually held in a somewhat flexed position, especially in view of the fact that the distal sloping of the trochlear notch would have rendered the joint unstable in positions of more than partial extension. In addition, the olecranon process is bent forward to give considerable leverage to the triceps only in flexed positions. These appear to be similar to conditions seen in living lemuriforms and are even possibly primitive for eutherians. As stated before, the spheroidal articulation in *Plesiadapis* had been modified from a primitive eutherian state in which the articulation was more ovoid or spindle-shaped. This modification has the effect of centering the axis of rotation through the articulation, thereby facilitating freer mobility in pronation and supination of the antebrachium. The paromomyiforms share this specialization with all other known primates. The forelimb, including the manus, was then capable of achieving a high degree of mobility comparable to those discussed for the foot below. These characters might reasonably be interpreted as a set of adaptations to adjust the limbs to a new substrate, the branch environment of an arboreal habitat, more precarious and geometrically different from those for which primitive Eutheria were adapted.

Because of the incongruence of corresponding articular facets the area of contact of two joint surfaces at any given moment may be smaller than the area of overlap. The increase in the size of the hip articulation in *Plesiadapis* probably relates to an increase in the area of contact reducing the stress per unit area of forces transmitted here larger than those in contemporaneous comparably-sized Eutheria; these may correspond to a change in substrate and thus of locomotor pattern. But whatever the reason for it may be, the relatively large size of the hip articulation in *Plesiadapis* probably constitutes a difference of significant functional consequence.

Since all specimens to hand are largely devoid of muscular rugosities, it is difficult to assess muscular dispositions along the femur of *P. tricuspidens*. Nevertheless, some suggestions can be made from the size and dispositions of the major bony features. The uniform strong development of almost all features of the proximal end of the femur suggests a very powerful hindlimb. A capacity for strong retraction of the upper hindlimb is indicated by an extensive area for the insertion of the gluteus minimus and medius on the anterolateral portion of the greater trochanter, and by the great enlargement of the third trochanter, the site of insertion of gluteus maximus.

We have already noted the discrepancy in proportions of the distal femoral end of *Plesiadapis* compared, for example, to *Hemiacodon, Notharctus* and living lemuriforms. In *Plesiadapis*, this area is less pronounced anteriorly than in the latter. Moreover, the tibial tuberosity in *P. gidleyi* is less pronounced than in *Notharctus* or Malagasy lemuriforms. These characters affect the moment arm of the quadriceps femoris which passes via the patella and ligamentum patella over the knee to the tibial tuberosity, increasing it in the non-paromomyiforms. These primates also exhibit high patellar cresting, especially laterally, in order to maintain the alignment of the patella. A similar contrast occurs between the closely related galagines and lorisines, the former showing the leverage for knee extensors that the latter do not. We do not suggest that *Plesiadapis* was necessarily a slow climber or even arboreal, on the basis of this character, but simply that it did not emphasize certain characters of the knee apparently arising as leaping specializations in some primates. In fact, the knee of *Plesiadapis* does not appear to be modified to the extreme seen in lorisines.

In addition to the foregoing, we can briefly summarize here some inferences concerning function of the tarsal complex of *Plesiadapis*, which have been discussed elsewhere in more detail [SZALAY and DECKER, 1974]. Reference to the previously presented list of apparent synapomorphies with primitive higher primates in this region may prove useful. The majority of these characters are consistently related to free mobility of the foot, permitting its inversion and eversion to adjust to arboreal branch substrates as, for example, in lemuriforms and other primates. The tarsus distal to the astragalocalcaneal pair may tilt superiorly-inferiorly, this mobility provided largely by the calcaneocuboid articulation. Associated with these motions via the astragalonavicular articulation are the rotations between astragalus and calcaneum analogous to radioulnar rotations on sets of anterior and posterior gliding articulations. The associated tilt of the distal tarsus and astragalocalcaneal rotation permit the inversion and eversion of the foot. Since the navicular slides medially on the astragalar head during inversion, the pronounced medial development of the navicular facet on the astragalar head and its slight dorsomedial orientation favor the reception of stresses in inverted orientations of the foot. As indicated by its enlarged facet on the astragalus, the plantar calcaneonavicular ligament also plays an increased role in medially butressing this articulation. Tilting of the distal tarsus approximates and draws distant, during inversion and eversion, respectively, the navicular and calcaneal aspect of the lower ankle joint. This would ordinarily crowd the astragalus and force separation of the lower ankle joint or, barring this, inhibit inversion and eversion to any great degrees. The problem is resolved by the helical configuration of the posterior astragalo-calcaneal articulation, which permits helical motion. This provides the rotation necessary to invert and evert the foot, and the pitch required to adjust astragalocalcaneonavicular relationships. Thus, extreme orientations of the pes may be achieved, since the articular surfaces retain the ability to receive stresses in these positions. The calcaneal plantar groove for the flexor fibularis tendon can be explained as a feature designed to maintain the alignment of the tendon and, therefore, the efficiency of the muscle in the various positions.

The characters of the upper ankle joint are associated with an axis of rotation which is not very perpendicular to the lateral or fibular side of the trochlea. BARNETT and NAPIER [1953] have associated these developments in Primates with mobility of the fibula and its capacity to resist inversion and eversion strains in various positions of flexion and extension of the ankle joint.

We have largely avoided the traditional, and usually somewhat speculative, inferences from fossil material concerning the development of the muscles themselves. Such comparative information on the structure and properties of muscles in living mammals is broadly lacking, and the study of the relation of shape of bones to stresses resulting from muscle forces and loads is still in its early stages.

IV. Summary and Conclusions

We believe then, that *Plesiadapis* should not be conceived as bizarre, a point of view expressed in recent primatological literature. It is, in fact, in some respects still a primitive eutherian; this accounts for most of its differences from undisputed primates, which have specialized in new directions.

We have emphasized the value of the distinctive radiohumeral articulation and a tarsal complex of characters in arguing the relationship of *Plesiadapis* and, therefore, of other, archaic primates, the paromomyiforms, to the undoubted primates. Other similarities with the latter do occur but are presently of less value since we cannot conclusively decide whether these were primate synapomorphies rather than eutherian symplesiomorphies. Nonetheless, our conclusion is in agreement with known dental and basicranial evidence.

Not only do a great many of the postcranial characters constitute synapomorphies, we argue, but so do the basic adaptations to an arboreal habitat from an ancestry that was probably habitually terrestrial. Classification, therefore, of the Paromomyiformes with higher categories other than the Primates is unwarranted and it appears to us to be a direct violation of the tenets of evolutionary systematics which dictate emphasis on both monophyly and adaptive similarity in constructing informative evolutionary classifications.

V. References

BARNETT, C. H. and NAPIER, J. R.: The rotary mobility of the fibula in eutherian mammals. J. Anat., Lond. *87:* 11–21 (1953).

CARTMILL, M.: Arboreal adaptations and the origin of the order Primates; in TUTTLE The functional and evolutionary biology of Primates, pp. 97–122 (Aldine-Ahterton, Chicago 1972).

CLARK, W. E. LeGROS: The antecedents of man (Harper & Row, New York 1959).

DECKER, R. L. and SZALAY, F. S.: Origins and functions of the pes in the Eocene Adapidae (Lemuriformes, Primates); in JENKINS: Primate locomotion (Academic press, New York 1974).

JENKINS, F. A., jr.: Limb posture and locomotion in the Virginia opossum *(Didelphis marsupialis)* and in other non-cursorial mammals. J. Zool., Lond. *165:* 303–315 (1971).

JENKINS, F. A., jr.: Tree shrew locomotion and the origins of primate arborealism; in JENKINS: Primate locomotion (Academic Press, New York 1974).

MARTIN, R. D.: Adaptive radiation and behaviour of the Malagasy lemurs. Philos. Trans. (B) *264:* 295–352 (1972).

MCKENNA, M. C.: Classification, range, and deployment of the prosimian primates. Coll. Int. Cent. Nat. Rech. Sci. – Probl. act. Paléont. *163:* 603–610 (1967).

RUSSELL, D. E.: Les mammifères Paléocènes d'Europe. Mém. Mus. nat. Hist. nat. *13:* 1–324 (1964).

RUSSELL, D. E.: Sur *Menatotherium* et l'âge Paléocène du gisément de Menat (Puy-de-Dôme). Coll. Int. Cent. Rech. Sci. Paris. – Probl. act. Paléont. *136:* 483–489 (1967).

RUSSELL, D. E.; LOUIS, P., and SAVAGE, D. E.: Primates of the French early Eocene. Univ. Calif. Publ. Geol. Sci. *73:* 1–46 (1967).

SIMONS, E. L.: Primate evolution: an introduction to man's place in nature (Macmillan, New York 1972).

SIMPSON, G. G.: The Tiffany fauna, upper Paleocene. II. Structure and relationships of *Plesiadapis*. Amer. Mus. Novit. *816:* 1–30 (1935a).

SIMPSON, G. G.: The Tiffany fauna, upper Paleocene. III. Primates, Carnivora, Condylarthra and Amblypoda. Amer. Mus. Novit. *817:* 1–28 (1935b).

SIMPSON, G. G.: Studies on earliest primates. Bull. Amer. Mus. nat. Hist. *77:* 185–212 (1940).

SZALAY, F. S.: The beginnings of primates. Evolution *22:* 19–36 (1968).

Szalay, F. S.: Paleobiology of the earliest primates; in Tuttle The functional and evolutionary biology of primates, pp. 3–35 (Aldine-Atherton, Chicago 1972).

Szalay, F. S. and Decker, R. L.: Origins, evolution, and function of the tarsus in late Cretaceous eutherians and Paleocene primates; in Jenkins: Primate locomotion (Academic Press, New York 1974).

Tattersall, I.: Man's ancestors: an introduction to primate and human evolution (Murray, London 1970).

Authors' addresses: Dr. Frederick S. Szalay and Mr. Richard Lee Decker, Department of Anthropology, Hunter College, 695 Park Avenue, *New York, NY 10021*, and Dr. Ian Tattersall, Department of Anthropology, The American Museum of Natural History Central Park West at 79 St., *New York, NY 10024* (USA)

In Szalay: Approaches to Primate Paleobiology
Contrib. Primat., vol. 5, pp. 167–217 (Karger, Basel 1975)

Evolutionary History of the Cercopithecidae

Eric Delson

Department of Anthropology, Lehman College, CUNY, N.Y.

Contents

I. Introduction

The reconstruction of the evolutionary history of a group of organisms requires a combination of data bearing first on the relationships or phylogeny of the living members, second an understanding of the group's paleontology and zoogeography, and finally, if possible, a model or hypothesis which attempts to 'explain' or interpret the obtained or observed facts. The current state of knowledge about the Cercopithecidae, the Old World monkeys, allows at least a good approximation to such a history to be developed.

Several other recent studies [e.g., Maier, 1970; Simons, 1970, 1972] have considered this group in terms of paleontology especially, but their results suffered in part from incomplete data and in part from the lack of evolutionary hypotheses which attempted to unify the diverse data into a complete picture. Obviously, many data are still lacking, both in terms of an incomplete fossil record and as regards the inter-relationships of the living genera, much less species, but the results of recent studies have led to some conclusions which may be reported here. A consideration of modern systematic methodologies will be followed by the results of applying these methods to the study of cercopithecid cranial and dental anatomy. The expected ancestral morphologies, or morphotypes, thus derived, will then be considered in light of the known fossil record of the Old World monkeys and their possible ancestors, along with paleoecological hypotheses, in a sequential time perspective. The modern subgroups of the family will be considered in turn, by geographical region rather than following a direct temporal sequence beginning with the Late Miocene.

II. Morphology and Evolutionary Reconstruction

The discipline of systematic zoology has benefited recently from the interaction of three differing philosophies of classification and phylogenetic reconstruction. The classical methodology has involved the morphological comparison of supposedly homologous structures in the several specimens or species under study. Shared characters or character states are considered to link the groups possessing such characters, and the greatest phyletic weight given to those features which have proven valuable in previous studies of this group or its nearest relatives. The numerical or phenetic systematists have advocated the use of numerous measurements and character-state rankings of many features, not only those which might be phyletically meaningful, and the sophisticated mathematical treatment of all this data as items of equal weight. Groups of samples (specimens) sharing an equal number of character states are considered as taxa of equal rank, although the formal names (genus, family) assigned to such groups is of lesser importance. Finally, the 'phylogenetic' or cladistic school depends less on measurement of any type and more on qualitative observation of characters and states, from which a direction or 'polarity' of morphological change may be determined, leading in turn to the grouping together of only those samples, sharing derived or 'advanced' states. The common presence of 'primitive' or ancestral features does not indicate special relationship within the larger group under study. A further major tenet of this philosophy is that the cladogram resulting from this method of grouping taxa may be directly translated into a classification, and that no additional information (such as relative morphological divergence or 'success') may be input to such a classificatory scheme.

Both the phenetic and cladistic methodologies, as well as those of classical systematics, have valuable features as well as possibly detrimental ones. The application of multivariate numerical analyses allows for arrival at a holistic picture including normal variation within samples; on the other hand, the absence of any weighting of the data may yield a result which does not necessarily answer biological questions, but instead finds *some* explanation for the observed pattern, although some items may be irrelevant or linked functionally in an unknown manner. The Hennigian or 'phylogenetic' approach makes a major addition to method in its formal reliance on the common presence of only derived characters and the search for 'sister groups', but some of its proponents have damaged their case and its acceptance through an insistence on dichotomous separation of sister groups, the one-to-one correspondence between phylogeny and classification, a disregard for morphological divergence and possibly some functional aspects, and a partly argumentative rejection of much fossil evidence. Each of these points can be (and has been) argued at great length [see recent volumes of *Systematic Zoology*], but such discussion will not be pursued here. Suffice it to say that these matters are more questionable than the search for derived characters, although the means of determining the latter are still not always clear-cut.

The positive features of the newer and more extreme approaches can be integrated with the body of 'classical' systematic methodology, as has often occurred in its past development. Thus, both multivariate and character-state analyses can yield information which leads to the determination of derived versus ancestral states within the studied group. Such results have been valued in past 'classical' work, but never sought out to the exclusion of groupings based on shared 'primitive' or ancestral features. An attempt to utilize such a combined approach in the study of Old World monkey morphology yielded a series or results which can be summarized here. Further consideration of the several methodological philosophies may be found in MAYR [1969], ELDREDGE and TATTERSALL (this volume) and especially SCHAEFFER et al. [1972] as regards paleontological applications.

Certain systematic results must be presented in advance in order that terms used and taxa discussed have an unambiguous interpretation. By comparison with other primate groups and especially following Simpson [1945], it is my interpretation that a single family is taxonomically sufficient to receive all modern Old World monkey taxa, as well as those fossils clearly referable to the group. On the basis of modern forms, two subfamilies may be recognized, the Colobinae or leaf-eaters and the Cercopithecinae or cheek-pouched monkeys. The latter taxon may in turn be subdivided into two tribes, the Cercopithecini to receive *Cercopithecus* and its allies (vernacularly termed cercopithecins) and the Papionini (papionins) for the more typical cercopithecines. The classification which follows is taken to the level of subgenus; novel features will not in most cases be explained here but are documented in Delson [1973, and in preparation] and are derived from the observations and morphologies discussed below. The morphological systems or regions which have yielded the most data bearing on cercopithecid evolutionary relationships, as well as being of use in the study of fossil forms, are the cranium (especially the facial skeleton) and the dentition, which will be considered in turn. In both cases, following the outline suggestions of Schaeffer *et al.* [1972], fossils were only included in the analysis to broaden the range of character-combination variations, and relative geological age was not of interest. The special value of fossils will be discussed below.

A. Classification

Order Primates Linnaeus, 1758[1]
  Infraorder Catarrhini E. Geoffroy, 1812
    Superfamily Cercopithecoidea Gray, 1821
      Family Cercopithecidae Gray, 1821
        Subfamily Cercopithecinae Gray, 1821
          Tribe Cercopithecini Gray, 1821
            Cercopithecus Linnaeus, 1758
            [*C. (Miopithecus)* I. Geoffroy, 1842][2]
            Erythrocebus Trouessart, 1897
            Allenopithecus Lang, 1923
          Tribe Papionini Burnett, 1828
          [Subtribe Papionina Burnett, 1828]
            Papio Müller, 1776
              P. (Chaeropithecus) Gervais, 1839

1   Bibliographic references to original taxonomic descriptions are not included due to space limitations.
2   Taxa wholly enclosed in square brackets are not formally recognized although possibly distinct.

      †3 *P. (Parapapio)* Jones, 1937
      Cercocebus E. Geoffroy, 1812
   † *Dinopithecus* Broom, 1937
     † *D. (Gorgopithecus)* Broom and Robinson, 1949
[Subtribe Macacina Owen, 1843]
     Macaca Lacépède, 1799
    + *Procynocephalus* Schlosser, 1924
    + *Paradolichopithecus* Necrasov, Samson and Radulesco, 1961
[Subtribe Theropithecina Jolly, 1966]
     Theropithecus I. Geoffroy, 1843
    + *T. (Simopithecus)* Andrews, 1916
Subfamily Colobinae Blyth, 1875
[Subtribe Colobina Blyth, 1875]
     Colobus Illiger, 1811
       C. (Procolobus) Rochebrune, 1887
       C. (Piliocolobus) Rochebrune, 1887
   † *Libypithecus* Stromer, 1913
   † *Cercopithecoides* Mollett, 1947
   † *Paracolobus* R. Leakey, 1969
† (Subtribe Presbytina Gray, 1825)
     Presbytis Escholtz, 1821
     Pygathrix E. Geoffroy, 1812
       P. (Rhinopithecus) Milne-Edwards, 1872
     Nasalis E. Geoffroy, 1812
       N. (Simias) Miller, 1903
(Subtribe unnamed)
   † *Mesopithecus* Wagner, 1839
   † *Dolichopithecus* Deperet, 1889
Subfamily incertae sedis
   † *Prohylobates* Fourtau, 1918
   † *Victoriapithecus* von Koenigswald, 1969

B. Cercopithecid Cranial Morphology

The two most perceptive of recent studies on the cranium of catarrhines have been those of Verheyen [1962] and Vogel (1966). Many of their results were confirmed by my work, which in addition included facial and cranial factor analyses in an attempt to determine if subgroups of cercopithecids could be distinguished on the basis of cranial morphology and if derived conditions could be separated from ancestral ones. The primary distinction within the Cercopithecidae is between longer-faced cercopithecines and

3   A '†' preceding a genus-group name implies a wholly fossil known range.

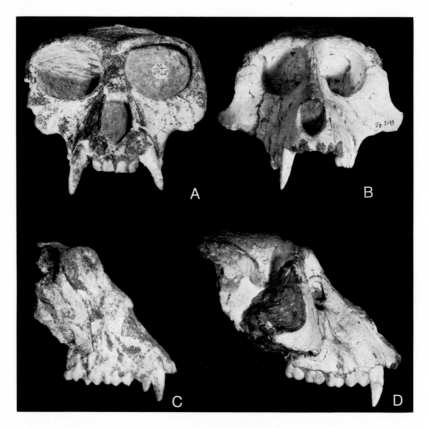

Fig. 1. Comparison of facial form in Colobinae and Cercopithecinae. A, C = Colobinae, *Mesopithecus pentelici*, Pikermi, male skull BMNH M8945; frontal, right lateral; B, D = Cercopithecinae, *Macaca 'majori'*, Capo Figari, male skull, NMB Ty 5199; frontal, right lateral. Specimens illustrated slightly less than natural size.

shorter- more upright-faced colobines, although there is much overlap in proportions (see fig. 1).

The interorbital distance tends to be larger in colobines, while the nasal bones are shorter, although in long-faced colobines (especially *Nasalis larvatus*) the elongated nasals may decrease interorbital width. Facial height is generally greater in cercopithecines (for a given overall size), especially in the suborbital zygomatic region. The lacrimal fossa is partly formed by the maxilla in colobines (and hominoids), but is restricted to the enlarged lacrimal bone in cercopithecines. As a result of facial shortness, landmarks such as the zygomatic root and the maximum palatal width are relatively more anterior in colobines. On the other hand, the Colobinae are relatively wider in the face as a whole, perhaps in conjunction with a wide interorbital region, especially in a smaller animal. The variation observed in choanal shape does not segregate along taxonomic lines. In general, cercopithecines and nonhylobatine hominoids (including *Aegyptopithecus*) have high and narrow openings, as do some colobines. A low and wide pattern is found in some *Cercopithecus*, most *Presbytis*, *C. (Colobus)* and *C. (Procolobus)*, and gibbons, while other *Presbytis*, *Pygathrix* and *Mesopithecus* have an intermediate high and wide shape. It is expected that this feature would be related to facial size, but no pattern is as yet discernable. Within Colobinae, *Colobus* and *Nasalis* species tend to possess a relatively low skull vault by comparison to face height, while the vault is higher in Presbytis and Pygathrix of similar or smaller size. A low skull is also found in the cercopithecines, suggesting inverse proportionality to size, but as hominoids of even larger size have rather high vaults, trends in the two main groups may be opposed.

In the mandible, colobines (as well as *Theropithecus* and *Cercopithecus*) have a relatively upright ascending ramus. Although certainly related to masticatory functions, the extremely back-tilted ramus of most larger papionins and *Erythrocebus* is most probably a derived condition for cercopithecids (and for catarrhines in general). The median mental foramen of cercopithecines is a similarly derived feature, occurring only rarely in other groups. Most cercopithecines, however, share with hominoids a mandibular corpus which increases in depth mesially; this contrasts with the range of states seen in colobines, in which the corpus may be of relatively constant height (*Nasalis*, most *Colobus*, *Mesopithecus*) or may shallow mesially. All colobines show some combination of expanded gonial region, 'bulging' under the rear molars and possibly overall deeper corpus, features seen individually in some cercopithecines and which again are probably related to masticatory processes.

Metrical comparisons as well as character-state observation pointed to these features as important in discrimination, and factor analysis was especially interesting in this way. Not only were separations possible into cercopithecines versus *Colobus/Nasalis* versus *Presbytis/Pygathrix* among modern forms, but individual specimens or genera could be seen to show close links. Most importantly, it was found that the African fossils *Cercopithecoides* and *Paracolobus* are close to each other and to African *Libypithecus* and *Colobus*, and that *Mesopithecus* and *Dolichopithecus* are rather intermediate between the three colobine 'poles' of *Colobus*, *Nasalis* and

Pygathrix. Further, it could be suggested that the morphological relations between *Dolichopithecus* and *Mesopithecus* are similar to those between *Nasalis larvatus* and *Nasalis ('Simias') concolor*, at least in the facial skeleton.

On the basis of these preliminary results, an attempt may be made to reconstruct an ancestral morphotype for the cercopithecid skull; in large part, this is an extension of the work of VOGEL [1966], who argued that colobines and gibbons shared many ancestral cranial patterns, while the Cercopithecinae were the most derived or specialized group. Thus, the early cercopithecid facial skeleton may have been much as in modern colobines, rather orthognathous (perhaps most like *Colobus* or larger *Presbytis* species) with widely spaced orbits, short and wide nasal bones, only moderate height of face below the orbits, small lacrimal bone with fossa extending onto the maxilla and upper incisors small and not widely divergent. The skull vault was probably small, partly as a result of the smallish brain, with vault height moderate to low by comparison with face height. Even less certain are the conditions of the choanae (possibly wide and low or intermediate in all catarrhines ancestrally) and the mandible: the ramus was probably slightly back-tilted, the gonion small, and the corpus as in cercopithecines, of low to moderate depth, possibly deeping slightly mesially. In general agreement with this morphotype is RADINSKY's [in press, 1974] view that the sulcal pattern of the colobine brain is ancestral by comparison to that of cercopithecines, and that gibbons may be even less changed from the early catarrhine conditions.

C. Dental Morphology and an Ancestral Morphotype

This summary of results is based on the extensive examination of dentitions of all genera and most species of living and fossil cercopithecids. As part of this study, a terminology was developed to describe important regions of especially the cheek teeth. JOLLY [1972] made use of an earlier version of this study and the terminology: figures 2 and 3 are modified slightly from those he has published, illustrating the right M_3 and left M^3 of *Theropithecus* with features labeled. Two sets of terms may require slight additional definition. The tooth margin, divided into four continuous segments (mesial, buccal, etc.) is the one-dimensional line which connects the cusp apexes around the rim of the occlusal surface. It may be formally defined as the set of points such that the neighboring points are either also on the margin or are closer to the roots or alveolar plane. The margin is unambiguously determined only on unworn teeth; with wear, its enamel edge is planed flat, revealing the dentine within, as when the crest of an anticline is eroded. The terms notch, cleft and groove are all used here for excavated regions, differentiated as follows: a notch may only occur where a tooth wall has been completely cut through, essentially in the sense of intercusp notches; a groove is defined as a narrow, linear depression, generally separating two cusps or other basic developmental features; a cleft, finally, is deeper and often broader than a groove but does not penetrate through the tooth wall. Essentially, any depression larger than a fine line or groove and not on the occlusal surface is a cleft; on the occlusal surface, an equivalent feature would be termed a basin or fovea, or more broadly a fossa.

The molariform teeth of Cercopithecidae are all based on a single ground plan, consisting of a high crown with four marginal cusps linked by transverse ridges or loph(id)s, and three foveas separated by the two ridges. Upper teeth are in general mirror images of their lower serial homologues, with buccal and lingual features reversed. The teeth widen or 'flare' outward laterally from the cusp apexes to the cervix (enamel line), especially on the buccal face of lowers and the lingual face of uppers. This flare is greatest in papionins (most in *Cercocebus* and *Papio*, least in *Theropithecus*), less in cercopithecins and least of all in colobines. Exceptions to this plan are found in M_3 and dP_3^3: in M_3, a hypoconulid is developed on the distal shelf in most forms; a paraconid is present on dP_3 mesial to the trigonid basin, often joined to the protoconid by a paralophid; the dP^3 has more rounded corners than other upper teeth and often a mesiobuccal extension as well. Cercopithecid premolars are far more heterogeneous than are molariform cheek teeth. The P_4 consists of a small trigonid with subequal protoconid and metaconid and molariform metalophid and a large talonid basin. There is only one cusp on P_3, preceded by a sloping mesial 'flange' for sharpening C' and with a distal fossa homoplastic but probably not homologous to the molariform talonid. In the female, the flange does not project far beyond the alveolar plane (if at all), but in males it sinks deeply below this level. The upper premolars are somewhat 'D'-shaped, with a straight buccal face. There is a prolongation of enamel onto the mesiobuccal root of P^3, forming a flange as in P_3, although it would not appear to be functional and may be considered as an 'overflow' effect from a canine/premolar honing morphogenetic field (as also for some P_4). Cercopithecid canines are large stabbing weapons which show high sexual dimorphism; uppers, especially those of males, present a deep, compressed cleft or sulcus on the mesial face which continues through the cervix onto the root. These features are typical of all Old World

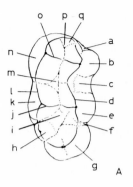

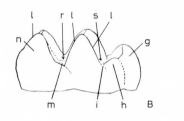

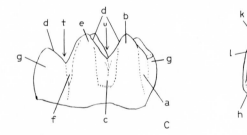

Fig. 2. Identification of cercopithecid dental landmarks on the right lower third molar of *Theropithecus* [modified after JOLLY, 1972, with permission of the author]. A = Occlusal aspect; B = lingual aspect; C = buccal aspect; D = mesial aspect (stippling indicates contact facet with M_2); E = distal aspect. a = Mesial buccal cleft; b = protoconid; c = median buccal cleft; d = buccal margin; e = hypoconid; f = distal buccal cleft; g = hypoconulid; h = 6th cusp (tuberculum sextum); i = distal fovea; j = hypolophid; k = entoconid; l = lingual margin; m = talonid basin; n = metaconid; o = metalophid; p = trigonid basin (mesial fovea); q = mesial shelf; r = median lingual notch; s = distal lingual notch; t = distal buccal notch; u = median buccal notch. In this and succeeding dental illustrations, elevated features (crests, ridges, outlines) are represented by solid lines, depressed features (grooves, clefts) by dotted lines.

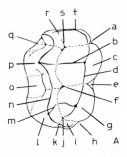

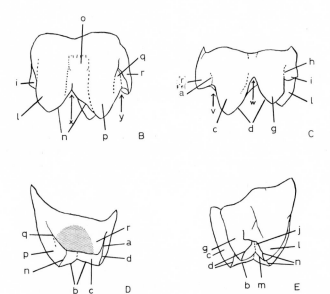

Fig. 3. Identification of cercopithecid dental landmarks on the left upper third molar of *Theropithecus* [modified after JOLLY, 1972, with permission of the author]. A = Occlusal aspect; B = lingual aspect; C = buccal aspect; D = mesial aspect (stippling indicates contact facet with M²); E = distal aspect. a = Mesial buccal cleft; b = paraloph; c = paracone; d = buccal margin; e = median buccal cleft; f = trigon basin; g = metacone; h = distal buccal cleft; i = distal shelf; j = distal fovea (talon basin); k = distal lingual cleft; l = hypocone; m = metaloph; n = lingual margin; o = median lingual cleft; p = protocone; q = mesial lingual cleft; r = mesial shelf; s = mesial fovea; t = mesial margin; u = distal buccal notch; v = mesial buccal notch; w = median buccal notch; x = median lingual notch; y = mesial lingual notch.

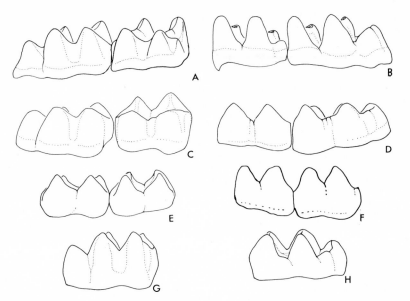

Fig. 4. Lateral views of right M(2-)3 in major cercopithecid morphologic groups. A, C, E, G = Buccal aspect; B, D, F, H = lingual aspect. A, B = Colobinae, M₂-₃ *(Dolichopithecus);* C, D = typical Papionini, M₂-₃ *(Macaca);* E, F = Cercopithecini, M₂-₃ *(Erythrocebus);* G, H = *Theropithecus,* M₃.

monkeys, but another series of characters distinguish between four morpho-logic-taxonomic groups, or morphotypes.

The Colobinae are characterized by moderate to high crown relief, especially visible in the deep median lingual notches of lower teeth (for this and other comparisons, see the drawings of 'typical' teeth in figures 4–6 and also the illustrations of actual fossils in figures 11–16, 18, 19). The trigonid basin (mesial fovea) is mesiodistally short in lowers, though it may be rather large in uppers; the distal margin of molariform teeth is asymmetrical, one of the few distinguishing features for isolated upper teeth. In the dP₃, there is a tendency to reduce or lose the distal fovea behind the hypolophid, and the mesial lingual notch is rather shallow. The maximum mesial width is less than the distal in M₁₋₂, but greater in M₃ except in *Colobus guereza* (and *Cercopithecoides* and *Paracolobus*). As in other groups, the mesial width is generally greater in upper teeth, but there is less tendency for reduction of the M³ distal loph; instead a fifth, midline cusp may be present, although other accessory cuspule development is slight in colobines. The median

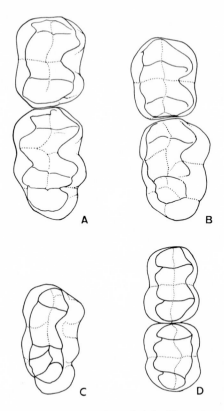

Fig. 5. Occlusal view of right $M_{(2-)3}$ in major cercopithecid morphologic groups. A = Colobinae, M_{2-3} *(Dolichopithecus);* B = typical Papionini, M_{2-3} *(Macaca);* C = *Theropithecus*, M_3; D = Cercopithecini, M_{2-3} *(Erythrocebus).*

lingual cleft of colobine uppers tends to continue nearly to the cervix, but the median buccal notch is not especially deep. There is almost always a hypoconulid on M_3, except in some populations or individuals of smaller-sized species (in *Presbytis* and *Mesopithecus*). The colobine P_3 is relatively wide with only moderate development of the talonid basin (which may, however, be expanded lingually) or the mesial flange. On the other hand, the mesio-buccal region of P_4 is more flange-like in colobines than in cercopithecines, the tooth may at times be turned obliquely to the molar row axis, and the metaconid is less wide and less tall than the protoconid, or at most subequal to it. In *Colobus*, especially, there is a tendency to greater reduction of the metaconid, although this appears to be variable among samples of the same

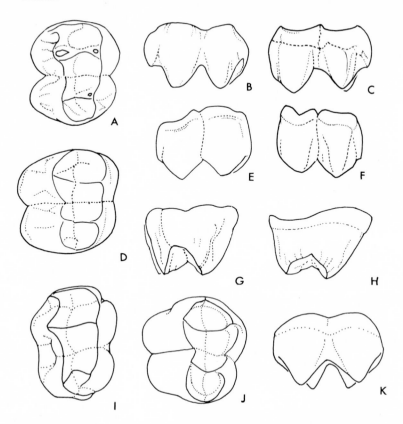

Fig. 6. Occlusal, lateral and distal views of left M² or M³ in major cercopithecid morphologic groups. A, D, I, J = Occlusal aspect; B, E = buccal aspect; C, F, K = lingual aspect; G, H = distal aspect. A, B, C, G = Colobinae, M²? *(Dolichopithecus)*; D, E, F, H, = typical Papionini, M²? (cf. *Paradolichopithecus* sp.); I = *Theropithecus*, M³ (see also fig. 3); J, K = Cercopithecini, M² *(Erythrocebus)*.

species or subgenus. The protocone of P³ is often reduced in colobines, extremely so in *Colobus* spp.

The dentally 'typical' Papionini (excluding *Theropithecus*) possess somewhat more 'bunodont' teeth, with less relief and especially with shallow median and distal lingual notches in lowers. The mesial foveas are subequal in length to the distal, accessory cusps are more common, and the distal margins are nearly symmetrical in both uppers and lowers; the distal fovea is normally developed in dP₃, but the mesial lingual notch is deeper than expected. The mesial width is greater than the distal in M_2 and in the M_1 of some species measured (*Macaca sylvanus* and *Cercocebus torquatus*); the usual pattern of mesial width smaller is seen in M_3 and most M_1. There is a tendency to reduce the distal loph of M^3, and no fifth cusps were seen on that tooth, but a sixth cusp is often found on M_3; hypoconulids were lacking only in rare *Papio* and *Cercocebus albigena* individuals. As in all cercopithecines, the P_3 has a well-developed distal 'talonid' basin and is relatively long and narrow. The P_4 is relatively wide for its length, and the mesiobuccal area is inflated, not flange-like, especially in larger cercopithecine species; the metaconid is always greater than or equal to the protoconid in both width and height. A most important but as yet only partly studied feature is the striking reduction or total lack of enamel on the lingual surface of lower incisors in all papionins. This developmental condition leads to the production of a self-sharpening, nearly gliriform chisel edge on these teeth.

The dental pattern of *Theropithecus* is distinct enough to warrant separate treatment: it preserves most papionin features, such as accessory cuspules, lower molar width relationships, moderate 'flare' and reduced lower incisor enamel, but shows several special features as well. The most important of these is great crown height and especially relief, with deeply excavated foveas as well as deep notches. In some ways, the pattern is one of convergence to the colobine condition. with a somewhat short (but deep) trigonid basin and high cusps. Other distinctive features include a large distal accessory cuspule on M_2 (and some M_1) which projects toward M_3; the mesial buccal cleft of lower molars is 'pocketed' and does not blend smoothly onto the buccal surface; the lower molar lophids and trigonid are turned mesiolingually oblique to the tooth's long axis; the buccal margin at the notch bases is nearly on the midline, producing the semblance of a longitudinal 'lophid' or crest; and partly as a result of this, the high relief and a delayed eruption pattern, *Theropithecus* has a distinctive wear pattern compared to that shared by other monkeys.

The Cercopithecini share the fourth dental morphotype, characterized by loss of M_3 hypoconulid and concomitant reduction of M^3. In most forms, expecially *Erythrocebus*, the teeth are elongate and flare is low, but in *Allenopithecus* there is highly developed flare, leading to very wide teeth like those of *Cercocebus*. In the lower molars, mesial width and trigonid size increase from M_1 to M_3. Other features are similar to those described for Papionini, except for the scarcity of accessory cuspules and the unreduced (colobine-like) character of lower incisor enamel.

In attempting to reconstruct the ancestral cercopithecid dental morphotype, obviously derived conditions can be found and rejected first. Such would be the loss of most lingual enamel on lower incisors in Papionini and the loss of M_3 hypoconulids in Cercopithecini. VOGEL [1966] has further noted that I^2 is conical and I^1 trapezoidal in gibbons, gorillas and colobines, and he has suggested that this condition is ancestral with regard to the derived homomorphy in other hominoids and enlarged I^1/tilted I^2 in many cercopithecines. In these features, the colobine condition is again ancestral, as was seen above for the facial region. For the lower premolars, it seems likely that the P_4 cusps were of subequal development in early cercopithecids (and ancestral catarrhines?), while there was a rather strong development of C'/P_3 honing with a P_3 flange extending below the alveolar plane, a derived state by comparison to the early catarrhines (as reconstructed). The mesial groove of the upper canine passing continuously onto the root, as well as the bilophodont molar pattern and the absence of dP_4–M_2 hypoconulids, are further shared derived features of all Cercopithecidae not to be expected in earlier ancestral catarrhines.

The most difficult problem is that of determining or at least estimating the ancestral conditions within the cercopithecid range for the main features of the molariform teeth, such as flare, crown relief and trigonid size. Comparison with a 'sister group' is not really feasible, because the hominoid fovea anterior and other structures are rather differently constructed. Based on consideration of the total range of patterns throughout the modern (and fossil) Cercopithecidae and other catarrhines as well, it would appear that a relatively large trigonid, perhaps as in macaques, was present in the ancestral cercopithecid. It is even more certain that the high relief and especially deep lingual lower-molar notches of colobines and *Theropithecus* are derived within Cercopithecidae. Although basal or lateral 'flare' is certainly derived within Catarrhini, it is less clear how strongly such flare was developed in the ancestral monkeys. This flare is generally greatest in just those areas where cingulum is present in other primate groups, and as I have argued elsewhere [DELSON, 1974 c], it appears that cercopithecid flare is in fact derived from cingular developments which, instead of being localized as a shelf or ledge, graded smoothly over the tooth face. From an intermediate degree of development, again perhaps as in macaques, the flaring would decrease (independently) in colobines, most Cercopithecini and *Theropithecus*, while possibly increasing in other African Papionini. *Allenopithecus* may have retained a more ancestral condition than other cercopithecins, or it may even have paralled papionins in this trend but not others.

In these molariform-tooth characters, therefore, it appears that colobines are rather derived, while macaques and other papionins are closest to the hypothesized ancestral conditions. In total, macaques may represent a good approximation to many ancestral cheek-tooth patterns: large but shallow trigonids, moderate flare, low rounded cusps, shallow notches, mesial width greater than distal in all molars (like hominoids), subequally long M^2 and M^3, and well developed M_3 hypoconulid. Combined with these features would have been the colobine incisor pattern (conical I^2 and lowers fully sheathed in enamel), as well as a colobine-like facial structure.

It will be noticed that no mention has been made of aspects of post-cranial or 'soft-part' morphology in this discussion of ancestral morpho-types. The former is generally accepted to be more reflective of habitus than heritage traits, and my studies [as well as those of JOLLY, 1967] have supported this view while finding that adaptations to terrestriality in both subfamilies lead to similar but not identical mosaic modifications of a presumably arboreal ancestral condition in catarrhines. On the other hand, further dis-cussion of soft parts, as well as possible hypotheses about the selection pres-sures leading to all of these modifications, seem better treated after a consi-deration of the fossil evidence of the two earliest stages in the evolution of the Cercopithecidae: early catarrhine ancestors and the first true cerco-pithecids. Following these discussions, a brief overview of the history of the main subgroups of Old World monkeys will draw upon later fossil evidence as well as the morphotypes described above, in an attempt to determine the relationships between and among all known forms. For this phase of the study, the fossil evidence is of greatest importance, not by revealing the 'true' ancestors of living or other fossil populations, but by indicating the scope and diversity, as well as perhaps the age, of evolutionary radiations within the group. For example, if only the 'main lines' of cercopithecid descent were to be reconstructed, i.e., those which led to living forms, we would have no knowledge of the several parallel trends to increased size in many taxonomic-geographic groups, and the especially fascinating lineage(s) of terrestrial European colobines would be a complete unknown, as would the past dominance of *Theropithecus* over *Papio* in Africa.

III. Early Evolution of Cercopithecidae

Any discussion of fossils which emphasizes their temporal placement must of necessity rest on a stratigraphic framework. Modern concepts of chronostratigraphy and geochronology are becoming more precise, with the result that some changes have been made in previous usage and correlation

AGE IN M.Y.	EPOCHS AND SUB-EPOCHS	STANDARD STAGES/AGES	EUROPEAN LAND MAMMAL AGES	SELECTED MEDITERRANEAN ASSEMBLAGES	N. AMERICAN LAND MAMMAL AGES	SIWALIK FORMATIONS	SOUTH AND EAST ASIAN ASSEMBLAGES	AFRICAN MAMMALIAN ASSEMBLAGES
0	P L E I S — LATE, MID-DLE	TYRRHENIAN	?	CAPO FIGARI, HEPPENLOCH	R. LABREAN, IRVING-			GAMBLIAN, KANJERA, OLORGESAILIE
1	EARLY	"MILAZZIAN", SICILIAN (EMILIAN)	BIHARIAN	MOSBACH, FARNETA				
2		CALABRIAN		SENEZE	TONIAN	TAWI, PINJOR	C 18 TRINIL, K HONAN, T 12 DJETIS	OLDUVAI — SK,KA ?BF?
3	P L I O — LATE	VILLA-FRANCHIAN / PIACENZIAN	VILLA-FRANCHIAN	ST. VALLIER, ROCCANEYRA	BLANCAN	TATROT	YUSHE	VOGEL R. ?JM 90? STS,M,T OMO KANAM
4		TABIANIAN		ETOUAIRES, VILLAFRANCA, PERPIGNAN		?		EKORA, LAETOLIL, KANAPOI, KAISO
5	O C E N E — EARLY	RUSCINIAN (ZANCLIAN)	RUSCINIAN	MONTPELLIER, POLGARDI				LOTHAGAM
6		MESSINIAN		WADI NATRUN, ARQUILLO				
7				HATVAN, MARAGHA	HEMP-	DHOK		MPESIDA
8	M — LATE		TUROLIAN	MARGEAU, VOSENDORF	HILLIAN	PATHAN		ONGOLIBA ?
9	I	TORTONIAN		SAMOS, MANSUETOS, PIKERMI				
10	O			MOLLON				
11	C — MID-DLE		VALLESIAN	MONTREDON, CAN PONSIC, HOWENEGG	CLAREN-DONIAN	NAGRI		NGORORA
12	E	SERRAVALLIAN		ST. GAUDENS, BACCINELLO-1				
13	N			LA GRIVE, NEUDORF-SDB	BARSTO-	CHINJI		
14	E — MID-DLE	LANGHIAN	"VINDO-BONIAN"	GORIACH, SANSAN				FT. TERNAN
15				NEUDORF-SPL, PONT-LEVOY, V. COLLONGES	VIAN	KAMLIAL		
16	— EARLY			LA ROMIEU				MABOKO, OMBO ? ?
17		BURDIGALIAN	"BURDI-GALIAN"	ORECHOV, BONREPOS		?		
18	N — EARLY			WINTERSHOF	HEMING-			LOPEROT, RUSINGA
19	E			ESTREPOUY, LAUGNAC				NAPAK, SONGHOR, KORU
20								

Fig. 7. Correlation chart of Oligocene-Pleistocene sequences. Abbreviations: in Mediterranean column, Neudorf-Spl. = Neudorf an der MarchSpalte (fissure), Sdb. = Sandberg (younger stratified deposit); in Asian column, CKT = Chou-Kou-Tien localities 1, 18 and 12; in Africa column, BF = Bolt's Farm; JM 90 = locality J.M. 90–91 in the Chemeron Formation; KA = Kromdraai; M = Makapan; SK = Swartkrans; STS = Sterkfontein; T = Taung. The time ranges of the Olduvai and Omo deposits are indicated by straight vertical lines; the East Rudolf deposits probably overlap those of the Omo valley and extend younger in time to an undetermined summit in the main series.

(see also Savage, pp. 2–27). Aspects of the calibration and paleoecology of Neogene (Miocene-Pleistocene) deposits containing fossil cercopithecids have recently been reviewed elsewhere [see Delson, 1974 a, b and references therein], and thus only a summary correlation chart is included here as figure 7. All localities, as well as faunal and chronological units mentioned in the text, are shown on this chart which attempts to provide accurate representation of temporal relationships, both relative and radiometric.

A. Oligocene Catarrhines and Ancestral Cercopithecids

Present knowledge of Oligocene Old World Anthropoidea is essentially restricted to the sample of some 7–10 species recovered from the Egyptian Fayum, mostly through the expeditions led by E. L. Simons. Since the first of these taxa was recovered over 60 years ago, opinions have varied widely about their relationships with modern forms. Most recently, Simons [1967, 1970, 1972, etc.] has argued that *Parapithecus*, and less clearly its relative *Apidium*, are the known forms closest in morphology to Old World monkeys. His latest papers have advocated the placement of these two genera as the subfamily Parapithecinae of the Cercopithecidae. This interpretation appears to be based on the presence in *Parapithecus* species of high and somewhat narrow lower molar crowns with slightly reduced hypoconulids, as well as the anthropoid, 'simian' or 'monkey' *grade* characters of frontal and symphyseal fusion found in these genera [Simons, 1972, p. 191]. Such grade features may certainly be developed in parallel by groups sharing a distant common ancestor and do not imply close propinquity of descent, as Simons has implied. In another study [Delson, 1974 c], I have discussed this problem specifically and shown that other characters of *Parapithecus* are quite distinct from those found in modern monkeys or in the postulated ancestral morphotype. These include small P_4 metaconids, short M_3, lack of cingulum, incorporation of large conules into possibly functional lophs and especially the derived development of C'/P_2 honing, which is unlikely to have been lost in descendants. Simons' comparisons have been especially with *Cercopithecus ('Miopithecus') talapoin*, of admittedly similar size but a probably rather specialized member of a group with quite derived dental pattern. It is possible that *Parapithecus* (and *Apidium*) may have been the ecological vicar of cercopithecids (or at least of '*Miopithecus*') in the Fayum region, but it seems unlikely that they were phyletically related to modern species.

Granting that the Fayum in fact represents only one small area and one environment of Oligocene Africa, it is nonetheless possible to examine other taxa known from there by comparison with the postulated dental morpho-type for early cercopithecid ancestors. Specimens allocated to *Oligopithecus* and *Propliopithecus* species variably present such characters as: evidence of hypoconulid reduction by fusion with entoconid on dP_4–M_2;[1] subequal size of P_4 metaconid and protoconid; well-developed P_3/C' honing; narrow M_3; M_2 wider and longer than M_1; and loss of paraconids. In addition, the lower molars possess labial cingulum in most cases, which would produce cerco-pithecid-type flare if smoothly incorporated into the tooth wall; bulging is in fact observed when cingulum is not clearly present. Finally, these spec-imens share the high crowns (but low relief) seen in the molars of all Fayum anthropoids. It may be suggested that out of this known variability of early catarrhines, one lineage may have combined the mosaic of features ancestral to cercopithecids, perhaps in an environment different from that of the Fayum. Such a cercopithecid ancestor would be termed a 'hominoid' on overall grounds of grade similarity, but a cercopithecid in terms of its de-scendants.

B. The Oldest Known Cercopithecids

Fossil Old World monkeys of early and middle Miocene age have so far been reported only from the northeast quarter of Africa, at one isolated locality and from several geographically and temporally close ones. A de-tailed study of these specimens is in preparation, but some preliminary com-ments can be reported. *Prohylobates tandyi*, as recognized by SIMONS [see 1969], is undoubtedly a cercopithecid, although its affinities are uncertain. The three fragmentary mandibles of this species are known from the Wadi Moghara local fauna of Egypt, which now seems to be of rather early Mio-cene age [HAMILTON, 1973; C. MADDEN, personal commun.]. Although the teeth are worn, they appear to have been less completely bilophodont than in modern or other fossil monkeys; the crowns are relatively high, but the lingual notches shallow; the trigonid basins are badly worn but not large; there may have been a small distal cuspule on M_2, perhaps as in some *Theropithecus* rather than a true hypoconulid; and the third molar was prob-

---

1   VORUZ [1970] has noted this reduction in *Oligopithecus* and '*Moeripithecus*', now recognized as a species of *Propliopithecus*; her observations, however, did not consider other characters nor did she evaluate the position of *Parapithecus*.

ably rather short but with a strong hypoconulid. Most of these features are as expected in an early cercopithecid, or slightly colobine-like, as may be the deep corpus. Until more complete and less worn specimens are available, this taxon will continue to be enigmatic and tantalizing.

The majority of African Miocene monkeys (and of course hominoids) are derived from deposits in the Lake Victoria region. From Napak, PILBEAM and WALKER [1968] reported a frontal bone said to resemble platyrrhines or immature colobines and an M^- of cercopithecine aspect. The tooth is morphologically of papionin type, with approximated cusps and moderate to high flare. RADINSKY [in press, 1974] has suggested that if the frontal is cercopithecid, it is 'primitive' in its lack of an indication of the arcuate sulcus on the endocranial cast. Recalling that early cercopithecids are expected to combine cercopithecine-like molars with colobine- (or gibbon-) like crania and brains, these specimens merely conform to the postulated morphotype and do not clearly indicate separation of the subfamilies as claimed by PILBEAM and WALKER [1968].

Possibly more convincing evidence as to this separation is provided by the collection partly studied by VON KOENIGSWALD [1969] from Maboko and perhaps Rusinga and Ombo; further specimens have now been identified from Loperot and possibly Songhor. Essentially on the basis of size, VON KOENIGSWALD recognized two species of the then new genus *Victoriapithecus*, but the actual variation may be better understood in terms of sexual dimorphism, the incorrect allocation of hominoid specimens and a rare second 'morph'. The five most complete mandibular fragments (fig. 8 A–E) demonstrate a fair range in size, but four of these (and most of the isolated teeth) differ from the fifth (fig. 8 C) in proportions and some morphological features. For example, the ranges of M_2 mesial maximum width, maximum length and width/length ratio in eight specimens are 60–78 mm, 67–81 mm, and 86–96%, respectively (note KNM MB-37 smallest, MB-36 largest, MB-1 narrowest). The M_2 of KNM MB-34 falls within these ranges for absolute size, but is significantly narrower, with values of 62 and 77 mm and 80%. A similar situation holds for the M_3 of this jaw as opposed to most others. Moreover, it may be seen from figures 8 G, I that while the median lingual notches of both MB-34 and MB-1 (type of *Victoriapithecus macinnesi*) are relatively shallow, the trigonid basins of the former are large as compared to the small ones of MB-1 and most other specimens. Moreover, despite its larger tooth size, MB-34 has a shallower corpus and possibly a more backwardly-tilted ascending ramus, although this area is damaged. All upper teeth are isolated, but most are of similar, typically cercopithecid aspect.

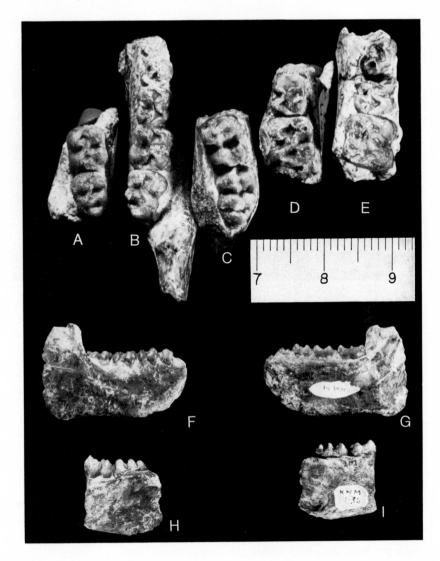

Fig. 8. Victoriapithecus mandibles, occlusal and lateral views. A = Left M_{2-3}, KNM MB-37, occlusal; B, F, G = right P_4–M_3, KNM MB-1, holotype *V. macinnesi*, occlusal, buccal, lingual; C, H, I = left M_{2-3}, KNM MB-34, occlusal, buccal, lingual; D = left M_{2-2}, KNM MB-33, occlusal; E = right P_4–M_3, KNM MB-36, occlusal. Occlusal views at approximately $1.9 \times$ natural size (see scale) lateral views $\times$ 0.95 (approx.).

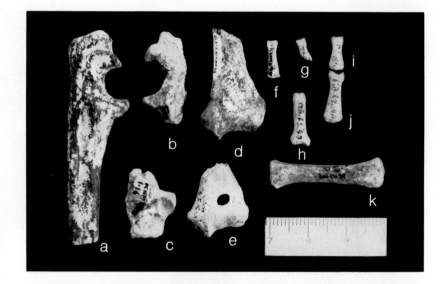

Fig. 9. Postcranial bones from Maboko, attributed to *Victoriapithecus* spp. a = Right ulna, KNM MB-32, lateral aspect; b = left ulna, KNM MB-2, lateral aspect; c = left calcaneum, KNM MB-16, proximal aspect; d = left humerus, KNM MB-19, anterior aspect; e = left humerus, KNM MB-3, anterior aspect; f = middle phalanx, KNM MB-93, dorsal aspect; g = middle phalanx, KNM MB-21, dorsal aspect; h = proximal phalanx, KNM MB-12, dorsal aspect; i = middle phalanx, KNM MB-13, dorsal aspect; j = middle phalanx, KNM MB-22, dorsal aspect; k = caudal vertebra, KNM MB-20. All natural size.

The most distinctive is KNM MB-75, an $M^{2?}$ which preserves a crista obliqua and which was designated by VON KOENIGSWALD [1969] as the type of *V. leakeyi*. If it is indeed cercopithecid, it probably should be associated with MB-34, with which it articulates reasonably well. The many isolated canines are all of male type, and no morphological or metrical distinctions have yet been discerned within the series.

A similar dichotomy of form is revealed by the postcranial fragments, illustrated in figure 9. The ulnae indicate a straight shaft and large olecranon, partly retroflexed as in the most terrestrial colobines or semiterrestrial cercopithecines (macaques). Both humeral fragments are too small to articulate with the ulnae, but the larger is both more cercopithecine-like and more terrestrially adapted in its distal morphology [compare JOLLY, 1967], especially the greater posterior reflection of the medial epicondyle. A caudal vertebra and a partial calcaneum do not provide much information, but the several phalanges indicate that at least some Maboko monkeys were quite terrestrially adapted in their extremities, with short and stout phalanges, while others may have had more elongate, arboreally-suited digits.

The sum total of this evidence is inconclusive but suggestive. Again recalling that cercopithecine teeth are postulated as ancestral, it appears that the reduced trigonids seen in most of the specimens indicate a first step toward the development of a colobine dentition. The rather square form of these teeth, as compared to the more elongate MB-34, agrees with this hypothesis, although the retention of the ancestrally shallow notches indicates a mosaic situation of evolution in progress. If it may be suggested further (and most hesitantly) that the two ulnae represent larger, and the large humerus a smaller individual of the same species as the robust phalanges, this animal would have been a relatively terrestrial, cercopithecine-like monkey of moderate size. The smaller and more gracile humerus (and phalanx?) might be more arboreal, possibly more colobine-like in the humerus at least. One step further allows for the possible association of the larger teeth, of cercopithecine aspect, with the larger bones, and the smaller, more colobine (derived) teeth with the more arboreal (ancestral?) limbs. Such associations, if held up by further evidence from Maboko, do suggest that the separation of the modern cercopithecid subfamilies was at least in progress by the Middle Miocene. With the consideration of possible ecological or functional hypotheses to explain the differentiation of cercopithecids and their subgroups, this observed situation makes even more evolutionary 'sense'.

C. Nonskeletal Characters and Early Cercopithecid Paleobiology

JOLLY [1966] has suggested that the original adaptive niche of the cercopithecids was leaf-eating, and he has proposed that the colobine molar pattern is most 'primitive'. NAPIER [1970] has argued that the colobines are morphologically ancestral in all character states, essentially because they are more arboreal, could not be descended from already terrestrial cercopithecines, and thus must be 'primitive'. Although I have agreed with VOGEL [1966] and others that the colobine (and gibbon) cranium is probably representative of the ancestral condition for catarrhines, the cercopithecine dentition appears least derived and the colobine rather strongly so; limb structure, as noted above, is probably more indicative of function than of ancestry.

Moreover, the modern colobines are further equipped to process large quantities of foliage through an enlarged and sacculated stomach and modified digestive tract [see KUHN, 1964]; KUHN's work further does not indicate any differences in the construction of this system between African and Asian colobines, suggesting that this clearly derived complex (with regard to cercopithecids as a whole) may be ancestral for Colobinae and is perhaps to be expected in the earliest forms with clearly colobine teeth (see below). On the other hand, modern cercopithecines possess buccal pouches for temporary storage of food items prior to mastication, but a digestive apparatus similar to that of hominoids. These cheek pouches may be of value to a semi-terrestrial animal which is hesitant about feeding on the ground; in times of stress, flight would be up into the trees, and the ability to gather and store food, and then to process additional material when there is time, would permit the greatest freedom of action. Another aspect of monkey anatomy which could not have fossilized but which still may provide information about relationships is karyology. Most gibbons share with some colobines a diploid number of 44 chromosomes, which may be the ancestral number for catarrhines; *Nasalis* has 48, as one of its several derived features; all papionins share a diploid number of 42 and very similar karyotype, indicating a probably closely-knit ancestry; finally, the diploid number ranges between 48 and 72 among (and within) species of Cercopithecini, another of the strongest specializations of this group.

Considering these facts and the postulated morphotypes discussed above, the most plausible explanation for the origin of Old World monkeys as an adaptative unit (rather than as a grade) is NAPIER's [1970] hypothesis that in seasonal forests of deciduous and even some evergreen types, the ability to subsist on leaves rather than fruits when necessary would have been a great selective advantage. Thus, the ancestors of cercopithecids may not have been dependent on leaves, but were able to supplement the basic primate diet of fruit (and some protein) with foliage in those habitats or times of year when it was most plentiful. Such an environment may not have occurred in the Fayum region, and thus none of the known Fayum forms 'became' cercopithecids, although it can be suggested that the actual ancestors were more closely related to *Oligopithecus* and perhaps *Propliopithecus* than to *Parapithecus* and *Apidium*. By the late Oligocene to early Miocene, proto-cercopithecids probably had given rise to the first true Cercopithecidae, of which *Prohylobates* may represent one lineage whose relationships with modern forms is unclear. The early cercopithecids which actually did lead to modern forms probably had colobine-like skulls, macaque-like teeth, subequally long limbs with joints and digits of 'arboreal' type; less certainly, they may have had a diploid chromosome number of 44, a hominoid-type digestive apparatus (with neither cheek pouches or a sacculated stomach) and perhaps discontinuous ischial callosities. They were likely to have been arboreal (or semiarboreal) quadrupeds who ate fruits when possible but supplemented their diet with leaves under certain ecological conditions.

By the Middle Miocene, the ancestors of colobines may have begun to concentrate on a folivorous diet with concomitant changes in dental morphology (reduction of trigonid basins followed by an increase in crown relief) and digestive system, as well perhaps as a relative reduction of the thumb and tarsus, possibly in connection with development of arm swinging or leaping behaviors and invasion of the forest canopy. At the same time, ancestral cercopithecines may have begun experimenting (in an evolutionary sense) with semiterrestrial behaviors and a more eclectic, omnivorous diet, leading in turn to cheek pouches for temporary food storage and perhaps larger size and allometrically longer faces (and related bony modifications) as a corollary of defense activities. An intermediate level in this reorganization may have been sampled at Maboko Island with the local fauna containing the two species (?) of *Victoriapithecus*.

Unfortunately, there is as yet no decent fossil evidence to document the next period of cercopithecid evolution. No monkeys are certainly known from Fort Ternan, despite the published statements of the late L. LEAKEY [1968]. The tooth from Ongoliba (Zaire, ex-Congo) described by HOOIJER [1963, 1970] is of cercopithecine form, and may be of Late Miocene age, as is the as yet unpublished tooth from Ngorora [BISHOP and CHAPMAN, 1970]. In the period between perhaps 15 and 10 million years ago, several major changes and splits took place in the cercopithecids as I interpret them, events which can only be hinted at here. It may be expected that the early cercopithecines may have undergone some form of chromosomal fusion, resulting in a diploid number of 42. A more arboreally inclined population or set of populations of these animals may have diverged in this time interval to become the ancestors of Cercopithecini, losing their M_3 hypoconulid, narrowing their teeth in general, and apparently undergoing wholesale duplication of chromosomes or the even more unlikely fission thereof. Knowledge of whether chromosome number in this group has increased or decreased with more recent speciation would be of great importance in deciphering relationships, especially of *Erythrocebus*, *Allenopithecus* and '*Miopithecus*', with rather low numbers. The remainder of the cercopithecines may have been represented across Africa in the Late Miocene by one or more polymorphic, intergrading species which probably were at least as terrestrial as most macaques and mangabeys and which had at least begun to reduce the enamel layer on the lingual surfaces of their lower incisors. The population of Turolian age from Marceau, Algeria represents a cercopithecine which is part of this group of species. ARAMBOURG [1959] referrred to it as *Macaca flandrini*, but his type and some other specimens are in fact colobines, and the cercopithecine specimens are at present best left unnamed or termed ?*Macaca* sp.

There was probably less diversification among the colobines of 10–15 million years ago. They may already have been in possession of the main features of modern colobines, and pollicial reduction is likely to have begun but not progressed very far. Paleontological and paleoecological evidence suggests that there may have been a division between a more arboreal group which remained in Africa and a somewhat more terrestrially inclined section which crossed into Eurasia via an open woodland 'corridor' in the early Late Miocene. This view is rather different from that offered by NAPIER [1970], who suggested that cercopithecines arose in Eurasia during the Early Miocene from an essentially colobine ancestor through specialization first of the limbs and locomotion and then of the skull and teeth. African and Asian colobines were seen as independent local developments paralleling each other in thumb reduction, stomach enlargement and other features. NAPIER concluded his article with the thought that audacity such as his stimulates new research through controversy; this in fact has been the case, but I hope that the novelty and details of my reconstruction have been well grounded in data produced by this research. The final sections of this paper will be a brief review of the later history of the groups whose origins I have suggested here. The phylogeny of figure 10 is an attempt to represent the discussion up to this point by combining fossil evidence with inferred relationships of higher taxa.

IV. Phylogeny and Dispersal of Later Cercopithecidae

The fossil Colobinae, especially of the circum-Mediterranean area have been the subject of a recent study [DELSON, 1973a], and thus their relationships and distribution are now probably better known than those of the cercopithecines, which are in general more common as fossils. In order to retain a sense of the unity of the taxonomic groups, they will be discussed in turn, using a geographic and temporal perspective, rather than following the

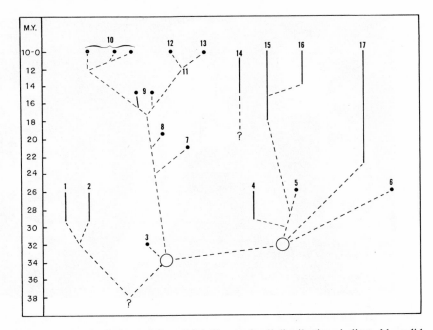

Fig. 10. Early radiations of Catarrhini. Known fossil distributions indicated by solid lines (or dots for single-spot occurrences), inferred relationships by dotted lines; inferred times of lineage separation calibrated to time scale. M.Y. = Millions of years before the present. Note that the family rank has been used for taxa of the 'Hominoidea', but this is more a matter of convention than conviction. Populations and taxa indicated by numerals as follows: 1 = *Apidium* spp.; 2 = *Parapithecus* spp.; 3 = *Oligopithecus savagei;* 4 = *Propliopithecus* (including *Moeripithecus*) spp.; 5 = *Aegyptopithecus zeuxis;* 6 = *Aeolopithecus chirobates;* 7 = *Prohylobates tandyi;* 8 = Napak cercopithecid(s); 9 = *Victoriapithecus* 'species'; 10 = Colobinae; 11 = common ancestor of (modern) Cercopithecinae; 12 = Cercopithecini; 13 = Papionini; 14 = Oreopithecidae; 15 = Pongidae (including *Dryopithecus* spp.); 16 = Hominidae (including *Ramapithecus* spp.); 17 = Hylobatidae (including *Pliopithecus* and *Limnopithecus* spp.).

solely time-sequential approach of the preceding section. The several figures are intended to illustrate not only the fossil taxa themselves, but also the dental morphotypes which they typify.

A. Colobinae

The oldest known colobine, *Mesopithecus pentelici*, ranges from the late Vallesian or early Turolian into the late Turolian (about 11–6 million years)

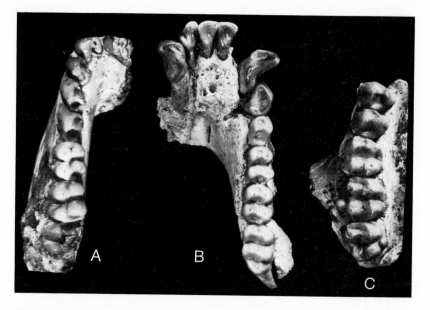

Fig. 11. Dentition of *Mesopithecus* species in occlusal view. A = *M. pentelici* female juvenile mandible (L I$_2$–M$_3$, erupting), Pikermi, NHMW uncatalogued; B = ?*M. monspessulanus*, male mandible (L C$_1$–R M$_3$), Villafranca d'Asti, NMB VJ 130; C = *M. pentelici*, male maxilla (R P^3–M^3), Pikermi, MNHN-P PIK 011. All natural size.

in southern and central eastern Europe, but it is still known essentially from the large population (on the order of 100 individuals, most fragmentary) from Pikermi, the type site. In cranial and dental characters, it is clearly colobine with few special features, despite MAIER's [1970, p. 201] suggestion that the teeth are of cercopithecine type (see fig. 1 A, C, 11, 12). The choanae are high and wide, the mandible of even depth with moderately large gonion, the face rather short and upright, and there may have been a low sagittal crest far back on the male skull. Postcranially, *M. pentelici* was rather terrestrially adapted, at least as much so as modern *Presbytis entellus*, as evidenced by its long-bone robusticity, elbow joint morphology, stoutish phalanges and upheld by its association with open-country local faunas. The thumb was relatively longer than in any modern colobine, but shorter than in cercopithecines, while the tarsus was short as in most colobines [see also GABIS, 1960]. In terms of its skull and teeth, *M. pentelici* corresponds well with a postulated early colobine morphotype, and given its age and geo-

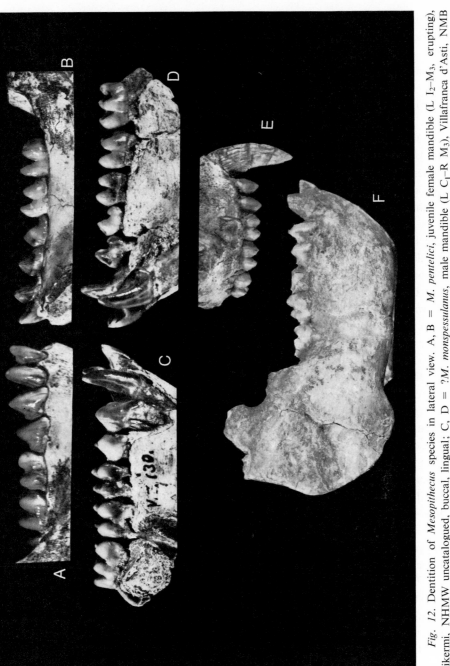

Fig. 12. Dentition of *Mesopithecus* species in lateral view. A, B = *M. pentelici*, juvenile female mandible (L I₂–M₃, erupting), Pikermi, NHMW uncatalogued, buccal, lingual; C, D = ?*M. monspessulanus*, male mandible (L C₁–R M₃), Villafranca d'Asti, NMB VJ 130, buccal, lingual; E = *M. pentelici*, male maxilla (R C¹–M³), Pikermi, BSM AS II 6, buccal; F = *M. pentelici*, male mandible, Pikermi. MNHN-P PIK O34, lateral. All natural size.

graphic position could be typical of ancestral Eurasian leaf-eaters, if not African ones as well.

A younger colobine species of similar size and distribution (Ruscinian to mid Villafranchian of Europe) has previously been placed in '*Semnopithecus*' but is better considered as ?*Mesopithecus monspessulanus*. Its teeth are significantly narrower (and slightly smaller overall) than those of *M. pentelici* (fig. 11 B, 12 C, D), and the few postcranial remains studied or previously reported suggest a possibly less terrestrial adaptation. Referral of this species to *Mesopithecus* is questionable, but less so than would be allocation to any other fossil (or recent) genus. Two Italian populations of apparently latest Miocene age (almost all the specimens of which have been lost) are included in this species and may show some evidence of transition from *M. pentelici*.

Dolichopithecus ruscinensis has a pan(southern)-European range during the Ruscinian and early Villafranchian, but it too is known mostly from a moderately complete collection from the type locality at Perpignan (or Roussillon). This rather large colobine is dentally typical but evidences facial lengthening and concomitant changes at least to the degree seen in *Nasalis larvatus* (fig. 13, 14). Partial postcranial material reveals a locomotor adaptation to life on the ground more pronounced than in any other known colobine and within the range of 'baboons' as defined behaviorally by JOLLY [1967]. The long bones are robust, the humeral trochlear flange strong and medial epicondyle reflected posteriorly, the ulnar shaft is slightly concave and the olecranon prolonged, and the phalanges are quite stout. No humerus is complete, but estimates of the length of a male specimen suggest it to have been about as long as male femora, a situation otherwise found only in some large and extremely terrestrial populations of fossil *Theropithecus* studied by JOLLY [1972]. A mosaic combination of features in part determined by its colobine heritage brought *Dolichopithecus* to a level of terrestrial adaptation similar to that found in modern mandrills *(Papio sphinx)* and *Macaca sylvanus*, which share a rather different character mosaic. The presence of *D. ruscinensis* during a time characterized by humid forest [see DELSON, 1974b] suggests further eco-ethological comparisons to mandrills and some of the larger macaques of southeast Asia.

The earliest specimen attributed to *Dolichopithecus* is an ulna from the latest Miocene of Hungary, similar in all comparable respects to less complete Perpignan specimens; two teeth referred to *M. pentelici*, but of rather larger size than usual, are known from the same spatiotemporal region. In light of the lack of observed change in morphology of *Dolichopithecus* populations with time, these data support a 'punctuated equilibrium' or allopatric model

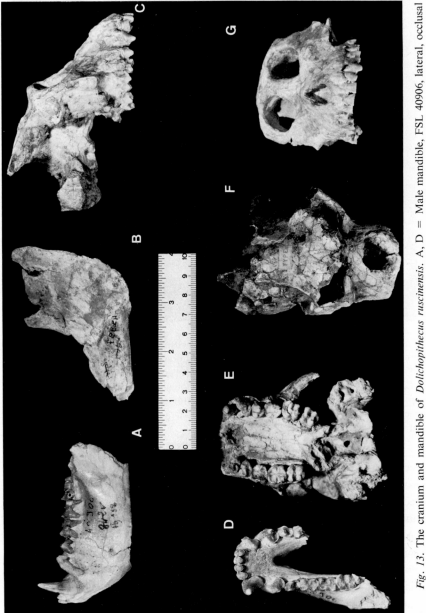

Fig. 13. The cranium and mandible of *Dolichopithecus ruscinensis*. A, D = Male mandible, FSL 40906, lateral, occlusal (R M$_2$–L M$_3$); B = male (?) mandible, posterior portion, FSL 49997, left lateral view (photographically reversed); C, G = female partial cranium, FSL 41327, right lateral and frontal views, occlusal plane horizontal; E, F = female partial cranium, MNHN-P PER 001, ventral and dorsal views, occlusal plane horizontal. All specimens from Perpignan (Roussillon), illustrated one-half natural size.

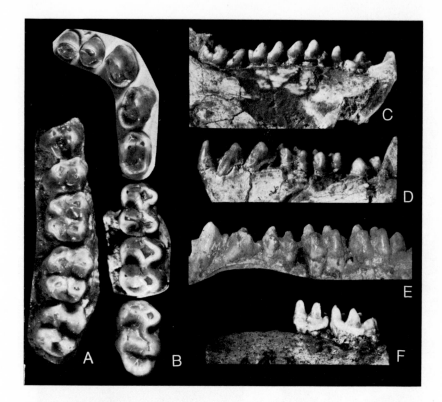

Fig. 14. The dentition of *Dolichopithecus ruscinensis*. A = Left maxilla of female cranium FSL 41327 (P³–M³), occlusal; B = right dentition of female mandible MNHN-P PER 004, occlusal – isolated teeth are placed in a rough approximation to the dental arch; C, D = female mandible ML Pp (6), L C_1–M_3, lingual, buccal; E = male mandible ML Pp 1, L P_3–M_3, buccal; F = male (?) mandible fragment ML Pp 2, R M_{2-3}, lingual. All specimens from Perpignan (Roussillon). A, B at one and one-half times natural size; C–F at natural size.

of speciation [see ELDREDGE and GOULD, 1972; ELDREDGE and TATTERSALL, this volume]: a semi-isolated population of Pannonian Basin *M. pentelici*, unaffected by latest Miocene Mediterranean desiccation, developed large size and increased terrestriality, then spread rapidly over the previous range during Ruscinian reforestation. It is even possible that the effect of competition with early *Dolichopithecus* might have led to character displacement in *M. pentelici*, resulting in ?*M. monspessulanus*. No derived feature of *M. pentelici* is absent from *D. ruscinensis*, and several trends (e. g., facial lengthening, choanal narrowing and especially increased terrestriality) are continued from the earlier to the later species. There is little doubt that *Mesopithecus pentelici* shared a recent common ancestor with *Dolichopithecus ruscinensis* (and with ?*M. monspessulanus*) – it may be further suggested that the relationships were in fact those of actual ancestor and descendants.

There are of course no living colobines in Europe, and the two other fossils from the Mediterranean region appear to have closer affinities with African forms, but the extinct European species may be specially related to the several modern and fossil Asian colobines. Only one important extinct population is known from Asia, represented by six gnathic fragments from the Dhok Pathan Formation of the Siwaliks. SIMONS [1970] has correctly suggested that these fossils, previously termed *Macaca sivalensis* and ?*Cercopithecus* or *Semnopithecus asnoti*, probably represent a single small colobine species which may be called ?*Presbytis sivalensis*, using *Presbytis* here in the sense of a form genus. The open-country aspect of the Dhok Pathan fauna and its rough age equivalence to Turolian *Mesopithecus* suggests that these species may have been descended from an African emigrant of Vallesian/Nagri age. No Pliocene or Early Pleistocene fossil colobines are known from Asia, and the Middle to Late Pleistocene forms are clearly related to modern species groups. GROVES' [1970] interpretation of the modern genera, can be generally supported, especially in linking *Pygathrix* with *Rhinopithecus* and *Nasalis* with *Simias*, rather than indiscriminately grouping (or recognizing) them all. However, it seems that the components of each pair are about equally distinct and both *Rhinopithecus* and *Simias* are here recognized as subgenera, the latter possibly even more distinct cranially than the former, despite GROVES' views to the contrary. Cranial analyses suggest that *Presbytis* and *Pygathrix* are most similar to one another, while GROVES noted that *Nasalis* and *Pygathrix* share a high intermembral index and other long-bone features not found in other modern colobines; unpublished protein studies by SARICH do not clearly support either view and may be most important in helping to decipher their relationships. At present, *Nasalis* does seem to have been a Plio-Pleistocene offshoot from a possibly 'central' *Presbytis/Pygathrix* stock.

Modern African *Colobus* species are linked by the derived reduction of the thumb and such dental characters as P³ protocone reduction, while each of three subtaxa is distinguished by other derived features of the dentition, skull and postcranium, as well as soft parts [see especially VERHEYEN, 1962]. These may best be ranked as subgenera, with *C. (Colobus)* for the 'black-and-white' forms, *C. (Procolobus)* for the 'olive' and *C. (Piliocolobus)* for the 'red' group. Only two species, *C. guereza* and *C. polykomos* need be

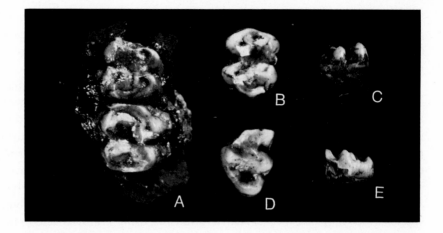

Fig. 15. Selected specimens of ?*Colobus flandrini* [ARAMBOURG, 1959] from Marceau, Algeria. A = Holotype maxillary fragment, R M$^{1-2}$, occlusal; B, C = left M2?, occlusal, lingual; D, E = right M$_3$, mesial portion broken away, occlusal, lingual. All specimens MNHN-P uncatalogued. A, B, D at one and one-half times natural size; C, E at natural size.

recognized in the first subgenus, not four as is often done recently [e.g., RAHM, 1970; THORINGTON and GROVES, 1970]. No fossil *Colobus* has yet been published in any detail, but small specimens probably referable to this genus although not yet identifiable to subgenus are now known from the Pliocene of East Africa (Kanam, Omo, East Rudolf). Moreover, as was noted above, several fossils from Marceau, Algeria, represent a colobine of moderate size and Turolian (Late Miocene) age. Only some nine teeth can be recognized as colobine, but these include rather wide upper molars, as in *Cercopithecoides*, and the M$_3$ distal width may have been greater than the mesial (see fig. 15). Because ARAMBOURG [1959] applied the name *Macaca flandrini* to one of these specimens, and as the relationships of the population would seem to be African, it may be termed ?*Colobus flandrini*, again using the modern genus as a form genus for early African fossils without the implication of special relationship [DELSON, 1973].

Three other African fossil colobines are of great interest. *Cercopithecoides* has been described from a number of South African localities [see FREEDMAN, 1957, 1965; MAIER, 1971 a]; this moderate-sized form is relatively well known from cranial and dental remains, but postcranial elements have not yet been separated from those of contemporaneous cercopithecines, so

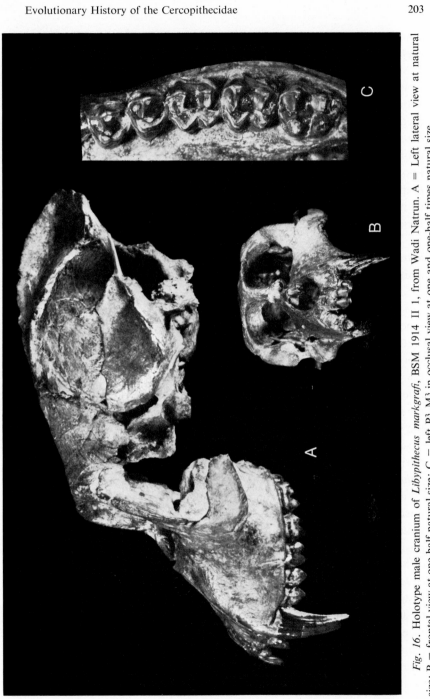

Fig. 16. Holotype male cranium of *Libypithecus markgrafi*, BSM 1914 II 1, from Wadi Natrun. A = Left lateral view at natural size; B = frontal view at one-half natural size; C = left P^3–M^3 in occlusal view at one and one-half times natural size.

that its locomotor adaptation is unknown. The somewhat larger *Paracolobus chemeroni* is known from a nearly complete skeleton collected at locality J. M. 90–91 in the Chemeron Formation [Leakey, 1969]. Despite some elbow-joint features similar to terrestrial colobines such as *Dolichopithecus* and *Presbytis entellus*, the foot of *Paracolobus* is elongate and the overall picture may have been similar to *Colobus* or *Nasalis*, enlarged in size; further detailed study of the holotype skeleton is necessary. Additional material probably referable to both *Cercopithecoides* and *Paracolobus* species is now known from East Rudolf, the Omo deposits and Laetolil [M. Leakey, in preparation; Delson, in preparation], and one specimen labeled as being from 'Rawi' may indicate the persistence of large colobines into the middle Pleistocene.

A third group of African Colobinae may be represented by *Libypithecus markgrafi*, of the latest Miocene of Wadi Natrun, Egypt, which was correctly assigned to this subfamily by Stromer [1913], the original describer. Succeeding authors have in most cases been unsure of its affinities, and all that Simons [1972, p. 201] could add from his observations was that the holotype skull is smaller than was thought. In addition to the skull, the only Natrun specimen which can probably be assigned to this species is one very worn colobine-type lower molar, probably an M_1 – many other teeth and jaw fragments belong to a cercopithecine of larger size (see below). The dentition and lacrimal configuration of the cranium clearly places this genus in the Colobinae (see fig. 16B, C), and Radinsky [in press, 1974] has restudied the endocranial cast which also appears colobine-like. Factor analyses and other comparisons suggest that the skull is morphologically closest to other African fossils (and somewhat farther from *Colobus*, *Nasalis* and *Mesopithecus*). The middle and lower face of this specimen is hafted to the neurocranium without certain alignment, and it appears probable that a slight posteroventral rotation of the face would produce a more accurate reconstruction. The large sagittal and nuchal crests, and especially the posterior position of the greatest sagittal-crest height (fig. 16A), suggest that the mass of M. temporalis was placed rather far posteriorly, even by comparison with *Colobus badius*, recognized as cranially similar by Jolly [1967]. By analogy with the mandrill-baboon-gelada continuum in cranial features described by Jolly [1970], *Libypithecus* would be expected to have a dentition dominated by anterior elements, with the masticatory forces most efficient mesially, but in fact the incisors are small as in all colobines, while the posterior molars are largest; the resultant functional complex is not comparable to any seen in modern cercopithecids and cannot yet be fully interpreted.

Tooth size in *L. markgrafi* is similar to that in *Mesopithecus pentelici*, but dental and especially cranial (facial) shape differ strongly. JOLLY [1967] suggested relationship to *Dolichopithecus* in cranial form, and SIMONS [1970, 1972] has accepted this idea uncritically, but even though the one known *D. ruscinensis* male skull is badly crushed, facial shape and cranial crest positioning, and especially dental proportions, are quite different in the two. Instead, it seems most likely that *Libypithecus* represents an extinct group of (northern?) African colobines, roughly equally distinct 'from modern *Colobus* species and the also extinct *Cercopithecoides/Paracolobus* group. As the latter two sections are apparently arboreal in adaptation, it is possible that *Libypithecus* was also, and it may be suggested that when the thumbs of these three genera are recovered, they will be rather reduced, along the lines of *Colobus* and more than in Eurasian species. The three African groups may have shared a common ancestor during the middle of the Late Miocene; *?Colobus flandrini* would not appear to represent that ancestor, and its relationship to these groups is still uncertain.

In terms of formal taxonomy, the three main subgroups of Colobinae (African, European and Asian) appear certainly less distinct than the tribes of Cercopithecinae; they could be recognized as subtribes, but I do not now consider such a step necessary or worthwile. No further grouping of the genera within any geographic 'subtribe' is required either. Figure 17B is a cladogram representing the inferred relationships of the genus-group taxa of Colobinae, without regard to the time dimension; in figure 17A, this dimension is added, and the possible ancestor-descendant relationships of the European forms are illustrated.

B. Cercopithecinae

The fossil record of the two cercopithecine tribes is heavily weighted in favor of the papionins. At present, the only definite fossils of Cercopithecini are a mandible and perhaps some isolated teeth from the Omo deposits [ECK and HOWELL, 1973] and a juvenile mandibular fragment from Kanam. Relationships among the many recognized species groups of *Cercopithecus* and between these and the more distinct forms are perhaps the least understood of any part of the family. Karyological and biochemical evidence could be of great help in the study of this tribe. If cercopithecins invaded the high forest early in the Late Miocene after a period of semi-terrestriality shared by all early cercopithecines [see also NAPIER, 1970], their

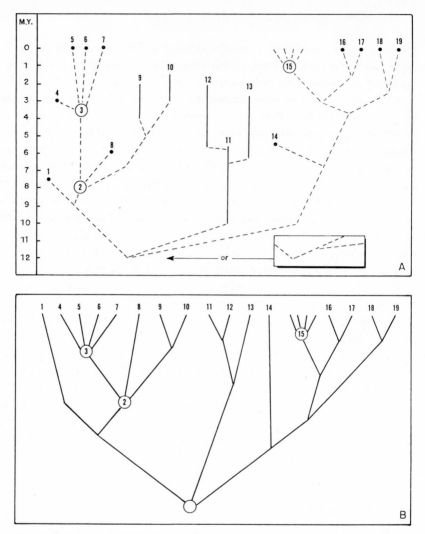

Fig. 17. Phylogeny of Colobinae, with taxa and populations in time-equivalent positions and inferred lineage patterns indicated by dotted lines (A), and cladogram indicating relationships among taxa and populations of Colobinae, with no time dimension, absolute or relative (B). 1 = ?*Colobus flandrini,* Marceau; 2 = common ancestor of three main groups of African colobines; 3 = common ancestor of *Colobus* spp.; 4 = *Colobus* sp(p)., Kanam, Rudolf, Omo, etc.; 5 = *Colobus (Colobus)* spp; 6 = *Colobus (Procolobus) verus;* 7 = *Colobus (Piliocolobus)* spp.; 8 = *Libypithecus markgrafi;* 9 = *Paracolobus* spp.; 10 = *Cercopithecoides* spp.; 11 = *Mesopithecus pentelici;* 12 = ?*Mesopithecus monspessulanus;* 13 = *Dolichopithecus ruscinensis;* 14 = ?*Presbytis sivalensis,* Dhok Pathan; 15 = modern *Presbytis* spp.; 16 = *Pygathrix (Pygathrix) nemaeus;* 17 = *Pygathrix (Rhinopithecus)* spp.; 18 = *Nasalis (Nasalis) larvatus;* 19 = *Nasalis (Simias) concolor.*

major chromosomal rearrangements might have been a means of rapid diversification of the genotype in order to take advantage of new niches.

It was suggested above that the Papionini may have been represented in the earlier Late Miocene by one or more widespread semiterrestrial species in Africa. Elsewhere (DELSON, 1974b], I have suggested that these animals may not have left Africa when some colobines did because they were already too committed to the ground to take advantage of partly forested corridors. Further, a drying of the region of the present Sahara at the end of the Miocene, when the Mediterranean became desiccated and southern Europe was rather arid, would account well both for the first presence of fossil cercopithecines in Europe in the earliest Pliocene and for the known fragmentation of the tribe. From a common ancestor of possibly macaque-like morphology, a northern group which retained the ancestral characters of lateral facial profile, general lack of maxillary and mandibular fossae and moderate molar flare would have been able to cross into Europe, and thence Asia; a similar group, possibly represented by specimens assigned to *Parapapio* species, formed a sub-Saharan source group for the dentally typical African Papionini; and a third population or set of populations may have become adapted to isolated wetland pockets and given rise to the *Theropithecus* group as suggested by JOLLY [1970, 1972]. The evolutionary history of each of these sections will be briefly reviewed here – as with the colobine geographic units, these may be recognized at the subtribe level [not as tribes, cf. MAIER, 1970], and although such recognition is not here advocated for either set of evolutionary units, consistency suggests that if one set were so recognized, the other should be also.

The earliest known fossil cercopithecine is the population from Marceau noted above as indeterminate or cf. *Macaca;* of more clearly macaque (or simply ancestral) morphology is the Wadi Natrun population represented by some 20 specimens found with *Libypithecus*. Macaque is known from Montpellier in southern France in the early Pliocene, alongside *Dolichopithecus ruscinensis* and ?*Mesopithecus monspessulanus*, at the same time that certain rodents are also found on both sides of the Mediterranean. While colobines dominated in the Ruscinian, macaques became the most common and then the only cercopithecid form in Europe through the Villafranchian and later Pleistocene. Most if not all European and later North African *Macaca* (fig. 18) are referable to the modern *M. sylvanus*, perhaps as temporal-geographic subspecies; their distribution is reviewed by DELSON [1973, 1974a]. An exception may be a relatively small-sized population of questionably 'Holocene' age from Capo Figari, Sardinia, named *M. majori* by AZZAROLI [1946]

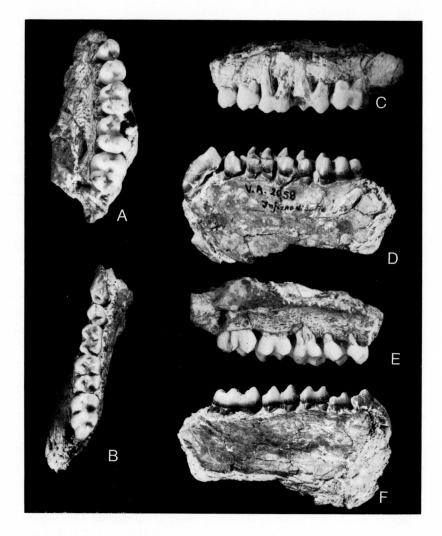

Fig. 18. European Pleistocene macaques. A, C, E = Left maxilla of uncertain sex, P3–M3, SMNS 7683a, late Middle Pleistocene of Heppenloch Cave, Germany, holotype of *Macaca* 'suevica', occlusal, labial, lingual; B, D, F = male left mandible, P3–M3, NMB VA 2058, earliest Pleistocene (late Villafranchian) of Upper Valdarno, Italy, topotype of *Macaca* 'florentina'. All at natural size.

(see fig. 1 B, D). Cercopithecines are not known as fossils in Asia until the Plio-Pleistocene (Tatrot Formation): the two mandibular fragments known as '*Semnopithecus*' *paleindicus* [and often ignored by reviewers, e. g., by SIMONS 1970, 1972] are in fact cercopithecine and may be generically termed ?*Macaca*. The larger and somewhat younger populations named '*Cynocephalus*' *falconeri* (Pinjor Fm.), *M. anderssoni* (Honan Villafranchian), *M. robusta* (North China) and *M. speciosa subfossilis* (Tung-Lang, ?mid-Pleistocene of Viet-Nam) are all closely similar in morphology and probably linked to the modern *M. thibetana* species group, rather than the later-invading *M. mulatta* group.

Cercopithecines of even larger size are also known from Eurasia. At the late Villafranchian (earliest Pleistocene) localities of Seneze, France and Graunceanu, Romania, cranial materials have been published as '*Dolichopithecus*' *arvernensis* and *Paradolichopithecus geticus*, respectively. Smaller but more fragmentary material is known from the earlier Villafranchian sites of Cova Bonica, Spain [DELSON, 1971], Vialette, France and Malusteni, Romania. It seems likely that the younger populations may best be called *Paradolichopithecus arvernensis* (and *P. a. geticus* if necessary) and the older termed cf. *Paradolichopithecus* sp. They may represent stages in an *in situ* transition from a macaque ancestor, as evidenced by cranial similarities to Asian macaques rather than African papionins; further cranial and post-cranial material from Graunceanu, now under study, should clarify the relationships of this population. A similarly large-sized form in Asia has been named *Procynocephalus wimani* (Honan and Choukoutien), to which long bones associated with isolated teeth have been referred. JOLLY [1967] has shown that these bones indicate a highly terrestrial habitus and has further suggested that '*Cynocephalus*' *subhimalayanus* (Pinjor), the first fossil monkey specimen ever reported, may be an Indian member of this genus. Study of the type specimens involved supports JOLLY's views, but despite the assertions of SIMONS [1970, 1972], there is no evidence of a special relationship with *Paradolichopithecus*. VERMA [1969, received after this paper had been completed] has named a new species of *Procynocephalus*, based on a recently-found mandible from the Pinjor horizon. Comparison with '*C.*' *subhimalayanus* was limited to the observation that the latter is known only from a maxilla, which rendered comparison difficult; I would expect that the two specimens may well prove conspecific and thus provide further information as to the morphology of *Procynocephalus*.

The paleobiology and relationships of *Theropithecus*, both fossil and extant, have been discussed in detail by JOLLY [1970, 1972], to which only few comments need be added. The oldest known fossil specimen is a tooth from a 4-million-year-old level at Lothagam, Kenya, which is indistinguishable from those of modern *T. gelada*. Another ancient *Theropithecus* population is that from the (?)early Villafranchian of Ain Jourdel, Algeria, represented by a single lower molar originally called '*Cynocephalus' atlanticus*. JOLLY has postulated that the geladas are essentially adapted to graminivorous feeding habits, originally in low-altitude edaphic grasslands while other papionins inhabited the surrounding woodlands, although *T. gelada* today is secondarily restricted to the highlands of Ethiopia. At least two extinct species or lineages are now known, including one with distinctive facial structure from the Omo region, and one which attained great size and presents derived conditions of anterior tooth reduction by comparison with 'primitive' living geladas (fig. 19). As SIMONS [1970, pp. 117–118] has noted, this phyletic pattern suggests and the older fossils document a longer separation of *Theropithecus* from other Papionini than would be expected from SARICH's [1970] finding of no protein differences between *Theropithecus* and some *Papio* species; this indicates that, if SARICH is in error on this point (by a factor of greater than four), some of his other estimated divergence times might be equally inaccurate.

    Not enough is known of the early history of the dentally typical African Papionini to support more than a 'guesstimate' of genus-group relationships. If macaque-like cranial and dental features are indeed ancestral for the tribe, then *Parapapio* species from Plio-Pleistocene deposits in South Africa may relate to the common ancestor of this section of the tribe, as suggested also by MAIER [1971 b]. *Cercocebus*, the most arboreal of the modern forms, is poorly represented in the fossil record, in part because of preservation difficulties in acid forest soil; three teeth from Kanam, if associated, may suggest the presence of a small species similar to *C. albigena*. The population(s?) represented by the named species *Cercocebus ado* and *Papio (Simopithecus) serengetensis* from Laetolil, as well as newly collected material from Omo, East Rudolf and Olduvai, might relate to *Cercocebus* but could more easily fit with *Parapapio* [studies in preparation by M. G. LEAKEY and R. E. F. LEAKEY and by DELSON]. The evolutionary meaning of the named species of *Parapapio* from South Africa is still uncertain, despite the recent studies of MAIER [1971 b] and FREEDMAN and STENHOUSE [1972]. The few good skulls of this 'genus' appear to have a different facial morphology than those of true *Papio*, living and fossil. Of the latter, South African *P. robinsoni* Freedman, 1957, and East African *P. baringensis* R. Leakey, 1969, seem unequivocally related to the modern savannah/hamadryas group,

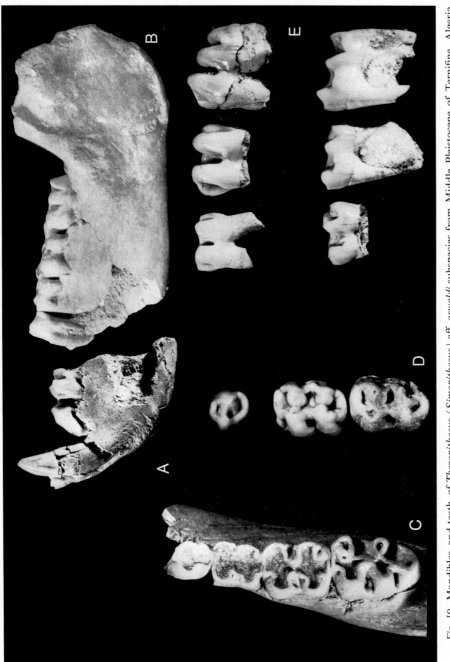

Fig. 19. Mandibles and teeth of *Theropithecus* (*Simopithecus*) aff. *oswaldi* subspecies from Middle Pleistocene of Ternifine, Algeria. A = Male mandibular symphysis with R and L I₁–P₄, left lateral view (photographically reversed); B, C = male (?) partial mandible with L P₄–M₃, left lateral and occlusal views; D = occlusal view of unassociated upper teeth of large size, L P⁴, M²?, M³?; E = buccal (above) and lingual (below) views of unassociated left and right M₂₋₃ respectively. All specimens MNHN-P uncatalogued. A, B at two-thirds natural size; C–E at natural size.

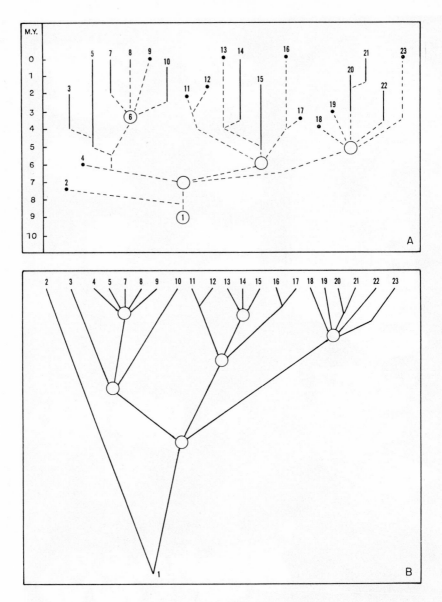

Fig. 20. Phylogeny of Papionini, with taxa and populations in time-equivalent positions and inferred lineage patterns indicated by dotted lines (A), and cladogram indicating relationships among taxa and populations of Papionini, with no time dimension, absolute or relative (B). 1 = Common ancestor of all papionins; 2 = ?*Macaca* sp., Marceau, Algeria; 3 = *Paradolichopithecus* spp.; 4 = ?*Macaca libyca*, Wadi Natrun;

while the smaller *P. angusticeps* and *P. wellsi* are indistinguishable from each other and not clearly separate from the enigmatic *P. izodi* [see FREED-MAN, 1957, 1961, 1965]. Preliminary considerations suggest that by analogy with other cercopithecid genera, the differences in facial morphology, dental proportions, postcranial function and behavior in the complex of species referred to *Papio*, *Parapapio* and '*Mandrillus*' are not greater than would be expected in a large genus with three subgenera. Because of the rules of nomenclature, the term *Papio* must be used with reference to mandrills, while *Chaeropithecus* is available for the savannah/hamadryas baboons [HOPWOOD, 1947; DELSON, in preparation]. As in Europe and Asia, very large cercopithecines are known in the fossil record of at least southern Africa. *Gorgopithecus* is represented only by a few gnathic fragments and one male skull which is heavily 'reconstructed' in plaster, while *Dinopithecus* fossils, including cranial parts, are more common. A brief comparison suggests that they are not as distinct from each other as they may be from the *Papio* complex, and they may be tentatively ranked as subgenera; they are certainly not specially related to *Theropithecus*, as suggested by FREED-MAN [1957]. Further study is necessary, especially including comparison of skulls with the range of variation in *Papio* as here understood, as well as with *Paradolichopithecus* and *Procynocephalus*. These relationships, as well as others which can be inferred among genus-group taxa of the Papionini, are presented in the cladogram and phylogeny of figures 20A, B.

As little as 10 years ago, there were no meaningful discussions of cerco-pithecid phylogeny or evolutionary history except the outline review by FREEDMAN [1957]. Several overviews have appeared in the last decade, of which it is hoped that this is the most complete. Paleoprimatologists are realizing that valuable studies must be based upon observation of the original specimens, and this trend combined with the increasing tempo of fossil

---

5 = European and North African Plio-Pleistocene *Macaca* spp.; 6 = possible common ancestor of Asian cercopithecines; 7 = *Macaca* aff. *thibetana* (= *Macaca anderssoni*, *M. robusta*, *M. speciosa subfossilis*, etc.); 8 = *M. fuscata*; 9 = other modern macaque species and species groups; 10 = *Procynocephalus* spp.; 11 = *Dinopithecus (Dinopithecus) ingens;* 12 = *Dinopithecus (Gorgopithecus) major;* 13 = *Papio (Papio)* spp. (mandrills); 14 = *Papio (Chaeropithecus)* spp. (savannah/hamadryas); 15 = *Papio (Parapapio)* spp.; 16 = modern *Cercocebus* species; 17 = ?*Cercocebus* spp., Kanam, Laetolil, etc.; 18 = *Theropithecus* sp., Lothagam-3; 19 = *Theropithecus* 'atlanticus', Ain Jourdel, Algeria; 20 = *Theropithecus (Simopithecus) darti* [*sensu* JOLLY, 1972]; 21 = *Theropithecus (Simopithecus) oswaldi* [*sensu* JOLLY, 1972]; 22 = *Theropithecus* cf. *brumpti*, Omo deposits; 23 = *Theropithecus (Theropithecus) gelada*.

finds in Africa especially will surely make the present analysis obsolete well
before the end of the decade.

V. Summary

The application of cladistic methods (especially a concentration on shared derived
rather than ancestral characters) permits the distinction of four dental and two cranial
morphotypes among the Cercopithecidae. Comparison with fossils suggests that the Fayum
parapithecids are not specially related to the Old World monkeys, while other undoubted
Fayum 'hominoids' may share derived features with monkeys. Miocene *Victoriapithecus*
'species' may document a stage very close to the split between Colobinae and Cercopithecinae. Later African colobines appear to form a monophyletic group, more arboreal than
the extinct European branch; Asian forms may be specially related to the latter. Among
the Cercopithecinae, after a Miocene separation from the high-forest Cercopithecini, the
Papionini divided into three groups: geladas, sub-Saharan *Papio*-related 'baboons' and
Mediterranean-Eurasian macaque relatives; each of these underwent Plio-Pleistocene
adaptive radiations and subsequent taxonomic diminution.

VI. Acknowledgements

I thank Dr. Malcolm McKenna and Dr. Clifford Jolly for discussions of and
comments on an earlier version of this paper. Many curators of paleontological and zoological collections graciously permitted me to study specimens in their care and provided
technical facilities as well. For additional discussions of the topics covered here and related
stratigraphic problems, I thank other colleagues, too numerous to name. Dr. Jolly and
Prof. J.-P. Lehman provided assistance in the preparation or obtaining of illustrative
materials (figures 2–3 and 13, respectively). Figures 4, 5 and 6 were prepared by Miss
Robin Ingle; figures 10, 17 and 20 by Mr. Raymond J. Gooris; all photographs except
those of figure 13 were taken by me. The research reported here was supported in part by
financial assistance from Columbia University, the Wenner-Gren Foundation for Anthropological Research (grant No. 2018), the National Geographic Society and the Uni
versity of Pittsburgh.

VII. References

Arambourg, C.: Vertébrés continentaux du Miocène supérieur de l'Afrique du nord.
    Publ. Serv. Carte géol. Algérie (n.s.), Paléont., Mém. No 4, pp. 5–159 (1959).
Azzaroli, A.: La scimmia fossile della Sardegna. Riv. Sci. preist. *1:* 168–176 (1946).
Bishop, W. W. and Chapman, G. R.: Early Pliocene sediments and fossils from the
    northern Kenya rift valley. Nature, Lond. *226:* 914–918 (1970).
Delson, E.: Estudo preliminar de unos restos de simios pliocenicos procedentes de
    «Cova Bonica» (Gava) (Prov. Barcelona). Acta geol. hisp. *6:* 54–57 (1971).

DELSON, E.: Fossil colobine monkeys of the circum-Mediterranean region and the evolutionary history of the Cercopithecidae (Primates, Mammalia); diss. Columbia Univ., Ann Arbor, University Microfilms (1973).

DELSON, E.: Preliminary review of cercopithecid distribution in the circum-Mediterranean region. Bur. Rech. géol. min., Mém. No. 78 (in press, 1974 a).

DELSON, E.: Paleoecology and zoogeography of the Old World monkeys. Proc. 9th Int. Congr. Anthrop. Ethnol. Sci., Chicago; World Anthropology (Mouton, the Hague, in press 1974 b).

DELSON, E.: Toward the origin of the Old World monkeys. Actes Coll. Int. C. N. R. S., 218. Probl. Paléont., Evol. Vert., Paris (in press, 1974 c).

ECK, G. and HOWELL, F. C.: New fossil *Cercopithecus* material from the lower Omo basin, Ethiopia. Folia primat. *18:* 325–355 (1973).

ELDREDGE, N. and GOULD, S. J.: Speciation and punctuated equilibria: an alternative to phyletic gradualism; in SCHOPF Models in paleobiology (Freeman, San Francisco 1972).

FREEDMAN, L.: The fossil Cercopithecoidea of South Africa. Ann. Transvaal Mus. *23:* 121–262 (1957).

FREEDMAN, L.: New cercopithecoid fossils, including a new species from Taung, Cape Province, South Africa. Ann. Sth. Afr. Mus. *46:* 1–14 (1961).

FREEDMAN, L.: Fossil and subfossil primates from the limestone deposits at Taung, Bolt's Farm and Witkrans, South Africa. Palaeont. afr. *9:* 19–48 (1965).

FREEDMAN, L. and STENHOUSE, N. S.: The *Parapapio* species of Sterkfontein, Transvaal, South Africa. Palaeont. afr. *14:* 93–111 (1972).

GABIS, R.: Les os des membres chez les singes cynomorphes. Mammalia *24:* 577–602 (1960).

GROVES, C. P.: The forgotten leaf-eaters and the phylogeny of the Colobinae; in NAPIER and NAPIER Old World monkeys, pp. 555–586 (Academic Press, New York 1970).

HAMILTON, W. R.: North african Lower Miocene rhinoceroses. Bull. brit. Mus. nat. Hist. Geol. *24:* 349–395 (1973).

HOOIJER, D. A.: Miocene mammalia of Congo. Ann. Mus. roy. Afr. cent., Sér. 8vo, Sci. géol. *46:* 1–77 (1963).

HOOIJER, D. A.: Miocene mammalia of Congo, a correction. Ann. Mus. roy. Afr. cent., Sér. 8vo, Sci. géol. *67:* 163–167 (1970).

HOPWOOD, A. T.: The generic names of the mandrill and baboon, with notes on some of the genera of Brisson, 1762. Proc. zool. Soc. Lond. *111:* 533–536 (1947).

JOLLY, C. J.: Introduction to the Cercopithecoidea, with notes on their use as laboratory animals. Symp. zool. Soc. Lond. *17:* 427–457 (1966).

JOLLY, C. J.: The evolution of the baboons; in VAGTBORG The baboon in medical research, vol. 2, pp. 427–457 (Univ. of Texas Press, Austin 1967).

JOLLY, C. J.: The large African monkeys as an adaptive array; in NAPIER and NAPIER Old World monkeys, pp. 141–174 (Academic Press, New York 1970).

JOLLY, C. J.: The classification and natural history of *Theropithecus (Simopithecus)* (Andrews, 1916), baboons of the African Plio-Pleistocene. Bull. brit. Mus. nat. Hist. Geol. *22:* 1–122 (1972).

KOENIGSWALD, G. H. R. VON: Miocene Cercopithecoidea and Oreopithecoidea from the Miocene of East Africa. Foss. Vert. Afr. *1:* 39–51 (1969).

KUHN, H. J.: Zur Kenntnis von Bau und Funktion des Magens der Schlankaffen (Colobinae). Folia primat. *2:* 193–221 (1964).

LEAKEY, L. S. B.: Upper Miocene primates from Kenya. Nature, Lond. *218:* 527–530 (1968).

LEAKEY, R. E. F.: New Cercopithecidae from the Chemeron beds of Lake Baringo, Kenya. Foss. Vert. Afr. *1:* 53–69 (1969).

MAIER, W.: Neue Ergebnisse der Systematik und der Stammesgeschichte der Cercopithecoidea. Z. Säugetierk. *35:* 193–214 (1970).

MAIER, W.: New fossil Cercopithecoidea from the Lower Pleistocene cave deposits of the Makapansgat Limeworks, South Africa. Palaeont. afr. *13:* 69–108 (1971a).

MAIER, W.: Two new skulls of *Parapapio antiquus* from Taung and a suggested phylogenetic arrangement of the genus *Parapapio.* Ann. Sth. Afr. Mus. *59:* 1–16 (1971b).

MAYR, E.: Principles of systematic zoology (McGraw-Hill, New York 1969).

NAPIER, J. R.: Paleoecology and catarrhine evolution; in NAPIER and NAPIER Old World monkeys, pp. 55–95 (Academic Press, New York 1970).

PILBEAM, D. R. and WALKER, A.: Fossil monkeys from the Miocene of Napak, northeast Uganda. Nature, Lond. *220:* 657–660 (1968).

RADINSKY, L.: The fossil evidence of anthropoid brain evolution. Amer. J. phys. Anthrop. (in press, 1974).

RAHM, U.: Ecology, zoogeography and systematics of some African forest monkeys; in NAPIER and NAPIER Old World monkeys, pp. 589–626 (Academic Press, New York 1970).

SARICH, V.: Primate systematics with special reference to Old World monkeys: a protein perspective; in NAPIER and NAPIER Old World monkeys, pp. 175–226 (Academic Press, New York 1970.)

SCHAEFFER, B.; HECHT, M. K., and ELDREDGE, N.: Phylogeny and paleontology. Evol. Biol. *6:* 31–46 (1972).

SIMONS, E. L.: The significance of primate paleontology for anthropological studies. Amer. J. phys. Anthrop. *27:* 307–325 (1967).

SIMONS, E. L.: Miocene monkey *(Prohylobates)* from northern Egypt. Nature, Lond. *223:* 687–689 (1969).

SIMONS, E. L.: The deployment and history of Old World monkeys (Cercopithecidae, Primates); in NAPIER and NAPIER Old World monkeys, pp. 97–137 (Academic Press, New York 1970).

SIMONS, E. L.: Primate evolution: an introduction to man's place in nature (Macmillan, New York 1972).

SIMPSON, G. G.: The principles of classification and a classification of mammals. Bull. amer. Mus. nat. Hist. *85:* 1–350 (1945).

STROMER, E.: Mitteilungen über die Wirbeltierreste aus dem Mittelpliocaen des Natrontales (Ägypten). Z. dtsch. geol. Ges. *65:* 350–372 (1913).

THORINGTON, R. and GROVES, C. P.: An annotated classification of the Cercopithecoidea; in NAPIER and NAPIER Old World monkeys, pp. 631–647 (Academic Press, New York 1970).

VERHEYEN, W. N.: Contribution à la craniologie comparée des primates. Ann. Mus. roy. Afr. cent., Sér. 8vo, Sci. zool. *105:* 1–247 (1962).

VERMA, B. C.: *Procynocephalus pinjori* sp. nov.: a new fossil primate from the Pinjor

Beds (Lower Pleistocene) east of Chandigarh. J. palaeont. Soc. India *13:* 53–57 (1969).

VOGEL, C.: Morphologische Studien am Gesichtsschädel catarrhiner Primaten. Bibl. primat., No. 4, pp. 1–226 (Karger, Basel 1966).

VORUZ, C.: Origine des dents bilophodontes des Cercopithecoidea. Mammalia *34:* 269–293 (1970).

Author's address: Dr. ERIC DELSON, Department of Anthropology, Lehman College, City University of New York, Bedford Park Boulevard West, *Bronx, NY 10468* (USA)

In SZALAY: Approaches to Primate Paleobiology
Contrib. Primat., vol. 5, pp. 218–242 (Karger, Basel 1975)

Evolutionary Models, Phylogenetic Reconstruction, and Another Look at Hominid Phylogeny

NILES ELDREDGE and IAN TATTERSALL

Departments of Invertebrate Paleontology and Anthropology, American Museum of Natural History, New York, N.Y.

Contents

I. Introduction

The belief that the fossil record alone can reveal the information required for thorough reconstruction of phylogenies is a pervasive one in modern paleontology. Invariably criticised for its incompleteness, the fossil record is nonetheless regarded as the final arbiter of all phylogenetic problems. Since this record is a matter of physical discovery, phylogeny has been viewed likewise, with the result that attention has rarely been paid to the existence of alternative ways of elucidating evolutionary history, or, indeed, to the basic methodologies underlying conventional approaches to phylogenetic reconstruction.

Such methodologies necessarily result, consciously or otherwise, from *a priori* models of the evolutionary process. In this paper, we will briefly characterize the two current models of evolutionary change, consider their merits, and attempt to show that each dictates its own strategy of phylogenetic reconstruction. In illustrating this discussion with examples drawn from hominid phylogeny we do not, of course, hope to be exhaustive. Rather we have chosen the hominids as appropriate in an illustrative context because of all taxa with a reasonable fossil record, Hominidae excites the greatest amount of general interest, and, more importantly, because the study of hominid phylogeny yields numerous excellent examples both of the methodology conventionally used by paleontologists in producing phylogenetic trees, and of the view of the evolutionary process from which this methodology stems. Paleoanthropology indeed, far from being, as sometimes claimed by those occupied with other groups, a special and inferior case of paleontological practice, actually epitomizes many of the more established concepts and methods of paleontology in general. Those deficiencies occasionally identified with the study of fossil man are in fact liberally shared with all branches of paleontology.

II. Evolutionary Models

Only two models of the evolutionary process have survived the century of post-Darwinian thought. They are, first, the splitting of a lineage, i.e. speciation, and second, simple phyletic evolution or phyletic gradualism, in which a lineage undergoes progressive modification in both gene content and gene frequencies, a process culminating eventually in the production of a 'new' species. The crucial difference between these two processes is that whereas under the speciation model new species arise as discrete entities, under phyletic gradualism they are no more than arbitrary subdivisions of an evolving continuum. To the gradualist the incomplete nature of the fossil record, while inconveniently reducing the frequency of 'intermediate' fossils, nevertheless serves a useful purpose in providing gaps at which such arbitrary horizontal boundaries may safely be drawn. Both evolutionary models have recently been examined in detail by ELDREDGE and GOULD [1972], who have characterized those of their properties which are particularly relevant to paleontology. We do not intend to repeat that discussion here, but will briefly contrast the two models since each has its methodological implications for phylogenetic reconstruction.

Despite the fact that most evolutionary biologists view speciation and phyletic gradualism as complementary, and would regard any theoretical consideration of evolution as incomplete without the inclusion of both models, in practice the study of speciation has rested almost entirely in the hands of neontologists. All speciation models, whether allopatric, sympatric or stasipatric, attempt to explain the splitting of a single parent species into two or more daughter species, and have generally fallen into the neontological domain simply because the few centuries or millennia involved in the establishment of a daughter species are, although a long time in terms of the human lifespan, virtually an instant in terms of geologic time. In addition, a sort of corollary to Walther's stratigraphic law, which states that within a section lateral facies changes mirror the vertical sequence, seems to apply in comparative studies of extant organisms. Thus, for example, the allopatric model, which involves the establishment of genetic incompatibility between populations during a period of genetic isolation, implies a general sequence of events which the Recent biota indeed reflects, containing as it does an entire spectrum of species types ranging from the relatively monomorphic to the highly polytypic.

In contrast, paleontological methodology has in the past been almost exclusively dictated by adherence to phyletic gradualism. Understandably, since those very properties of speciation, in particular its relative rapidity, which make it amenable to study by neontologists, have militated against its study, or even its 'documentation', by paleontologists. Quite naturally, paleontologists have chosen to adopt that model of the evolutionary process which requires lengths of time commensurate with those comfortably studied by geologists. Neontologists have not concerned themselves with phyletic gradualism for the simple and very good reason that they cannot; it takes too long. But neontologists, too, will readily admit its validity, and most modern textbooks present both models without discussing possible incongruities between them. We believe, however, that such incongruities are of critical significance, and that many difficulties experienced in reconciling current paleontological dogma with the fossil record stem from the fact that paleontologists have been concerned with fitting the fossil record into an exclusive framework of phyletic gradualism, a theoretical orientation which has directly influenced paleontological policies of phylogenetic reconstruction. Our aim in this paper is to elaborate an alternative methodology, one based on the speciation model.

Why do evolutionary biologists of all specialties tend to accept simultaneously both the speciation and gradualist models? Superficially, at least, the reasons are simple. We *know* speciation, or at least some process of lineage splitting, must have occurred; otherwise life could not have diversified. We *think* phyletic gradualism has been an active process because DARWIN viewed evolution largely as the gradual, long-term modification of lineages by natural selection. And the gappy nature of the fossil record has made it easy to 'document' the latter. In general, organisms change through the stratigraphic column; once a number of morphological stages are established in the sequence, it is easy to infer the existence of intermediates which are not known because the fossil record is incomplete. The degree to which theoretical beliefs can affect practical expectations is emphasized, for instance, by the recent debate in the paleo-anthropological literature over whether or not the International Code should be changed to facilitate the naming of largely hypothetical 'intermediate' forms [e. g. TOBIAS, 1969].

The problems inherent in applying phyletic gradualism to the fossil record are discussed at length by ELDREDGE and GOULD [1972] and will only briefly be referred to here. However, it is worth making the point that, at least in many cases, the fossil record is not compatible with the notion of phyletic gradualism, according instead far better with the speciation hypothesis. In no instance, of course, is the known stratigraphic distribution of a given species totally, or even nearly, complete. But it is generally found that individuals representing the first and last recorded occurrence of a fossil species generally do not differ very greatly from each other, and the conclusion has to be drawn in such cases that stasis has predominated through the life-span of the species. This phenomenon is most strikingly seen among marine invertebrates, which admittedly appear to evolve more slowly than terrestrial organisms; nonetheless, under the model of phyletic gradualism *some* change would be expected over the average 5- to 10-million year life-span of a Paleozoic benthic marine invertebrate species. It is noticeable in this context that in some areas of paleontology workers have recently begun explicitly to point to *lack* of change in the fossil record over long periods of time [e. g. ELDREDGE, 1971; ELDREDGE and GOULD, 1972; R. A. ROBISON, personal commun.]. It appears that the validity of phyletic gradualism as a general model of the evolutionary process is at last beginning to encounter some practical scepticism from members of a profession in which it was long held to be axiomatic.

Theoretically, too, phyletic gradualism, at least as a general model, is somewhat suspect. The concept of orthoselection enduring in its linear course over millions of years and thousands of generations is not an easy one to accept. This point is in fact one which has recently been raised by paleo-anthropologists [e. g. UZZELL and PILBEAM, 1971] in countering the arguments of SARICH [e. g. 1968], whose calibration of hominid phylogeny based on serum protein comparisons depends upon orthoselection on the albumens he studied. Selective forces arise from both the physiochemical milieu and from interaction with other organisms. It is true that biotic interactions can be highly complex, as in the relationships between insects and plants, and may lead to long-term trends over vast periods of time, although even here such trends do not appear to be gradual and strictly linear. But, for the reason given above, processes of adaptation to local edaphic conditions, whether biotic or abiotic, are virtually certain to be short-term events.

A further point of great importance is that evolutionary adaptation is not the only response available to organisms faced with strong and persistent selection pressures. Migration to more congenial environments, and extinction, either local or total, are generally easier options. Certainly, a certain amount of *in situ* adaptation is nearly always possible. But it is highly unlikely that such adaptation would persist over thousands of generations, since migration or extinction would eventually interrupt any process of linear change.

Further doubt is cast on the general applicability of the gradualist model by the concept of genetic homeostasis. The developmental conservatism of organisms is well established, and LERNER [1954] has discussed the parallel notion of resistance to change in terms of the genetics of populations. It is quite clear that evolutionary change is not inevitable in the sense that, given enough time, all species will undergo some conspicuous form of modification. On the contrary: evolutionary change is a difficult and costly process, and will occur only under certain conditions. Change is rare with respect to stasis, and is inevitable only in a probabilistic sense; only over the vast span of geological time will such evolutionary events become common.

Since we have included this brief discussion of phyletic gradualism in the belief that this model has almost exclusively dictated the approaches of paleontologists to phylogenetic reconstruction, it is only fair to point out that paleontologists do, of course, recognise lineage splitting. But such splitting is viewed not so much as a consequence of the speciation process as a special case of gradualism, i.e. as a bifurcation characterized by gradual phyletic modification of the two offshoots.

Our belief is that the claims of the speciation model are not satisfied by interpretation of this kind. In contrast, we would argue that in the speciation process most morphological change takes place in isolated populations prior to (in the form of geographic variation), during, and immediately following the speciation event. The remaining portion of life of a new species, the vast bulk of it, is then a time of relative stasis. The implications of each model for the formation of the fossil record are thus quite distinct, and because of this each necessarily dictates its own approach to the reconstruction of evolutionary events. The question of which strategy should be preferred depends on a single criterion: which model of the evolutionary process is more nearly general? We suggest that phyletic gradualism has not played the almost exclusive part in the evolutionary process with which it has been credited by paleontologists. Indeed, it is not unlikely that its role has been subordinate to that of speciation. And if it is accepted that speciation is preferable, in terms of being a more general picture of evolutionary change, an alternative method of phylogenetic reconstruction is indicated.

III. Phylogenetic Reconstruction in Paleontology

The phylogeny of a group of organisms is its evolutionary history; from the simple notion of organic evolution it is evident that phylogeny consists of ancestor-descendant sequences. This, we *know*, must be true. And in conjunction with the concept of evolution as a gradual, phyletic, phenomenon, this knowledge has provided paleontologists with a straightforward approach to the reconstruction of phylogeny: directly preserved in the fossil record, phylogenies are a matter of discovery, and have thus generally been subject to 'documentation' rather than to analysis. When difficulties arise in this procedure, refuge is readily available in the form of the 'woeful' incompleteness of the fossil record.

This general concept carries with it a corollary, subscribed to by many neontologists as well as by paleontologists: in the absence of a fossil record, discussion of phylogeny is fruitless. We would contend that this is not the case; the methodology we suggest here, one based ultimately on the speciation model of the evolutionary process, provides a uniform approach to phylogenetic reconstruction irrespective of whether the organisms under consideration are entirely fossil, exclusively extant, or a mixture of both.

Simply stated, then, phylogenetic reconstruction in paleontology has generally amounted to the *discovery* of ancestor-descendant sequences in the appropriate stratigraphic order. Series of fossil specimens, garnered from various areas, localities and stratigraphic horizons, are organized into such sequences; the main data inputs into such schemes are thus morphology, stratigraphy and geographical distribution. Of these, only the first two are of importance in a general consideration of paleontological practice; this is particularly so in the case of hominid paleontology, since *Homo sapiens* is a widespread polytypic species, and modern paleoanthropologists have usually seen no good reason to suppose that its ancestors were otherwise.

We are left, then, with morphology and stratigraphy as the main constituents in the consideration of phylogeny. These two aspects are combined in infinitely varying proportions. To take one extreme, some phylogenies are no more than lines drawn on a chart of biostratigraphic occurrence, connecting taxa all considered referable to the same monophyletic taxon of next higher rank. In such cases, morphology is of minimal importance, its use hardly extending beyond the provision of an indication of some kind of relationship; the nature of the relationship is determined by stratigraphy. This can be true even of studies which devote vast quantities of space to morphological description; the point is that, quite often, morphology only barely intrudes into the analysis of relationships.

Even if such extreme cases are not typical, it must be admitted that stratigraphy has profoundly influenced paleoanthropological practice. Each time the South African *Australopithecus* sites are redated, for instance, authors rush to modify the phylogenetic trees in their textbooks. And this is so despite the fact that, at least partly because of the uncertain stratigraphy associated with much of the hominid fossil record, paleoanthropologists have tended to indulge in a great deal of detailed morphological comparison.

The other end of the spectrum of such research reveals a greater emphasis upon comparative morphology. In such cases, although ancestor-descendant sequences are still sought, and expressed when 'found', more attention is paid to the fact that, while in the ideal case primitiveness is equatable with ancientness, in the real world it may not correspond to stratigraphic position. But the basic endeavor in either case is the same: recognition of ancestors and descendants concordant with stratigraphy. In general, it is true that when a system of phylogenetic relationships based on morphology conflicts with stratigraphy, no hypothesis of relationship is inferred, since such a hypothesis, to be acceptable, must be expressed in terms of ancestors and descendants.

Although cursory, this brief characterization does outline a dominant theme in the history of general paleontological practice. We do not pretend to have been comprehensive, or to claim that paleontologists invariably proceed in this manner. But it is nonetheless true that few paleontologists have seen fit openly to take issue with the assumptions we have described.

IV. An Alternative Strategy

It is thus not surprising that an alternative approach, one based on the speciation model and gradually gaining increasing acceptance by paleontologists and neontologists alike, was first mooted by neontologists, in particular by the entomologist HENNIG [1966]. Recent discussions of this cladistic methodology have been provided by BRUNDIN [1968, 1972], NELSON [1970], SCHLEE [1971], SCHAEFFER et al. [1972] and others. Our purpose here is to examine the extent to which the cladistic approach advocated by these authors may be applicable to the fossil record, and to contrast the method with standard paleontological procedures.

If in any given instance the assumption is made that the applicable model of the evolutionary process is that of lineage splitting after the speciation model, an immediate conclusion emerges. This is that, for every taxon (A) under consideration, there exists another taxon (B) to which, by virtue of common ancestry, A is more closely related than to any other such taxon. The statement holds true even if any number of other taxa more closely related than B to A is unknown for whatever reason. The only subsidiary assumption contained here is that not more than two taxa have resulted from a single speciation event; but although a more complex outcome than a simple bifurcation can theoretically result from such an event, the likelihood that this will in fact happen appears fairly remote.

This concept opens up a new avenue in phylogenetic reconstruction. Initially, instead of searching for ancestor-descendant relationships, we may first seek that taxon B which is most closely related to A in terms of recency of common ancestry. A cluster of two such taxa is referred to by HENNIG [1966] as a 'sister group'. Especially to the paleontologist, however, the sister group concept may appear at first glance to be somewhat paradoxical, since one species necessarily originates from another. Almost invariably the 'parent' species will persist essentially unchanged following the speciation event which produced the 'daughter'; why not, then, look for parent-daughter relationships rather than for sister groups? The answer lies in the generality of the sister-group concept. Unlike the highly specific parent-daughter group, a phenomenon with which paleontologists, in particular, are rarely confronted in real life, the sister group merely consists of those two species (or other taxa) *among all those known to the systematist* which most recently shared a common ancestry. It is not required that the two members of a sister group necessarily be the parent-daughter pair. Indeed, the probability that any two species known to us actually do bear such a relationship is remote, and would in any case be difficult to prove. A further constraint arises from the fact that, although one species in such a pair (usually the parent) generally does retain more of the ancestral character states while the other is, on the whole, more divergent, the jump from single character analysis to whole organism analysis introduces a variety of complexities.

Thus, for instance, a statement to the effect that one species in a sister group is relatively more primitive than another can be highly misleading; frequently, indeed, it is impossible to decide which of two species is the daughter and which is the parent. In any instance where two species under consideration actually do possess an ancestor-descendant relationship, it is generally best to regard them as sister species, since this is usually the maximum that may be claimed with assurance.

This approach to phylogenetic reconstruction thus reduces to the attempt to recognise sister taxa among fossil and living organisms. The hierarchy implied by the procedure is a real one in terms of the nature of the evolutionary process; in general, then, it must hold that for every sister group (I) there must be another sister taxon (II) most closely related to I by reason of recency of common ancestry. The sister group II may be a taxon at any level of the taxonomic hierarchy.

V. The Uses of Data in Systematics

We have already characterized the data available to systematists as falling into three categories: morphology, and temporal and spatial distributions. What proportions of this information should be used in the formulation of sister groups? We believe that only morphology, albeit defined in the very broadest sense, bears a clear imprint of evolutionary history. Spatial distribution is affected by a wide variety of adventitious influences, while it has as yet proven impossible to demonstrate the existence of any factor related to its phylogenetic, as opposed to adaptive, history which dictates a necessary distribution pattern for a species. The closest approach to such a factor lies in the observation that, in accord with the allopatric model, the early part of a new species' history must be spent in an area geographically discrete from that of its parent.

Similarly, the use of temporal distribution data should be approached with the greatest caution. A complete knowledge of the time ranges of all species under consideration in a particular study might, indeed, be useful, for if we could be sure that the known biostratigraphic range of a species coincides totally with its actual biochron, we would be able immediately to eliminate some taxa from the formulation of hypotheses of relationship. But unfortunately – and this is the most truly 'woeful' inadequacy of the fossil record – we can never know how complete the record is, or even which parts of an age range are represented. The possibility always remains that a given fossil form may be the daughter of one known only from later in the stratigraphic record. We propose, then, that of the three general categories of data available to the systematist, only morphological attributes, broadly defined to include behavior, biochemistry and so forth as well as anatomy, may confidently be used in the framing of theories of relationship.

Although the formation of sister groups may often not be easy to accomplish, the procedure by which it is done is quite easily explained. Since primitive characters can be shared by organisms which are only very remotely related, it follows that sister groups can be recognised only on the basis of uniquely shared morphological specializations. The essential distinction to be made is between shared primitive and shared derived characteristics. This is a simple, but radical, departure from conventional evolutionary systematics, under which evolutionary relationships are usually judged on overall resemblance, and where distinctions are rarely made between primitive and derived character states. Admittedly, the terms 'primitive' and 'specialised' are widely, albeit loosely, used throughout the paleontological literature; but they are usually employed in a descriptive sense, and seldom in the context of framing hypotheses of relationship.

Thus, although the term 'phenetic' is usually associated with numerical taxonomy, an essentially similar overall similarity/dissimilarity model underlies most attempts to reconstruct phylogeny. The notion on which this concept is based, i.e. that on the whole the more closely two taxa are related, the more closely they will tend to resemble each other, is, of course, a reasonable one. But it has defied all attempts (the books by SIMPSON [1961] and MAYR [1969] are perhaps the best-known recent ones) to broaden it into a methodological yardstick, and it has lent weight to the belief that systematics and phylogenetic reconstruction are forms of art rather than of science[1].

What is missing here is the necessary attention to different *kinds* of similarity. If two taxa are similar because of shared possession of primitive characters, we have no grounds for a statement upon the nature of their

1   Numerical taxonomists have also been irked at the suggestion that systematics is an art. But their attempts to introduce rigor into organismic biology by the liberal use of clever algorithms, while retaining the cherished but wholly inadequate underpinnings of phenetics, has resulted in something which, while it may not be art, is hardly better described as science. The problem with the apparent lack of rigor in evolutionary phenetics is not that it lacks hard numbers, but rather that it lacks precision in defining, among other things, what is meant by 'relationship', and that it fails formally to distinguish between shared primitive and shared derived characters. It is noteworthy that some numericist [e.g. FARRIS *et al.*, 1970] appear recently to have grasped this point and are essentially producing a parallel school of numerical cladistics.

relationship. It is not infrequently pointed out that similarities exist between the cranium of *Australopithecus africanus* (notably Sts 5), and those of various of the more gracile great apes. It is useless to deny that such similarities exist, particularly between *A. africanus* and *Pan paniscus*. But it does need to be pointed out that they are meaningless in phylogenetic terms; they represent shared primitive characters, and contain no specific information upon evolutionary relationships.

It is the derived character states, then, which contain the information essential to the reconstruction of phylogenies. In the following discussion, we examine the problems attendant upon distinguishing between primitive and derived character states. Although this distinction is one crucial to phylogenetics, there is no simple series of rules, strict adherence to which will automatically provide the 'right' answer. All too frequently, as in the case of fossil hominids, crucial comparative data are lacking. Sometimes, available criteria for determining whether a particular character state is primitive or derived are unsatisfactory. But whether or not the data will permit a ready solution, the process of detailed comparative analysis is intrinsically worthwhile, generating hypotheses which can be falsified or confirmed. We present no simple guidelines – we cannot – as a panacea to all our problems in phylogenetic reconstruction. But, whether 'successful' or not, the methodology outlined below invariably conduces to a more thorough understanding of the material to hand.

The comparison of (presumably) homologous attributes among a series of (presumably) related taxa inevitably discloses a great deal of variation among those taxa; frequently, this variation is non random, i.e. it falls into a spectrum. Such variation may intergrade, or there may be more or less large gaps between states of the same character in the taxa under study. Although not all between-taxa variation conforms to such a pattern, it is nonetheless true that some sort of order, in the form of a series, is most often present. Such spectra result from the nature of the evolutionary process. For despite the fact that the implications of the speciation model for the mode of formation of the fossil record differ from those of phyletic gradualism, it is nonetheless true that speciation itself still results as a rule in relatively small morphological changes between sister species. It is quite consistent with the speciation model that morphological change in the evolutionary process should approximate to a continuum, rather than produce large gaps. The critical difference between the two models is that while phyletic gradualism produces a smoothly continuous intergradation among successive taxa, speciation only approximates a continuum by producing a series of steady-state taxa separated, among true sister species, by rather small, yet discrete, morphological gaps. Unfortunately, the degree of resolution permitted by the fossil record is virtually never fine enough to allow a direct distinction to be made between the two situations, even on the basis of the best stratigraphic controls.

Character spectra of the kind outlined above have been termed 'morphoclines' [MASLIN, 1952], a word whose use was originally confined to the description of character states in a series of contemporaneous (specifically Recent) taxa. Since morphoclines reflect a spectrum based on the evolutionary history of the group in question, it follows that in the simplest case, that of a gradient in a single character, one *state* of that character is relatively most primitive, while another is the most derived from the primitive condition. 'Most primitive' here simply denotes, for the group under consideration only, that state which is, or most closely approximates to, the state of a given character in the common ancestor of a group. 'Derived' refers to *any* deviation from the primitive condition. MASLIN [1952] refers to the primitive-derived sequence inherent in a simple morphocline as its 'polarity'. The determination of the polarity of a morphocline lies at the core of the comparative procedure in phylogenetic reconstruction.

Perception of a morphocline is simple; determination of its polarity may not be. Essentially, there are two approaches to the resolution of the problem of morphocline polarity. The first lies in the communality of character states. If a given state is typical of a large number of taxa under study, and, in particular, if related taxa of roughly equal rank to the group being considered share it, then it is reasonable to regard that character state as being primitive. Thus, it is clear that a relatively small brain is a primitive state among the hominoids. Of course, this is also perfectly obvious from the conventional view of the fossil record, but the point here is that the fossil record is not absolutely necessary before the conclusion can be drawn. The fact that before a good hominid fossil record was available a large brain was expected to typify early hominids simply means that the comparative material was being ignored, or, more accurately, was looked at from the wrong viewpoint. In the case of brain size, morphology and stratigraphy agree in the indicated morphocline polarity. But apart from the fact that often fossil evidence is lacking (as in the case of the number of digits in the primitive eutherian manus, for instance, which we know only from comparative evidence must be five), does this mean that biostratigraphic data, where available, should routinely be used in assessing morphocline polarity? There are two primary reasons for doubting this, both of which are implicit in our previous discussion.

First, it frequently happens that the oldest known taxon of a series of fossil taxa is clearly specialized in one or more characters, and may even be more specialized in general than many taxa occurring later in the fossil record. Further, and perhaps more importantly, organisms are almost always mosaics of primitive and advanced character states. Determination of the primitive state in twenty characters of an organism will not guarantee that the twenty-first character examined will also be primitive. Yet the use of biostratigraphic data has to proceed on the assumption, whether or not overtly recognised, that within a group older taxa must be equally or more primitive in *all* characters than those occurring later in the stratigraphic record. That this is never the case is perhaps too strong an assertion; that it will be a relatively rare occurrence is, however, undeniable. The only response to this reality is to reject biostratigraphic data as reliable criteria for the assessment of polarity, at least in the first instance.

The second avenue of approach to the problem is provided by patterns of morphogenesis. Even if ontogeny does not recapitulate phylogeny in any real or useful sense, it still indicates, for instance, that the gill slits seen in tetrapod embryos are further evidence that such structures are primitive vertebrate features. This we known must be true despite the fact that they are not preserved in know Ordovician agnathans.

We fully realize that for many characters in a given group it will be difficult, and frequently impossible, to provide a strong, logical argument for polarity. But the lack in such circumstances of the hard information required does not result from any methodological inadequacy; rather, it is a reflection of the limitations imposed by the biological materials themselves.

VI. Formulation of Hypotheses of Relationship

As we have already stated, sister taxa can only be determined on the basis of the common possession of one or more unique character states. Usually, however, it is found that some taxa appear to be allied with certain others in one or more characters, but with yet others when different characters are considered. One example of this among the hominoids concerns the structure of the entoglenoid process [TOBIAS, 1969]. Among the African apes, the temporal bone alone contributes to the formation of this process. In modern *Homo sapiens* it lies entirely on the sphenoid. In gracile, and certain robust, *Australopithecus*, the situation is intermediate; both bones are involved in the structure of the entoglenoid process. But in Peking *Homo*

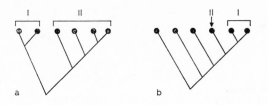

Fig. 1. For explanation, see text.

erectus, the process is situated firmly on the temporal. Such situations are invariably the result of parallelism, and perhaps the most profound generalization yet to have emerged from cladistic studies has been that parallelisms more rampant an evolutionary phenomenon than even its most ardent champions had previously claimed. The problem of parallelism is a very real one to the systematist, and appears, since it is related to the mosaic nature of organisms, when the transfer is made from single character analysis to 'whole-taxa' comparisons. A final theory of relationships must, therefore, be arrived at by consideration of the total number of derived character states, and by the assessment of the relative probabilities (or least likelihood) between any cases of conflicting characters revealed by the analysis. Often it will happen that most rival hypotheses of relationship can be rejected until only one remains; this will stand at least until new data are brought to bear on the problem. Those inevitable cases where this cannot be done are invariably due to an insufficiency of biological information.

After the initial sister group (I) is defined, it is compared as a whole with the other taxa under consideration. Thus, the sister group (II) of I is determined exactly as was the latter: on the basis of the possession of shared derived character states. Sister group II may be another species, or indeed, any or all of the other taxa as a whole (fig. 1). Of course, by the very nature of the comparative process only two items can be compared at once. To say, then, that sister group I, already an amalgam of taxa, is compared with other taxa, is really to say that a theoretical construct is being compared. This construct corresponds to the morphotype of the hypothetical common ancestor of sister group I. NELSON [1970] has provided the most recent discussion of the morphotype concept; the reader is referred to NELSON's paper for more extensive treatment.

By repetition of this procedure, the relationships of all the taxa under study may be approximated as far as the evidence allows. Even if it is based entirely or in part on fossil forms, the final theory of relationships produced

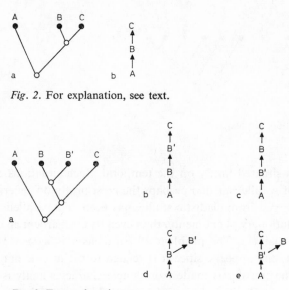

Fig. 2. For explanation, see text.

Fig. 3. For explanation, see text.

differs from the standard phylogenetic tree in that no ancestor-descendant relationships are postulated. Since this is indeed so, can we legitimately characterise such a theory as a phylogeny? The answer to this lies in the relative information content of this type of theory compared to that of the standard phylogenetic tree. Examples of each are given in figure 2; the same taxa, A, B and C, are represented in figure 2a (a theory of relationships) and in figure 2b (a typical phylogenetic tree)[2].

Both of these figures require modification if a new taxon, B', possessing certain derived characters in common with B, is encountered. Figure 2a could be modified in only one way (fig. 3a) to reflect this new information. But there are four ways in which B' could be incorporated into figure 3b (fig. 3b–e). The greater and more specific information contained in the phylogenetic tree is thus seen to be only apparent, since figure 3a summarizes all that can emerge from a comparative analysis. Maybe so; but a number of objections might be raised to a statement as bald as this. First of all, B may be considered more 'primitive' overall than B', in which case figures 3c and 3e can be eliminated. But we are still left with figures 3b and 3d

2   Figure 2b is only one of several which can be drawn on the basis of the relationships shown in figure 2a. Our discussion applies regardless of which of the possible phylogenetic trees is contrasted with figure 2a.

as alternative possibilities. Again, let us suppose that A, B and C are fossils, and occur in a progressive, non overlapping, stratigraphic sequence. In such a case, the standard practice would be to eliminate all the trees except figure 3b. But is this permissible? Strictly speaking, for figure 3b to be totally acceptable, the fossil taxa would have to conform, *in every detail* of the characters utilized in the analysis, to the hypothetical morphotypes depicted as open circles in figure 2a. We have already suggested that such a circumstance is highly unlikely to occur. And in that vast majority of cases where it does not, to change a hypothetical morphotype to embrace the morphological pecularities of a given fossil, purely because of its stratigraphic position, is to abandon the logic of the comparative procedure. Even if this procedure is not viewed favourably, it must be admitted that to take such a course is tantamount to assuming from the very beginning that which the paleontologist sets out to prove. To act thus is essentially to reject phylogenies as testable theories of relationship.

What we have tried to indicate, then, is that a theory of relationships amongst a group of organisms, whether all fossil, all living, or a mixture of both, usually contains all the legitimately derived testable conclusions about phylogeny which may be gained from any particular analysis. The phylogenetic tree, while appearing to be more precise and specific, is unlikely to contain any additional information on relationships beyond what would have to be described as untestable speculation. Thus, as far as useful information content on relationships is concerned, theories of relationship of the type which we have described fully qualify as phylogenies. This is not to say that the phylogenetic tree is not useful as a means of expressing informed speculation on relationships, and on framing new hypotheses which may be falsified or confirmed on the basis of new evidence. But its limitations in this regard should be explicitly recognized. This is not to deny, however, that the phylogenetic tree, which takes into account stratigraphic information, does contain one type of data which the hypothesis of relationships omits: only the fossil record can supply minimum ages for the appearance of actual taxa, and hence for evolutionary events.

VII. Comments on Hominid Phylogeny

The following brief discussion is not intended to be exhaustive. We have not in most cases been able to examine original specimens, and have, therefore, been strongly limited in the number of characters we have been able to compare. We wish to emphasize, then, that the purpose of this section is illustrative, although we believe that, at the level of resolution to which we have been restricted, our comments do carry some validity.

One of the largest problems facing the systematist in any attempt to form sister groups lies in the choice of characters to be used. Among the hominids, the problem is particularly acute, for two primary reasons. The first of these is the restricted diversity of the family: only one species extant today, and no more than three generally agreed genera throughout the family's entire evolutionary history. Inevitably, then, the distinctions between primitive and derived states of the same character are likely in most cases to be subtle, and not easily observed without detailed comparisons between all known members of the family: a difficult requirement for a single investigator to satisfy.

The second factor limiting character choice is the functional intercorrelation of many of those characters which are so easily described discretely. Thus, the large size of the cheek teeth in *Australopithecus robustus*, compared to those of *A. africanus*, is directly linked to such features as the robustness of the face and zygomatic arch and the possession of a sagittal crest, while reduction of the anterior dentition determines the shortness of the face, and, for mechanical reasons, the relatively anterior positioning of the sagittal crest compared with that of pongids. Characters begin to melt away when one realises the extent of the functional covariance of so many of the variations observed. Many, if not most, of the characters distinguishing gracile and robust South African *Australopithecus* may be traced to the influence of the dentition, if only as a starting point for understanding their functional interdependence, for in a complex interlocking functional system it is rarely possible to pinpoint with confidence an original adaptation to which all the others are concomitant. Nevertheless, observable variations do exist, and provide the basis for the procedure of sister group formation.

A further simplification here concerns the basic taxonomic groups which we have chosen to use. We have not been able to concern ourselves with hominid taxonomy, and our groupings are, therefore, frequently arbitrary at the specific level.

Figure 4 is a cladogram in which we provide a tentative hypothesis of the relationships between various hominid groups. It should be noted that in this type of diagram open circles at the branching points indicate hypothetical morphotypes, while solid circles represent known forms.

Any scheme which includes *Ramapithecus* can be based only on characters of the jaws and teeth, and, therefore, our branching point (BP) 1 is

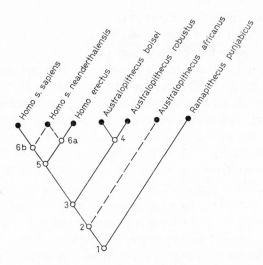

Fig. 4. Tentative cladogram of relationships within Hominidae. Solid circles represent known forms; open circles denote hypothetical morphotypes. It should be noted that this cladogram is based solely on cranial character states.

defined only by the following character states: incisor teeth relatively reduced; molar teeth foreshortened; occlusal surfaces of the molars expanded; cusps low; maxillae relatively deep; corpus of mandible relatively broad. All of these character states are derived within Hominoidea, but are primitive within Hominidae. Since all other hominids share one or more derived character states not possessed by *Ramapithecus*, they can be regarded together as comprising the sister group of the latter.

The location of the point we have designated as BP 2, i.e. that which expresses the relationships of *Australopithecus africanus*, is impossible to determine definitively on the basis of information available to us. It might, indeed, be placed almost anywhere, because cranially *A. africanus* resembles other hominids almost solely in primitive character states[3]. Indeed, this

3   It should be noted, however, that the picture may be somewhat different when the postcranial skeleton is considered. It is beyond the scope of this paper to examine the postcranial skeleton, but in any definitive treatment of hominid phylogeny the possibility that gracile *Australopithecus* does share derived postcranial character states with other hominids should be taken into account. The resulting phylogeny might then differ from the arrangement given here.

form possesses only one comparable derived character shared with all hominids other than *Ramapithecus:* homomorphic lower premolars. The placement of *A. africanus* within the entire group of non-*Ramapithecus* hominids is thus uncertain, and to this extent it is impossible to reject most of the numerous hypotheses as to its phylogenetic significance within Hominidae. On the basis of cranial and dental characteristics, *A. africanus* is potential ancestrally to any or all of the other hominids in its sister group, although it is impossible to say whether it is actually ancestral or not, and, if so, to what. We have placed BP 2 where it is because this is the least complex arrangement; but the inference must not be drawn from the clado-gram that the *A. robustus/A. boisei* group thus shares with *Homo* any de-rived characters which distinguish it from *A. africanus*. That this might be the implication of the cladogram is a reflection of the extreme difficulties inherent in classifying forms as primitive as *A. africanus* appears to be. To be quite blunt, then, *A. africanus* is so primitive that it tells us nothing, at least as far as cranial and dental morphology are concerned, about human evolution; nothing, certainly, which could not be deduced from a compara-tive analysis of extant hominoids.

We have chosen here to lump Olduvai hominid 7 and its lower Bed I paratypes into *A. africanus;* if a distinction were to be made, *A. africanus* and *A. habilis* would form a primary sister group divided by a single char-acter: possession by the latter of a brain greater than 600 cm^3 in size. Purely as a device, it would be possible to use this dichotomy to resolve the problem in forming the cladogram just discussed: *A. habilis* could, on the basis of its single derived character, be inserted at a point intermediate between BP 3 and BP 5, thus emphasizing the fact that something like *A. africanus/A. habilis* could serve as an ancestral morphotype for the entire group. We prefer, however, to retain an arrangement which expresses the uncertainty involved in hypothesising this relationship.

Those cranial characteristics of *A. africanus* which are primitive for its sister group include the following: cranial vault lightly-built and rounded; maximum cranial width bimastoid; external auditory meatus large; facial skeleton somewhat reduced; foramen magnum beneath cranial vault; rounded parieto occipital plane; slight basicranial kyphosis. Dental char-acteristics are, of course, those given for *Ramapithecus*, with the addition of bicuspid anterior lower premolars.

The derived characters shared by *A. robustus* and *A. boisei*, and used to define BP 3, include the following: substantial diminution of canine and incisor teeth; molarization of premolars; expansion and flattening of occlusal surface of molars, with a wear-pattern apparently reducing or eliminating shearing facets; possession of a sagittal crest, at least in males, albeit superimposed on a lightly-built cranial vault; considerable orthognathy; neurocranium hafted onto facial skeleton at a relatively low level; steep parieto-occipital plane; nasion and glabella almost coincident. Whether or not there should, in fact, be a bifurcation at BP 4 is highly debatable; the discovery of the cranium KNM-ER 406 in the Lower Unit at Ileret, East Rudolf [LEAKEY, 1970] has tended to obliterate the distinctions between the type of *A. boisei*, Olduvai hominid 5, and South African *A. robustus*. KNM-ER 406, although lacking tooth crowns, was evidently closely similar to O.H.5 in its dentition, while strongly resembling the robust South African skull Sk 48 in its comparable craniofacial morphology. If KNM-ER 406 is to be included in the hypodigm of *A. boisei*, BP 4 would have to be defined dentally, i. e. by extension of some of those derived character states already used in defining the branching from the primitive condition at BP 3. These would include: extreme diminution of canine and incisor teeth; extreme molarization of the molars, together with their almost total flattening; and extreme robusticity of the jaws. If, on the other hand, the much less likely course is taken of placing KNM-ER 406 together with South African *A. robustus*, BP 4 would depend almost entirely on the character states related to the extraordinary deepening of the face in O.H.5. Our choice of splitting here is, at any rate, at least as arbitrary as that of lumping lower Bed I *A. habilis* with *A. africanus*.

The traditional concern of paleoanthropologists with 'levels of organization', together with the fact that the chronocline and morphocline in the brain size of fossil hominids have tended to coincide, has attracted attention away from the observation that *Homo erectus*, when viewed in the entire context of hominid variation, does not present a good ancestral morphotype for *Homo sapiens*. Because *Homo erectus* is *there* in the middle Pleistocene, i. e. in the 'correct' stratigraphic position, it has been assumed that there is no difficulty in deriving *Homo erectus* from a species of gracile *Australopithecus*, or in deriving *Homo sapiens* from *H. erectus*. Purely on the basis of morphology, however, this is not the most parsimonious hypothesis. Indeed, when the problem is viewed in a comparative light, it becomes easier to understand why L. S. B. LEAKEY [e. g. 1966] preferred to derive *Homo sapiens* directly from '*Homo habilis*', in fact Olduvai hominids 13, 14 and 15, without the interposition of a 'stage' represented by O.H.9, 'Chellean Man', or of Neandertaloid forms. In the structure of its cranial vault, classical *Homo erectus* possesses a variety of derived characters not shared with *Homo sapiens;* in fact, the latter remains quite primitive in its upper neurocranial structure. *Homo sapiens* retains (or, at least, possesses) the round, lightly-built cranial vault of *Australopithecus africanus;* its derived osteological character states in this respect comprise its larger size, the shifting of the level of maximum width onto the parietals, and its more inflated frontal. All of these character states are related to a single factor: the enlargement, in a specific manner, of the brain.

In *Homo erectus,* on the other hand, the braincase is long, low, relatively flattened, terminates anteriorly in strong supraorbital tori, and, significantly, is thick-walled. All of these character states are derived, and none is shared with *Homo sapiens.* To state the case extremely crudely, and in conventional paleontological terms, it is easier to envision the derivation of *Homo sapiens* from a form such as that represented by Olduvai hominids 13, 14 and 15, by neural enlargement, than from classical *Homo erectus* by the same process. Admittedly, it is true that *Homo erectus* populations share derived dental features with *Homo sapiens,* although they do so hardly more than does, for example, O.H.13. It is plain that presently-known material is inadequate to permit resolution of this problem; in particular, it is unclear whether or not we are dealing with more than one taxon, and what the characteristics and ranges of variation of the real taxon/taxa are.

Of course, the question can always be raised at this point as to where, if not among the presently-known *Homo erectus* populations, the ancestry of modern *Homo sapiens* lies. After all, the species must have had a middle Pleistocene precursor. We have already suggested that available material is insufficient to characterize fully all the known hominid populations of this time period; but this aside, the possibility always remains that the most parsimonious hypothesis is not the correct one. Just conceivably, those characteristics of *Homo erectus* which differentiate it from both *Australopithecus africanus* and modern *Homo sapiens,* although without doubt specialized in the context of the whole of Hominidae, are primitive within the sister group of *Homo erectus* and modern *Homo sapiens.* The test as to whether or not this is the case must lie in the understanding of the functional and evolutionary significance of the two (or three, if the common features of *Australopithecus* and modern *Homo sapiens* crania are parallelisms) cranial types.

Similar problems apply in the consideration of the relative merits of points 6a and 6b (fig. 4) in expressing the relationships of modern *Homo sapiens* and the Neandertals, and the answers depend very much on those to the question of the relationships of *Homo erectus.* In one view, those features of the Neandertal skull which have generally been viewed (largely, one suspects, for reasons of stratigraphy) as 'primitive', for instance the long, low braincase, large supraorbital tori, angulated occipital and so forth, are in fact specializations which align their possessors with *Homo erectus* rather than with modern *Homo sapiens:* the *Gestalt* of the typical Neandertal (or Neandertaloid) skull is that of a *Homo erectus* with an inflated brain. In this view, a worthwhile question to ask of the Neandertals, even in phyletic terms, is not so much whether they gave rise to modern man, but whether *Homo erectus* and the Neandertals form a true sister group, or represent iterative offshoots. The alternative grouping of the Neandertals and modern *Homo sapiens* as a primary sister group would be dependent upon an evolutionary explanation of their cranial dissimilarities. We believe that available evidence is insufficient to permit a definitive choice between these alternatives, and the cladogram expresses this ambiguity.

No examination of a group containing fossils is complete without a consideration of stratigraphic distributions. We have recommended that temporal distributions should not be taken into account during the formation of an initial hypothesis of relationships; but to ignore such data at all stages would be to disregard that information which is uniquely contained in the fossil record. For only the fossil record can reveal the antiquity (at least in terms of minimum age) of actual evolutionary events.

Thus, we can say, for instance, that the *minimum* age of the hominid/pongid differentiation is in excess of 12.6–14.0 million years, since in *Ramapithecus* we have a form which is almost certainly hominid, at least in a strictly cladistic sense. Under conventional paleontological analysis, *Ramapithecus* emerges as a probable hominid, for the contention that this animal *is* the ancestor of all other hominids is not an objectively testable statement. But that *Ramapithecus* forms the sister group of all other hominids is such a hypothesis; and comparative analysis demonstrates that those characteristics which have been recently used to suggest that *Ramapithecus* is not a hominid are in fact primitive for Hominoidea, and are, therefore, meaningless in terms of phylogenetic relationships.

The stratigraphic distribution of *Ramapithecus* also introduces the question of stasis in hominid phylogeny, but examples of this may equally well be drawn from the *Australopithecus* group. *Australopithecus africanus*, for instance, as represented by Sterkfontein, provides a virtually perfect ancestral morphotype for *A. robustus*. But the latest stratigraphic orderings of the South African breccia cave sites, as well as the fact that it is becoming increasingly unlikely that only a single lineage is represented at each site, makes it impossible that any known South African population of *A. africanus* could be ancestral to *A. robustus*. From this it must be concluded that *A. africanus*, at least in South Africa, remained virtually unchanged from the condition of the ancestor represented at BP 2 in figure 4. How long a period of stasis is thus to be inferred is uncertain, but the mandibular fragment from Lothagam, Kenya, dated at *c.* 5.5 m.y., may perhaps best be referred to *A. africanus* among presently-named taxa, although the cranial morphology of the form it represents is unknown.

The evidence for stasis in the *A. robustus/boisei* group is particularly intriguing. In East Africa, it is clear, the robust form appeared later than the gracile one, but it was certainly well established before 2.6 m.y., when the earliest cranial evidence appears, and persisted until at least 1.1–1.2 m.y. (Chesowanja) with little, if any, appreciable change. Comparison of the East and South African robust populations, however, apparently demonstrates differential stasis: the Transvaal sites fall within the same time zone as some of the East African localities, yet while at the latter two types are quite obviously represented, at the former the gracile and robust forms are much less well differentiated. In any event, just as *A. africanus* presents a good ancestral morphotype for *A. robustus*, the latter provides an equally satisfactory ancestral morphotype for *A. boisei*, and it may well be that here we are confronted with some of the best evidence available from hominid phylogeny of the operation of the speciation model.

Evidence for overlap in the stratigraphic ranges of species elsewhere in the hominid fossil record is perhaps less clear-cut, although a list given by CAMPBELL [1972], in the most explicit recent exposition of current palaeo-anthropological practice, indicates very considerable overlap at almost all stages. The recent K/Ar dating of the Modjokerto locality of the Djetis zone in Java at *c.* 1.1 m.y. [JACOB, 1972] places *Homo erectus* from this site in the same time period as lower Bed I '*Homo habilis*', and considerably earlier than the latest occurrence in Africa of gracile *Australopithecus* (Chemeron); unfortunately, however, the only hominid fossil from Modjokerto is the calvaria of an infant. This specimen lacks the classic features of Java *Homo erectus*, but as an infant would necessarily do so, whatever its affinities. Its brain volume, however, has been estimated at 700 cm^3 [BOULE and VALLOIS, 1957], a figure well in excess of that of the considerably more mature O.H.7. The age of other hominid fossils ('Pithecanthropus IV', 'Meganthropus') is unknown; the Modjokerto date provides a minimum age for the base of the Djetis, but the zone's upper boundary is considerably younger.

More suggestive than this, however, is a discovery announced after this article was originally written. This is the cranium KNM-ER 1470 recently described by R. E. F. LEAKEY [1973] from the late Pliocene of East Rudolf. Reportedly some 2.9 m.y. old, and possessing an endocranial volume in excess of 800 cm³, this specimen provides perhaps the most spectacular evidence yet reported of a morphocline-chronocline conflict in the hominid fossil record. Specifically, it does considerable violence to the apparent concordance between the morphocline and chronocline in hominid brain volume to which we have earlier referred. And yet further confusion is introduced into conventional gradualist interpretations by HOLLOWAY's [1973] recent report that Olduvai hominid 9 (upper Bed II) possesses an endocranial volume (1,067 cm³) well above that of the stratigraphically much higher O.H.12 (Bed IV; 732 cm³).

In commenting on current palaeontological practice it has not been our wish simply to criticize the prevailing goals or methods of palaeontological research. Indeed, virtually all the traditional concerns of palaeoanthropologists, as of palaeontologists in general, are valid ones. What we have tried to point out, however, is that phylogenetic enquiries should be conducted on the basis of a hypothesis of relationships. Only when a minimal phylogeny of this kind has been established can further investigations be undertaken with confidence, and what is *known* be clearly separated from what is *believed*. And this points to a basic weakness in our present knowledge: we do not know precisely how many taxa we are dealing with in our discussions of the hominid fossil material, or what all their characteristics are. As we have suggested, we believe that this is subject to non arbitrary determination (although the possibility cannot be ignored that biocultural feedback may have increased the potential for phyletic evolution, at least during the later stages of hominid phylogeny); but, in any event, the question is only answerable through augmentation of the biological evidence. And, in this case perhaps more than in any other, augmentation of the biological evidence is synonymous with the augmentation of the fossil record.

VIII. References

BOULE, M. and VALLOIS, H. V.: Fossil men (Thames & Hudson, London 1957).

BRUNDIN, L.: Application of phylogenetic principles in systematics and evolutionary theory; in ØRVIG Current problems of lower vertebrate phylogeny. 4th Nobel Symp. pp. 473–495 (Wiley, New York 1968).

BRUNDIN, L.: Evolution, causal biology, and classification. Zool. Scripta *1:* 107–120 (1972).

CAMPBELL, B. G.: Conceptual progress in physical anthropology: fossil man; in SIEGEL, BEALS and TYLER Annual review of anthropology, pp. 27–54 (Annual Reviews, Palo Alto 1972).

DAY, M. H.: Omo human skeletal remains. Nature, Lond. *222:* 1135–1138 (1972).

DAY, M. H.: The Omo human skeletal remains, in BORDES The origin of *Homo sapiens,* pp. 31–35 (UNESCO, Paris 1972).

ELDREDGE, N.: The allopatric model and phylogeny in Paleozoic invertebrates. Evolution *25:* 156–167 (1971).

ELDREDGE, N. and GOULD, S. J.: Punctuated equilibria: an alternative to phyletic gradualism; in SCHOPF Models in paleobiology, pp. 82–115. (Freeman & Cooper, San
Francisco 1972).

FARRIS, J. S.; KLUGE, A. G., and ECKARDT, M. J.: A numerical approach to phylogenetic
systematics. Syst. Zool. *19:* 172–189 (1970).

HENNIG, W.: Phylogenetic systematics (Univ. of Illinois, Urbana 1966).

HOLLOWAY, R. L.: New endocranial values for the East African early hominids. Nature,
Lond. *243:* 97–99 (1973).

JACOB, T.: The absolute date of the Djetis beds at Modjokerto. Antiquity *46:* 148 (1972).

LEAKEY, L. S. B.: *Homo habilis, Homo erectus* and the australopithecines. Nature, Lond.
209: 1279–1281 (1966).

LEAKEY, R. E. F.: New hominid remains and early artifacts from northern Kenya. Nature,
Lond. *226:* 223–223 (1970).

LEAKEY, R. E. F.: Further evidence of lower Pleistocene hominids from East Rudolf,
north Kenya. Nature, Lond. *237:* 264–269 (1971).

LEAKEY, R. E. F.: Evidence for an advanced Plio-Pleistocene hominid from East Rudolf,
Kenya. Nature, Lond. *242:* 447–450 (1973).

LERNER, I. M.: Genetic homeostasis (Oliver & Boyd, London 1954).

MASLIN, T. P.: Morphological criteria of phyletic relationships. Syst. Zool. *1:* 49–70
(1952).

MAYR, E.: Principles of systematic zoology (McGraw-Hill, New York 1969).

NELSON, G. J.: Outline of a theory of comparative biology. Syst. Zool. *19:* 373–384 (1970).

SARICH, V. M.: The origin of the hominids: an immunological approach; in WASHBURN
and JAY Perspectives on human evolution, vol. 1; pp. 94–121 (Holt, Rinehart &
Winston, New York 1968).

SCHAEFFER, B.; HECHT, M. K., and ELDREDGE, N.: Phylogeny and paleontology; in
DOBZHANSKY, HECHT and STEERE Evolutionary biology, vol. 6; pp. 31–46 (Appleton Century Crofts, New York 1972).

SCHLEE, D.: Die Rekonstruktion der Phylogenese mit Hennig's Prinzip. Aufsätze und
Reden. Senckenberg. naturforsch. Ges. *20:* 62 (1971).

SIMPSON, G. G.: Principles of animal taxonomy (Columbia Univ. Press, New York 1961).

TOBIAS, P.: Bigeneric nomina: a proposal for modification of the rules of nomenclature.
Amer. J. phys. Anthrop. *31:* 103–106 (1969).

UZZELL, T. and PILBEAM, D. R.: Phyletic divergence dates of hominoid primates: a
comparison of fossil and molecular data. Evolution *25:* 615–635 (1971).

Authors' addresses: Dr. NILES ELDREDGE and Dr. IAN TATTERSALL, Departments
of Invertebrate Paleontology and Anthropology, American Museum of Natural History,
New York, NY 10024 (USA)

III. Evolutionary Morphology

In Szalay: Approaches to Primate Paleobiology
Contrib. Primat., vol. 5, pp. 244–292 (Karger, Basel 1975)

Allometry in Primates, with Emphasis on Scaling and the Evolution of the Brain

Stephen Jay Gould

Museum of Comparative Zoology, Harvard University, Cambridge, Mass.

Contents

I. Definition, Recognition and Treatment of Allometry

When Julian Huxley established the quantitative study of allometry in the 1920s, he pursued a strategy that proved excellent in the short run but rather restrictive in longer perspective. He tied both his term and his practice to a single form of mathematical expression – to the power function, $y = bx^a$. The short-term blessings were manifold: the function is easily fitted (as a straight line with logarithmically transformed data); its parameters have ready biological interpretations; it fits an extremely wide range of data and can be justified theoretically in many cases. But it had the most unfortunate effect of tying the study of a guiding concept in morphology to a particular equation. As multivariate techniques supersede bivariate studies in many areas, as other equations expressing the relation of size and shape come into greater use, the restriction of allometry to power functions relegates the subject to minor significance, if not to historical oblivion.

In a previous review [GOULD, 1966], I argued that allometry is simply the study of size and its consequences – more specifically, the study of departures from geometric similarity that scale regularly with changes in size, whether the change in size be expressed in ontogeny, phylogeny, or in a static series of standardized animals (adults within a species, adults within an order). The causal dependence of shape upon size has been recognized since GALILEO's great *Discorsi* of 1638, a work written while under house arrest for heresy. Allometry is perhaps the major principle regulating basic differences in form among related animals. It explains, among many other things, why large animals have relatively thick legs and small brains, why dachshunds can't be as large as elephants, why flies can walk up walls, and why large homeotherms metabolize so much more slowly than small ones [THOMPSON, 1942; WENT, 1968; GOULD, 1971a]. It is the key to what McMAHON [1973, p. 1204] has called 'the growing science of form, which asks precisely how orgaisms are diverse and yet again how they are alike'. To borrow HOWELLS' [1972, p. 126] assessment of multivariate analysis, 'it is an approach, not a technique'.

Yet allometry has suffered the common fate of unglamorous familiarity – to be recognized by all in the abstract, yet appreciated and applied rather rarely. Moreover, when applied, it is often applied badly. Thus, we encounter fallacies of recognition and of treatment.

1. Fallacies of Recognition

The allometry of brain size has long been known; CUVIER himself had commented that large vertebrates have relatively small brains. The primary implication of this pervasive negative allometry must be that neither absolute nor relative magnitude of the brain alone can stand as a criterion for cephalization. Small vertebrates must have maximal brain/body ratios, while the largest brains will reside in the largest animals (the brain increases more slowly than body weight, but it does increase). Any assessment of cephalization must take absolute body size into account; it must compare actual brain size with expected brain size for animals *of that body size*.

Nonetheless, we often encounter statements based upon absolute size alone. A recent and popular text states [CLARK, 1973, p. 577]: 'The original bipedal ape-man of South Africa had a brain scarcely larger than that of other apes and presumably possessed behavioral capacities to match.' Yet, for its small body size, *Australopithecus africanus* had a brain distinctly larger than any known pongid; a brain of equal size in a much larger gorilla does not challenge this assertion.

Nonetheless, SCHALLER [1964] feels compelled to apologize for the gorilla's failure to invent the australopithecine tools that must lie within its capacity: 'Why was australopithecus, with the brain capacity of a large gorilla, the maker of stone tools, a being with culture in the human sense, while the free-living gorilla in no way reveals the marvelous potential of its brain? I suspect that the gorilla's failure to develop further is related to the ease with which it can satisfy its needs in the forest.' TOBIAS [1971], though well aware of the allometric relation, virtually ignores body size in his long monograph of cranial capacity; this disembodied approach to the brain provides many figures, but little in the way of criteria for judging cephalization.

Statements based upon relative size alone are equally misleading. SCHEPERS [1949], writing at a time when the status of australopithecines as hominids was still in doubt, sought to establish the high cephalization of *A. africanus* by presenting simple ratios of brain weight to inferred body weight. On this fallacious criterion, *A. africanus* did distinctly better than gorillas, though SCHEPERS [1949, p. 93] was forced to admit that 'some lower [and smaller] primates yield much higher values for this index.' Even the master herself once claimed that a large camel was less cephalized than its smaller ancestor because the brain/body ratio had decreased [EDINGER, 1966, p. 160].

Intelligence testing has provided the chief arguments for 'scientific' racism in the 20th century, but craniometry was the refuge of the 19th century. When the 19th century arguments were not downright fudged or fallacious, they were based upon misunderstanding of allometric changes. TOBIAS [1970, p. 9] has shown that 'differences among various racial or population groups are negligible once allowance has been made for body size.' Some 'classic' studies of brain size in various black peoples did not take their low average body size into account. Using JERISON's [1963] 'extra neuron' criterion, TOBIAS [1970] estimates 8.5 billion for American whites and 8.7 billion for American blacks. The cephalic index (length/breadth of the skull) stood second only to cranial capacity as a racist argument. Yet BOAS showed as early as 1899 that its value is primarily a function of body size. 'Large bodies tend to yield long heads and heads which have absolutely the greatest lengths have the lowest indices' [1899, p. 448] as a result of the brain's negative allometry [for a modern treatment, see SCHAEFFER, 1962, p. 156; HEMMER, 1966, p. 201].

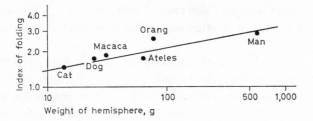

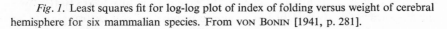

Fig. 1. Least squares fit for log-log plot of index of folding versus weight of cerebral hemisphere for six mammalian species. From VON BONIN [1941, p. 281].

In addition to its allometry with body size, the brain has intrinsic allometries of its own. Most notable, and a prime candidate for misinterpretation, is the relationship between surface convolution and brain size. LEGROS CLARK [1945, p. 2] explained convolution as a simple response to the necessity of keeping a constant ratio between the surface layer of grey matter (where neurons terminate) and the interior volume of white matter (where the same neurons arise) in large brains. 'The cerebral hemispheres of small animals are relatively or completely smooth (lissencephalous) and those of large animals are usually richly convoluted, irrespective of their zoological status... The proportion of white matter to grey matter remains approximately uniform, and to maintain this ratio the cortex must necessarily become folded.' Man's large brain for his body weight may be a criterion of cephalization, but the convolution that accompanies it is merely an automatic correlate of size itself. In fact, VON BONIN [1941] showed that the point for man is slightly below the line for a plot (admittedly based on very little data) of folding versus brain size (fig. 1). Nonetheless our textbooks continue to draw conclusions of phyletic 'advance' from allometric requirements of size. CAMPBELL [1966, pp. 224–226] writes: 'The actual area of the cortex in man has increased far more than might appear, since its folding is much more complex than in other animals. In man, 64 per cent of the surface of the cortex is hidden in the sulci (or fissures); in primitive [smaller] monkeys, only 7 per cent.'

2. Problems of Treatment

From an allometric viewpoint, the primary problem of many studies of shape in primates is a failure to use any quantitative technique beyond simple presentation and summarization. The voluminous compendia of SCHULTZ [e.g. 1933, 1940, 1953], despite their immense value as sources of data, suffer in this respect [see also, MOLLISON, 1910]. Moreover, when proportions alone are presented, it is often impossible to recover the original data.

The most bedevilling problem of allometric studies has long been the confusion among types of scaling. Size variation in nature is displayed in a variety of ways, and the predicted allometric scaling differs greatly among them. Consider, again, the brain. In the classic 'mouse to elephant' curve for adults of a related group of species, brain weight generally scales to the two-thirds power of body weight [an immense number of sources ranging from SNELL, 1891, to JERISON, 1973]; a relationship between brain weight and body surfaces is here implicated, but it has never been satisfactorily explained. On the other hand, brain weights in adults of a single species or of a group so closely related that they appear as different size-expressions of the same body plan (e. g. breeds of domestic dogs) scale at much lower powers of body weight, generally between 0.2 and 0.4 [an equally great number of sources, ranging from DUBOIS, 1898, to BAUCHOT and STEPHAN, 1964]. In short, intraspecific exponents are lower than interspecific values. Phylogenetic curves for adults of lineages increasing in size would, if they were known, be subject to no such generality, but I will argue later that many must have exponents near 1.0. Ontogenetic curves are more complex [COUNT, 1947; KERR et al., 1969]. Fetal exponents are high; later postnatal slopes are very flat, often approaching zero; the intermediate portion (late fetal through early infancy) often joins the two via a curved intermediate area. Figure 2 displays this variety in the brains's scaling with size. Ontogenetic data may be truly longitudinal (based on a single individual followed through life as in MATSUDA's [1963] study of limb proportions in white children); or, as is usually the case, inferred from samples based upon many individuals measured at different ages (as in MATSUDA's consideration of black children).

A confusion among type of scaling lay behind the exaggerated claims of the first important studies in quantitative allometry – the work of DUBOIS [1897] and BRUMMELKAMP [summarized in 1940b] on the vertebrate brain. In his first work, DUBOIS computed allometric slopes for seven pairs of related species, measured a mean exponent of 5/9 and claimed that it represented a universal constant (even though he himself had measured a lower slope for the intraspecific scaling of man). This dubious claim proved impervious to future test because DUBOIS, thenceforward, merely passed lines of 5/9 slope through single points (fig. 3). DUBOIS then computed the coefficients (y-intercepts) for his set of parallel lines at 5/9 slope and thought he had detected an even displacement by factors of 2 between curves for

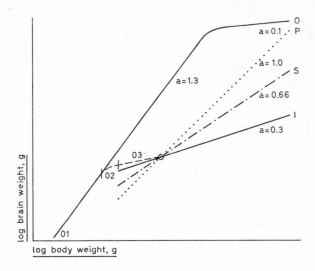

Fig. 2. An attempt to represent the various types of scaling for brain weight versus body weight (and a plea that they not be confounded). The circled point is an average adult for a given species. The curve 01–02–03 represents the complex ontogeny of that individual (an early prenatal period of high slope arbitrarily set here at 1.3, an intermediate curved segment, and a later postnatal period of very low slope, usually near 0.1). Curve I is the intraspecific regression, with its characteristic slope near 0.3, for variation among adults of a given species. Curve S is the static (mouse-to-elephant) regression for average adults of all contemporaneous animals of its broad taxonomic group (all mammals or all primates); it is shown with its characteristic slope of 2/3. The rest of the chart attempts to display a variety of phyletic possibilities. P is a phyletic curve; its slope can have any value (I have drawn the single possibility of 1.0 to illustrate a later point – pp. 281 – that brain increase in geometric similarity is still an 'improvement' relative to expectations of static scaling, i.e. a large descendant of the same brain weight/body weight ratio as its ancestor will have a relatively larger brain than a standard mammal of the same body weight). I illustrates what happens when animals increase or decrease in phylogeny along an extension of their own intraspecific curve (pp. 278). Phyletic increase yields a brain relatively small compared with a standard mammal on the static curve (I suggest that *Gorilla* and *Alouatta* have so evolved – p. 279); phyletic decrease yields a relatively large brain by the same criterion [BAUCHOT and STEPHAN have urged such an interpretation for *Miopithecus talapoin*] (p. 278). The ontogenetic curve is more complex, and O illustrates just one possibility for its evolutionary modification – the one implicit in the notion that neoteny is a major factor in human evolution (p. 284). By prolonging the rapid prenatal increase to larger sizes (O1 with its arbitrary slope of 1.3), we obtain a marked increase in relative brain size.

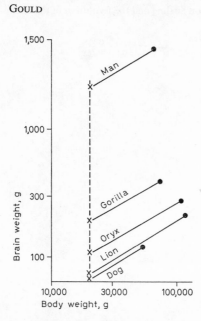

Fig. 3. DUBOIS' method for measuring 'degree of cephalization'. He simply passed a line of 'universal' slope 5/9 through the point for an adult of any given species. A scale of y-intercepts (or a scale of y-values at any x since the relative values are constant for a set of parallel lines) defines the levels of cephalization. From GOULD [1971 b], based on data of DUBOIS [1897].

primitive and advanced forms. Therefore, he claimed that brain evolution proceeded by occasional sudden jumps involving a doubling of the number of neurons at the outset of ontogeny in advanced descendants. 'The greatest importance should be attached to this geometrical progression, which is said to be founded on periodical, general divisions of the neurons. These divisions take place during embryonic development... and therefore cannot leave fossil traces of any stage of gradual transition' [BRUMMELKAMP, 1940 b, p. 3]. The confusion of allometric curves is legion. First of all, his lines represent the scaling of contemporary adults; they are not phyletic sequences display-ing discontinuities in cephalization. More important, his explanation invol-ved a doubling at the outset of ontogeny and this simply cannot be inferred from a set of interspecific curves. A modern, small adult simian with twice the brain weight of a small adult prosimian of the same body size is no model for a fetal *Ur*-simian of prospectively large adult size that has just suddenly doubled its supply of neurons.

In a far less serious example, TOBIAS [1971] has claimed that height rather than weight should be used in standard allometric plots for brain weight versus body weight among different species. As evidence, he cites the findings of PAKKENBERG and VOIGT [1964] that brain weight has a high partial regression with body height but none with weight for large samples of adult men and women. This is scarcely surprising: with height partialled out, differences in weight mark fat versus thin people and we would expect little correlation between nutritional habits and brain weight. The abscissal dimension of allometric studies is an estimate of that elusive quantity of general body size. There is nothing sacred about body weight as an estimator of this parameter. It is clearly a poor estimator for intraspecific studies of adult humans; but it is probably quite an adequate one for interspecific studies that compare not fat and thin people, but average individuals of small and large species. One would scarcely want to use body length in comparing hippopotomuses with boa constrictors.

As a final example, GILES [1956, p. 57] argued that 'allometric growth patterns in the chimpanzee would, given an over-all size increase, produce results quite similar to the exaggerated osteological morphology of the gorilla'. This claim was based on interspecific plots combining mountain and lowland gorillas with chimps and pygmy chimps. But straight lines through means for four separate species do not answer the question of whether a gorilla is an overgrown chimpanzee. We would need, rather, to follow two different approaches: (a) construct an ontogenetic curve for chimpanzees and extrapolate it to gorilla sizes; (b) compute an intraspecific curve for adult chimpanzees and extrapolate it to judge whether selection for larger chimpanzees might yield the proportions of a gorilla if the pattern of intraspecific variability were maintained during selection for increasing size. An interspecific curve cannot answer these questions. Similarly, the fact that an average adult Irish Elk lies on the interspecific curve for antler size versus body size in cervine deer does not prove that positive allometry produced the enormous antlers of the giant deer [GOULD, 1973, and in press]. The question is intraspecific. We must study the intraspecific allometry of adult stags in a single population to determine whether large deer tended to have relatively large antlers. They did.

In addition to these conceptual problems, we encounter a host of technical fallacies in the application of allometric equations, particularly of power functions [REEVE and HUXLEY, 1945; GOULD, 1966, pp. 594–600]. These include: (a) A failure to consider all parameters of allometric equations: GILES [1956] computed the allometric exponents for gorillas and chimps, found no significant differences and proclaimed the curves identical. But curves of equal slope can have different y-intercepts. Such sets of parallel

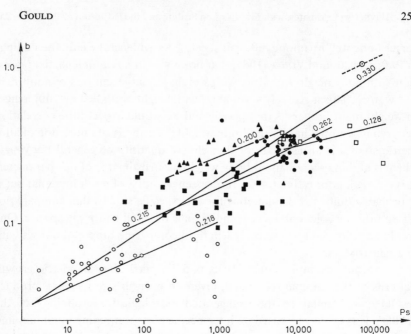

Fig. 4. BAUCHOT's plot of b-value versus body weight within five groupings of insectivores and primates to illustrate a problem in curve fitting [from BAUCHOT, 1972, p. 442]. The individual points clearly indicate that the variables are uncorrelated within a grouping; therefore, they can have no meaningful regression. BAUCHOT's values (near 0.2) are artifacts of his method; the reduced major axis computes a slope as the simple ratio of log standard deviations of the two variates. ○ = Insectivores; ■ = prosimians; ▲ = platyrrhines; ● = catarrhines; □ = anthropoids; ⊙ = man; Ps = body weight.

lines with slopes of 2/3 are particularly important in the analysis of interspecific allometry of the brain. (b) An overextended faith in power functions that leads to the fitting of several lines to curvilinear data: In the heyday of HUXLEY's power function, faith in log-log linearity was so great that curvilinear departure from expectations was usually met by approximation with two or three straight lines. We might defend this as a descriptive simplification, but its perpetrators generally imbued the points at which their straight lines met with special significance as 'critical stages' [ALCOBÉ and PREVOSTI, 1951, p. 17] marking sudden transitions from one type of growth to another. They are artifacts. NOBACK and MOSS [1956, p. 550], for example, plotted cerebellum versus cerebrum from the second fetal month to adulthood in man. They detected a 'critical point' at the end of the fourth fetal month, but admitted that they could find no morphological correlate for it. (c) Problems in curve fitting: Many authors, myself included, have advocated the use of 'regression' formulae that consider error in both variates. The most popular of these curves, the reduced major axis [IMBRIE, 1956; KERMACK and HALDANE, 1950], computes a slope for power functions as the ratio of log standard deviations of the two variates. This can lead to problems of the following type. BAUCHOT [1972] computed average slopes of 0.2 for reduced major axes of allometric coefficients (b-values) versus body weights in interspecific plots for 5 major groupings of insectivores and primates. He computed b-values for individual species by passing a line of slope 0.64 through the brain-body point for an average adult and computing the y-intercept. But since 0.64 is also his interspecific value for brain-body curves within each of the major groupings, there cannot possibly be any correlation between b-values so computed and body weight within a grouping. With correlation coefficients of 0.0, there can be no meaningful regression. Yet, since the reduced major axis is merely a ratio of standard deviations, you can always compute a slope, even though it is meaningless when the variates are uncorrelated. In BAUCHOT's case, a RMA of 0.2 only means that body weight is 5 times more variable than b-value (fig. 4). The standard least-squares regression of y on x avoids this dilemma since it is computed as the ratio of standard deviations multiplied by the correlation coefficient.

II. Basic Trends in Primate Allometry and Their Explanation

A. A Digression on Brain Size

Since most of this paper deals with the size of the brain and its parts in relation to body weight, I must note some of the controversy that has raged over the use of such measures. Can the gross size of the brain (or merely of the endocast in fossils) mean anything at all when we deal with such an immensely complicated computer of such subtle and variable structure? Is it not even a vulgarization to make such an attempt? HOLLOWAY [1969, 1970, 1972] has been particularly forceful in asserting these criticisms. He writes [1972, p. 188], for example: 'Each primate brain is reorganized according to the evolutionarily oriented adaptations of social, manipulative, and locomotor behaviors. Different primate brains are never simply smaller or larger versions of each other. Fossil endocasts reflect only poorly, if at all, the different internal organizations of neural structures... comparisons based on cranial capacities alone are not comparisons of equal units.'

The response to these critiques can be either apologetic or positive. The apologetic position simply laments our inability to do otherwise with fossils. It is this or nothing. 'This interest in brain size grows as much out of desperation in the face of limited data as out of a predisposition to work with gross measures ... Whether we like it or not we are faced with the problem: what does the brain size mean?' [JERISON, 1963, p. 263].

The positive position begins with an assertion of the strong order and utility that such data have provided: 'There is an obvious orderliness to brain: body data that cannot be ignored' [JERISON, 1970a, p. 228]. It continues with a statement that trends in cranial capacity are among the strongest and best documented in primate evolution: 'It is the most strikingly sustained trend shown in the fossil record and, hence, in the morphological evolution of the Hominidae... The trend towards increased brain size is the most continuous, long-lasting, and marked hallmark of hominization' [TOBIAS, 1971, pp. 114–115]. SACHER [1970, p. 246] has called brain/body allometry 'one of the important relationships of neurobiology'; he argues that HOLLOWAY's critique 'goes beyond the limits of valid and necessary criticism and becomes a classic instance of throwing out the baby with the bath water'. In fact, HOLLOWAY [1972, pp. 189–191] now seems more willing to entertain statements based upon cranial capacity.

I believe that the resolution of this controversy can be sought in a very insightful remark made recently by JERISON [1970a, p. 231]. He agrees that brain size *per se* may, as a parameter, measure very little of direct value. But it is more properly seen as a statistic for the estimation of important attributes strongly correlated to it. JERISON [1970a] mentions cortical volume, number of neurons and, possibly, degree of dendritic proliferation as characters best estimated from brain size [TOBIAS, 1971, p. 104, presents a similar list]. Such an argument, itself, rests upon allometric relations between brain size and its internal constituents. Of these, the negative allometry of neuronal density versus brain size is the most pervasive and predictable [JERISON, 1963; BAUCHOT, 1964].

B. Scaling in Bivariate Relationships

Primates, despite their evident diversity of external form, are no exception to the general observation that basic dimensions of internal organization scale with size in a remarkably regular manner for ordinal, interspecific relationships. STAHL [1965] compared the allometries (mainly negative) of primate organs (heart, lungs, liver, kidneys, adrenals, thyroid, pituitary, spleen, pancreas, and brain) with those of other mammals [see also FRICK, 1960, on primate hearts]. With one exception, primates are no different from other mammals. Power functions fit remarkably well and values of their parameters for primates are invariably within two standard errors of those for other mammalian groups (within one standard error for heart, lung, kidney, pituitary and spleen). STAHL [1965, pp. 1040–1041] comments: 'It could hardly be expected *a priori* that organ weights would be so well linearized by log-log plots, or that the characteristic parameters for mammals and primates would be identical within presently available statistical limits. These observations suggest that all mammals have in common a basic kind of physiological design.' The one exception, of course, is the brain. Its slope, in common with other mammalian orders, is about 2/3, but its y-intercept is unusually high, reflecting the large average brain size of primates with its equally positive percentage deviation at all body weights. I shall discuss various methods for assessing the special status of primate brain size in Section III.

STAHL and GUMMERSON [1967] studied the scaling of skeletal dimensions in a series of primates ranging from the tamarin at 0.27 kg to *Papio cynocephalus* at 22.0 kg. In accordance with GALILEO's principle, the midshaft diameter of long bones scales with positive allometry (slopes of 0.37–0.41 against body weight), while bone lengths scale isometrically (slopes of 0.31–0.37, spanning the isometric value of 0.33 for lengths plotted against weights). Although allometry of bone width is positive, it is not as strong as the 0.50 value predicted for length versus weight when the length involved reflects an area differentially increasing to support a volume. SCHULTZ [1953, p. 308] noted the same phenomenon and remarked: 'The absolute thickness of the long bones of the limbs is closely dependent upon the load they have to support, but a gain in body weight is followed by a much smaller relative increase in bone girth, as is to be theoretically expected.' Either small animals have larger margins of safety than large ones, or static considerations of bone strength are not the only determinants of bone thickness.

Other intriguing (if unexplained) ordinal allometries have been recorded. SACHER [1966] studied the logarithmic relationship between lifespan and brain and body weights in primates and other mammals. In primates, lifetime scales as the 0.38 power of brain weight and as the 0.23 power of body weight. Since brain and body are, themselves, so strongly correlated, SACHER considered the relationship of lifespan to each independently using partial correlation. With body weight, removed, lifespan correlated strongly with brain weight, but with brain weight removed, its correlation with body weight is slightly, though insignificantly, negative. SACHER obtained similar results for rodents and artiodactyls. Brain size seems to exert an effect upon lifespan that is independent of body size; the well-explored relationship of lifespan to body size [KURTÉN, 1953; STAHL, 1962] may be an artifact of the strong relation between brain and body. SACHER [1966, p. 13] sees some important implications for theories of aging: 'The differences in longevity, and therefore in rate of aging, between mammalian species are due for the most part to differences at the higher levels of coordinated behavior.' Lifespan may reflect the 'ability of the homeostatic systems to maintain the constancy of the *milieu intérieure*'.

LEUTENEGGER [1972] plotted fetal versus maternal weight for 15 simian species. Fetal weight scales strongly as the 0.70 power of maternal weight. I note that this is very close to the standard metabolic exponent of 0.75 and wonder if the similarity is more than coincidental. LEUTENEGGER used his results to estimate, via a long series of additional assumptions, the size of newborn australopithecines.

Other studies have considered ontogenetic and intraspecific allometries of single species. The strong positive allometry of facial length versus cranium length in baboons had been one of HUXLEY's [1932, pp. 18–19] prime examples of allometry. HUXLEY computed an exponent of more than 4.0 for a regression mixing ontogenetic and static adult data within single species; FREEDMAN [1962] has confirmed these results for several species. LUMER and SCHULTZ [1941] found differences in pre- and postnatal relative growth curves for limb dimensions in macaques.

C. Causes of Bivariate Brain Allometry

Since the overwhelming attention of allometric studies has been devoted to brain-body relationships, we should consider the causes of commonly-observed parameters of scaling.

The interspecific ordinal scaling of brain weight is so strong and general that we must regard it as a primary law of morphology. Values of the exponent are consistently found to be within 0.02 or 0.03 of 0.66. Moreover, the correlation is very strong; for all mammals, r is 0.95, indicating that 90% (r^2) of the variation in brain weight is related to differences in body weight. Astonishingly little attention has been devoted to the meaning of 0.66 scaling, despite the pervasiveness and importance of this relation. Many pious comments remind us that 0.66 implies a determination of brain weight by body surfaces rather than by volumes, but no one has really tried to explain why this should be so. We cannot even be sure that the presumed surface-dependency is more than coincidental; after all, surfaces scale to the 0.66 power of body weight only when geometric similarity is maintained with size increase. But this rarely occurs; that, after all, is what allometry is all about. I regard the 0.66 scaling of brain weight as a major, unresolved mystery in the study of morphology.

The ontogenetic curves seem easier to explain, at least in developmental terms: high prenatal exponents reflect the proliferation of neurons as the relatively enormous fetal head grows. The curved intermediate portion marks a slowing down of neuronal production, while the nearly flat later postnatal curve represents that portion of ontogeny during which few or no new neurons are added to the brain. KERR *et al.* [1969] emphasize the unusual nature of the brain's ontogenetic development compared with that of other organs. They computed body and organ weights at 50, 75, 100, 150, and 175 gestational days and at adulthood in *Macaca mulatta*. The allometric relation of most organs to body weight is a single straight line passing from pre- to postnatal growth without interruption. Yet the brain undergoes 'marked postnatal decrease in relative growth rate' [1969, p. 211]. Most organs have reached only 10% of their adult weight by 175 gestational days (20% for spleen and thyroid and 40% for adrenals); but the brain has, at that point, already reached 65% of its adult weight and must, thereafter, slow down in its development.

The static, intraspecific scaling of 0.2–0.4 for adults of a single population or of very closely related forms remains unexplained [SCHOLL, 1948; BAUCHOT, and STEPHAN, 1964]. Several proposals work for some cases, but cannot provide a general explanation:

(1) For intrapopulational curves, we can argue that most variation in weight represents differences between fat and thin animals, rather than intrinsic variation in size. Since there is no correlation between brain weight and the nutritional component of body weight, a potentially higher factor of scaling will be diluted and reduced by introducing this component. This argument will not work for interpopulation allometries in which, as in breeds of domestic dogs [WEIDENREICH, 1941], each point is an average adult of a different breed or subspecies.

(2) JERISON [1963] argued that 'progressive' mammals of the same body plan should possess the same number of 'extra neurons' and add brain weight with increasing body size only along the 'primitive' curve $y = 0.12 \times {}^{0.66}$.

If you add a constant to a regression scaling at 0.66 and then fit a new curve (to data admittedly somewhat curvilinear), the value of the exponent will be reduced. Although this explanation might apply to certain 'progressive' mammalian groups, it cannot work for reptiles and other groups on 'primitive' curves. PLATEL and BAUCHOT [1970] have computed an intrapopulational exponent of 0.33 for the reptile *Scincus scincus*.

(3) GEIST [1973] and JERISON [1973] have argued that brain size generally increases more slowly than 'expected' (meaning by expected, I assume, the interspecific exponent of 0.66) in evolutionary sequences leading to animals of larger size. This might explain some interpopulational allometries in which forms are arranged in ancestral-descendant sequences; it will not explain the intrapopulational exponents. In discussing the low cephalization of the giant Pleistocene Malagasy lemur, *Megaladapis edwardsi*, JERISON [1973] writes: 'As selection pressures towards gigantism were experienced, the evolutionary systems that responded did not have to include the brain and sensory systems to the same extent as, for example, skeletal and muscle systems. The increase in brain size necessary to accommodate a larger body in which the basic plan of neural and sensorimotor organization follows a species-specific pattern would be less than that necessary to accommodate a larger body of a randomly selected mammalian pattern.'

(4) I would argue that the interpopulational curve is a consequence of the intrapopulational pattern, i.e. selection operates upon the variability present in a parental population in order to produce breeds or subspecies of smaller or larger average size. The intrapopulational pattern, in turn, may be a consequence (in part) of the fundamental ontogenetic relationship. If small and large individuals within a population have, in general, travelled a smaller and greater distance along the postnatal part of the ontogenetic curve respectively, then intraspecific differences in adult size may scale along ontogenetic trends. However, the postnatal slope tends to be slightly flatter than the intraspecific slope.

D. Allometry in Multivariate Perspective

In keeping with the revolution in morphometrics spawned by the electronic computer, multivariate approaches to allometry have proliferated in recent years [JOLICOEUR, 1963; MOSIMANN, 1970; SPRENT, 1972; CORRUCCINI, 1972]. In precomputer days, multivariate approaches were limited either to pictorial representations of the type popularized by D'ARCY

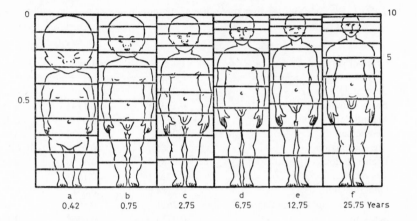

Fig. 5. The direction of human ontogeny as depicted by MEDAWAR [1945, p. 176]. MEDAWAR used these one-dimensional Thompsonian transformation grids to construct simple functions for the direction of overall shape in human growth.

THOMPSON in his method of transformed coordinates [KUMMER, 1951] or to simplifications portraying general changes in shape by elementary mathematical functions. In the second category, MEDAWAR [1945] tried to express the total allometry of external body form in human ontogeny as a single vector of change (fig. 5). He presented the following generalizations [1945, pp. 180–181]: (1) Shape is a continuous function of age. (2) The allometry in heights of various body parts is graded and monotonic, 'a simple tapered stretch', in MEDAWAR's terminology – basically, the postnatal head grows more slowly than the legs, with intermediate values in between. KROGMAN [1972, p. 4] has expressed this observation by stating an 'over all formula for the postnatal growth of the human body: 2–3–4–5' – i.e. in postnatal growth, the head and neck double their length, the trunk increases by 3 times, the arms by 4 and the legs by 5. (3) The rate of change of shape decreases progressively with time.

Both techniques have been recast in quantitative, multivariate perspective. SNEATH [1967] has quantified transformed coordinates as a variety of trend-surface analysis; he uses this method to compare the skulls of pongids with fossil and modern man. Attempts to express complex changes in shape with simple functions have yielded to the panoply of multivariate techniques that extract simplified patterns from a full account of raw data for many variables and samples.

Table I. Eigenvectors for first three principal components of factor analysis for 19 species of prosimians

Original variable	1st axis	2nd axis	3rd axis
Body weight	0.2651	−0.1486	−0.5271
Brain weight	0.2737	−0.1144	0.1945
Medulla oblongata	0.2717	−0.0330	−0.3822
Cerebellum	0.2727	−0.1429	0.0819
Mesencephalon	0.2685	−0.0799	−0.3315
Diencephalon	0.2735	−0.1230	0.0577
Telencephalon	0.2725	−0.1193	0.3152
Olfactory bulb	0.2184	0.8317	−0.0003
Paleocortex and amygdala	0.2611	0.4012	0.1291
Septum	0.2738	0.0568	−0.1268
Striatum	0.2731	−0.0270	0.3099
Schizocortex	0.2725	−0.0230	−0.0907
Hippocampus	0.2709	−0.1061	−0.0630
Neocortex	0.2691	−0.1979	0.4219

JOLICOEUR [1963] pointed out that the first principal component of the correlation matrix of logarithmically transformed data could serve as a general test for isometry in all variables simultaneously. In matrices for similarly designed animals differing widely in size, the first principal component is invariably a size factor expressing the changes of all variables with general increase in size [e.g., VAN GERVEN, 1972, on human femora]. Its eigenvector is a test for isometry; if all original variables have equal values upon it, then increase in size is occurring without corresponding change of shape. Table I, for example, shows the first three eigenvectors for 14 weights of the brain and its parts in 19 species of prosimians. These are based on the data of STEPHAN *et al.* [1970], the same set analyzed in a different way by SACHER [1970]; I have included *Tupaia* among the insectivores, rather than among the prosimians. The first principal component of logarithmically transformed data is, indeed, the expected size factor; all original variables project positively and nearly equally. General size increase of the brain is very close to isometric; only the olfactory bulbs have a distinctly lower projection, indicating that large prosimians do not have correspondingly large olfactory bulbs. It is a well-appreciated fact that the size of olfactory bulbs in primates does not vary nearly so much as that of most other parts of the brain. It is a special advantage of this technique for allometric studies that subsequent principal components are orthogonal to the first – i. e., they express differences in shape that are mathematically independent of the general size factor. The second eigenvector contains high projections only for olfactory bulbs and paleocortex, the most prominent of the 'ancient' parts of the brain [STEPHAN and ANDY, 1964]. There is, apparently, a joint component of their variation that is independent of general increase in brain size among prosimian species. The third eigenvector is equally interesting. Its high projections (negative in this case) are for body weight and medulla oblongata. The general size factor (first principal component) is one of brain size. Therefore, the third eigenvector expresses a component of body size that is independent of the brain. Only the medulla and, to a lesser extent, the mesencephalon are related to it in a positive sense. The study of allometry in primate brains has been hampered by the difficulty or impossibility of making accurate estimates of body size, especially in fossils. RADINSKY [1967] suggested that the area of the foramen magnum be used as a surrogate for body size, but (unfortunately) the primary correlate of this area seems to be with brain size itself (see pp. 263). We would greatly value any indication of body size that can be made on the cranium or its contents and that are not artifacts (as is the foramen magnum) of the entire brain's correlation with body size. Perhaps the medulla can serve this function. JERISON [1973] on other grounds, has suggested that it might.

III. Allometry as a Criterion for Judgment

In the last section, I discussed allometric trends for their intrinsic interest, particularly in terms of explanations that can be given for the parameters of power functions. But there is another, and probably more common, way in which allometric trends are used. Consider the following problem: two animals of the same general design but differing in size display numerous differences in shape. How are we to judge these differences in shape? Do they represent the mechanical necessities of increased size (relatively thick limb bones in the larger animal)? Do they reflect differences in adaptive strategies that bear no special relation to body size? We need criteria for judgment and the appropriate allometric curve offers the best starting point. It is, so to speak, a 'criterion of subtraction' that allows you to remove the pervasive influence of size from consideration. If both animals lie on the allometric trend, then their differences in shape may only express their differences in

size. If they do not, another type of adaptation is implied; even in this instance, the allometric component can be assessed and removed, thus identifying the nonallometric adaptation more clearly.

In the first category of differences in shape reflecting little more than
differences in size, WEIDENREICH [1941] argued, overenthusiastically, to be
sure (STARCK, 1953], that the relative size of the brain plays a dominant role
in the morphology of the skull. With the brain's negative allometry, absolutely small heads must contain relatively large brains and a host of com-
pensatory modifications as well. HEMMER [1966] attacked BROOM's distinction of Bushmen and Hottentots by the large skull and dolichocephaly of
Hottentots. The two features are both simple consequences of differences in
size (large heads are relatively long due to negative allometry of the brain);
moreover, many individuals are intermediate in size between supposed aver-
ages for the two groups. 'A division of Bushmen and Hottentots on the basis
of the length/breadth index is nothing more than a division according to
absolute size' [HEMMER, 1966, p. 202]. BIEGERT and MAURER [1972] plotted
the ratios of arm and leg length to skeletal trunk length versus skeletal trunk
length for anthropoids. Plots of such dimensionless ratios versus size display
allometry whenever any significant slope is measured on either linear or log
scales. They found strong positive allometry for both relationships, with arm
lengths increasing more rapidly than leg lengths with increasing size. They
conclude [1972, p. 144]: 'If *Gorilla*, in comparison with *Macaca*, has longer
arms and legs, this is only the expression of a general allometric tendency
among catarrhines, and not a specialization.'

In the second category of allometry as a criterion of subtraction,
SCHULTZ [1940] noted that nocturnal lemurs stand well above the standard
primate curve for strong negative allometry of eye weight to body weight.
SCHULTZ [1953] found that the relative thickness of lumbar vertebrae in bipedal man even surpasses that of the gorilla. STARCK [1953] noted the strong
positive deviation of *Daubentonia* and *Tarsius* from the brain/body curve for
lemuroids. BIEGERT and MAURER [1972] used their plots of relative arm and
leg length versus size as criteria for judging which primate species are
specialized for these characters. Above the line for arm length stand *Hylo-
bates*, *Symphalangus*, *Pongo*, and *Oreopithecus* (thus supporting a common
interpretation for this enigmatic fossil). *Homo* (as a response to requirements
of upright posture) and the two genera of modern gibbons stand above the
regression for leg length. Gibbons apparently have an unusually small body
with respect to their limbs. In a multivariate study, SACHER [1970] found that
factor scores for *Tarsius* placed it outside the prosimian envelope and nearer

to anthropoids for various measures of brain size. CORRUCCINI [1972] has presented a general method for removing allometric effects in multivariate studies: he illustrates his procedure with a study of ten dental measures in hominoids.

We return, inevitably, to the brain. Among the more persistent effects of anthropocentrism, we must count our desire to measure levels of cephalization, and thus to affirm in rigorous terms our obvious superiority. In no other area has such a diligent use of allometry as a criterion of subtraction been made; for we are convinced that once we learn how to partial out the influence of body size upon brain size (man, after all, is quite a large animal), man will stand alone on the pinnacle of cephalization. We will examine three approaches recently used to study cephalization in primates.

1. Radinsky's Measure of Foramen Magnum

RADINSKY [1967] proposed that the area of the foramen magnum might stand as a good surrogate for body weight in the study of brain allometry. He produced respectably log linear plots illustrating the relation of endocranial volume to foramen magnum area in several orders of mammals, including primates. In 1970, he applied his technique to assess the cephalization of fossil prosimians. He obtained the following results for the two oldest primate endocasts that also yielded data on foramen magnum area. *Smilodectes gracilis* of the North American middle Eocene lies below the curve, while *Adapis parisiensis* of the European late Eocene lies right on it. Of seven fossil species, including three giant Pleistocene lemurs from the Malagasy Republic, four lie below the curve for modern prosimians, two above and one right on (fig. 6). Thus, we obtain no evidence for progressive cephalization within the Prosimii; of the oldest fossils, only *Smilodectes* lies (very slightly) outside the range of modern values.

RADINSKY's measure has one great strength and one apparent problem. As its strength, it vastly increases the domain of data available for rigorous analysis, especially in paleontology. We only have cranial information for some species; for others, any estimate of body weight must be based subjectively on scraps of doubtfully associated postcranial material. As its problem, the area of the foramen magnum is probably not a 'pure' measure of body size. Partial correlation analysis [WANNER, 1971; JERISON, 1973] indicates that foramen magnum area is dependent *both* upon body size and absolute brain size. A plot of endocranial volume against foramen magnum is, in part, a plot of the brain against itself. This may explain the tightness of some of the regressions. It will also lead to strong underestimation of degrees of depar-

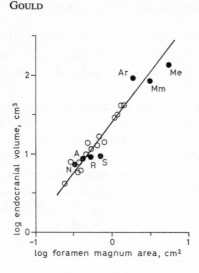

Fig. 6. Endocrinal volume versus foramen magnum area for fossil and Recent prosimians; from RADINSKY [1970, pp. 222]. A = *Adapis parisiensis*; Ar = *Archaeolemur majori*; Me = *Megaladapis edwardsi*; Mm = *Megaladapis madagascariensis*; N = *Necrolemur antiquus*; R = *Rooneyia viejaensis*; S = *Smilodectes gracilis*.

ture from general trends of brain-body allometry. Perhaps the fossil lemurs do not really lie so close to the modern prosimian brain-body curve. ASHTON and SPENCE [1958, p. 178] affirm JERISON's doubt on other grounds: 'As the growth pattern of the foramen magnum is practically indistinguishable from that of the endocranial cavity, it follows that age changes in the size of the brain may be deduced from a study of those in the foramen.' I can supply a further argument, founded upon dimensional analysis, to buttress this claim; I base this upon unpublished data kindly supplied by Dr. RADINSKY: If the area of the foramen magnum is primarily determined by brain weight rather than body weight, we would expect it to scale in the following way:

$$\text{brain weight } a \text{ (foramen magnum area)}^{1.50}.$$

For six groups (insectivores, rodents, prosimians, monkeys, artiodactyls, and carnivores) including 164 species, the value of the coefficient ranges from 1.37 to 1.58 with a mean value of 1.48. Therefore, foramen magnum area scales in geometric similarity with brain size and may be controlled by it; we may also assume that brain weight scales as the two-thirds power of body weight (RADINSKY's average is 0.67 for his 6 groups). From this we may predict a value for the regression of foramen magnum area on body weight:

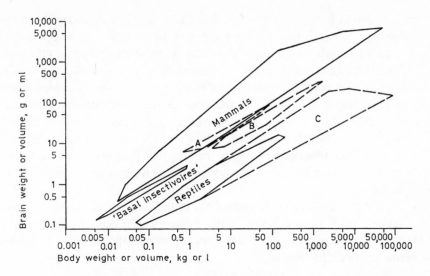

Fig. 7. JERISON's technique of 'minimum convex polygons' for the study of brain-body allometry; from JERISON [1970, p. 237]. Polygons have approximate slopes of 2/3, as expected. A and B represent archaic Tertiary mammals; C is the polygon for fossil reptiles. The 'basal insectivores' of BAUCHOT and STEPHAN lie below the mammalian polygon and, when extended, join the polygon for primitive fossil mammals. Living species: solid polygons; fossil species: dashed polygons. A = creodonts; B = condylarths and amblypods; C = therapsids and dinosaurs.

brain weight α body weight $^{0.66}$,
f.m. area α brain weight $^{0.66}$,
brain weight $^{0.66}$ α (body weight $^{0.66})^{0.66}$,
f.m. area α body weight$^{0.44}$.

RADINSKY's calculated value for a lumped sample of all 164 species is 0.42.

2. Jerison's Minimum Convex Polygons and Cephalization Quotients

JERISON [1970a, and 1973] has devised a semi-quantitative method for displaying brain/body allometries that avoids a reliance upon very precise (and probably incorrect) values for allometric exponents. He simply draws a polygon connecting extreme points and thus encompassing all brain/body points for the group under analysis (fig. 7). These interspecific polygons are invariably elongated with their major axes at slopes of approximately 2/3, thus affirming both the orthodox values and the primary orderliness of brain/body data. They provide many important insights in the use of allometry as a criterion of subtraction. *Tupaia* lies in the lower range of the prosimian polygon. *Australopithecus africanus* is well above the simian polygon; it would need a body weight of 60 kg (well above any conceivable estimate) to reach the uppermost margin of the simian polygon. Of the five oldest primate endocasts, *Adapis parisiensis, Smilodectes gracilis* and *Tetonius homunculus* lie below the polygon of modern prosimians; *Necrolemur antiquus* is within it, and *Rooneyia viejaensis* above. RADINSKY may well have underestimated his deviations by using foramen magnum areas as an estimate of body size. Though it avoids RADINSKY's problem of artifacts, JERISON's method for the treatment of fossils requires that body weights be estimated by such subjective procedures as comparison with modern forms of similar body design.

To attain a quantitative measure of cephalization that position within a polygon cannot supply, JERISON [1970b] devised an cephalization quotient (EQ) in the following manner. The general interspecific curve for all modern mammals (with its high correlation coefficient of 0.95) is:

brain weight $= 0.12$ (body weight)$^{0.66}$.

EQ is simply the ratio between the actual brain weight of any individual sample and the brain weight predicted by this basic equation for an 'average' mammal *of the same body weight*, i.e.

$$EQi = \frac{Ei}{0.12\,(Pi)^{0.66}},$$

where E is brain weight, P body weight and i the sample under consideration. This measure provides another set of important insights, this time in more rigorous formulation. The mean EQ for prosimians lies near 1: i.e. prosimians have the same brain size as average mammals. Other primates have distinctly higher EQs. Man's maximal value of 6.3 puts him within dolphin range. STEPHAN [1972, p. 174] cannot be supported in his claim that 'a high encephalization is not a specific characteristic of primates ... man is the only primate with an outstanding brain size.' Cebids, cercopithecids and pongids may not display astronomical cephalization, but their values of EQ are so consistently greater than 1 that some strong generality of higher cephalization must be asserted (fig. 8). JERISON [1973] has also analyzed the Paleogene primate endocasts with his EQ statistic. Since he has accumulated a great amount of data on brain/body allometry of fossil mammals in many orders, he can, for the first time, compare fossil primates directly with their contemporaries. 'Archaic' mammals of the first portion of the Paleogene had EQ values averaging about 0.20 (i.e., their endocasts were 1/5 the size of values for average modern mammals of the same body weight). Progressive carnivores and ungulates attained values of 0.5 by lower Oligocene times.

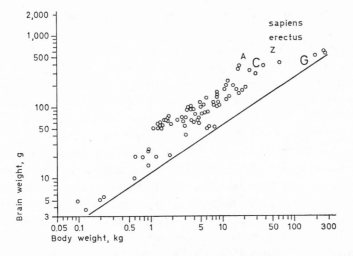

Fig. 8. The high general cephalization of primates; from JERISON [1963, p. 277]. The solid line is the mouse-to-elephant curve for all mammals. Primates also lie on a slope near 0.66, but their line is clearly above that for mammals in general. C is the domain of chimpanzees, G of gorillas. A = *Australopithecus africanus*; Z = '*Zinjanthropus*'.

Yet the early Eocene *Tetonius homunculus*, the oldest primate endocast, has an EQ of 0.68 – a value 'several times as great as that of any of the other mammals of that period' [JERISON, 1973]. Although 0.68 is low for a modern prosimian – it is at the 10th percentile of a cumulative frequency distribution of EQ for modern prosimians – it is a remarkably high value for a mammal of its time and indicates that high cephalization has been a hallmark of primates from the early part of their history. Three other Eocene prosimians, *Smilodectes gracilis*, *Adapis parisiensis*, and *Necrolemur antiquus* had EQ values of 0.55, 0.58, and 1.22 respectively. The Oligocene *Rooneyia viejaensis* has a remarkably high EQ of 1.75. This value is higher than that of *Tarsius* (1.53) to which it is presumably related; among modern prosimians, only *Daubentonia* exceeds 1.75 in EQ. JERISON [1973] remarks: '*Rooneyia* had actually achieved a relative brain size that was within the living anthropoid (simian) range... For a lower Oligocene species, this was something special. Its brain was actually 3 or 4 times the size typical for the mammals of its time.' For later Neogene endocasts, JERISON could detect no differences in EQ between fossil and modern nonhominid anthropoids. He concludes that there is no evidence for progressive increase of cephalization in cercopithecids during the Neogene.

3. Bauchot, Stephan and Their Colleagues on Comparisons with 'Basal Insectivores'

In a long series of excellent papers, BAUCHOT, STEPHAN and their colleagues use a technique conceptually similar to JERISON's EQ criterion for the judgment of cephalization [BAUCHOT, 1972; BAUCHOT and STEPHAN, 1964, 1966, 1968, 1969; PIRLOT and STEPHAN, 1970; STEPHAN, 1972; STEPHAN and BAUCHOT, 1965; STEPHAN et al., 1970; STEPHAN and ANDY, 1964, 1970; STEPHAN and PIRLOT, 1970]. Where JERISON used the mass curve for all mammals as his criterion, BAUCHOT and STEPHAN construct a brain/body curve for 'basal insectivores' – essentially shrews, tenrecs, and hedgehogs. In these forms 'not only the total brain but also all progressive structures ... are quantitatively the least developed ... They can be expected to be still comparatively similar to the early forerunners of the placental mammals, and therefore to represent a good base of reference for evaluating evolutionary progress' [STEPHAN, 1972, pp. 157–158]. Their index of progression (IP), the measure similar in principle to JERISON's EQ, is simply the ratio between brain size for a form under consideration and the predicted value for a basal insectivore of that body size. In theory, the 'basal insectivore' method raises many doubts and problems:

(a) Shrews, tenrecs, and hedgehogs are modern groups with modern specializations; they are not a series of models of Ur-mammals at different sizes. The brain/body curve of basal insectivores is largely a comparison between small shrews and larger tenrecs and hedgehogs. If this comparison reflects modern specialization rather than simple inheritance from an idealized primal mammalian design, then the regression is no reflection of any basic mammalian pattern.

(b) The largest basal insectivore is still quite a small animal. To compute the IP of a rhinoceros, we must compare its brain with a value on the basal insectivore curve extrapolated more than two orders of magnitude from any real data. Extrapolation is always a dangerous game; distant extrapolation is even worse.

In practice, however (and thankfully), the basal insectivore curve seems to serve its designated function remarkably well. It has a slope of 0.63, reassuringly near the predicted value of 2/3. Moreover, an extrapolation of the basal insectivore curve passes right through the polygon for fossil condylarths and amblypods [JERISON, 1970] (fig. 7), an 'archaic' group of large animals: 'A line through the basal insectivore polygon and the condylarth-amblypod polygon is readily drawn, and it can have a slope of 0.63 as required by Stephan's analysis' [JERISON, 1970, p. 238].

Yet BAUCHOT and STEPHAN carry their technique much further; for they have laboriously calculated the weights of major brain divisions (medulla, cerebellum, mesencephalon, diencephalon, telencephalon) and of subdivisions of the telencephalon (olfactory bulbs, paleocortex, septum, striatum, schizocortex, hippocampus, and neocortex) for all their species of insectivores and primates. Their valuable matrices of data are presented in full in STEPHAN et al. [1970].

Analysis by parts opens a large new domain of insights. The following figures provide a quantitative account of the importance of neocortical enlargement in primate evolution: Values of IP for the entire brain range from 2.4 to 7.0 *(Lepilemur* to *Daubentonia)* in prosimians, 4 to 12 *(Alouatta* to *Miopithecus talapoin)* in simians; man stands at 28.8. IPs for the neocortex alone average 14.5 in prosimians, 45.5 in simians, and 156 in man.

STEPHAN and ANDY [1964] have summarized the status of brain parts in their 'ascending primate scale' (a ranking based on IP values for the neocortex). Of the major divisions of the brain, the telencephalon (unsurprisingly) increases most strongly and consistently within this ranking. It is followed by the diencephalon and cerebellum, both with similar patterns of 'progression'. Parts of the telencephalon may be roughly divided into three groups. The striatum ranks most closely to the neocortex in the first group. Parts in the second group (hippocampus, schizocortex, and septum) increase slightly or irregularly on the 'ascending primate scale'. In the third group, olfactory bulbs tend to regress (decrease in IP) while the paleocortex shows no change.

The value of allometric regression upon body size, and of comparison with a 'basic' curve, is strongly affirmed in the following example: STEPHAN and ANDY [1970, p. 127] document the common assumption that the septum of the telencephalon undergoes reduction during primate evolution and especially in man. But measures of IP prove just the opposite: the septum undergoes a small but definite increase in the 'ascending primate scale' and in man himself. The impression of reduction was created by the more rapid enlargement of other parts of the brain, especially of the neocortex.

BAUCHOT and STEPHAN [1968, and several other papers in the series] use measures of IP to correlate the status of brain parts with evolutionary specializations. They found that five species of semi-aquatic insectivores share a large number of features with more advanced primates – large IP values for the total brain produced largely by neocortical enlargement and despite reduction of the olfactory bulbs (fig. 9). (There are, of course, several differences as well: the insectivores have poorly-developed vision and a large

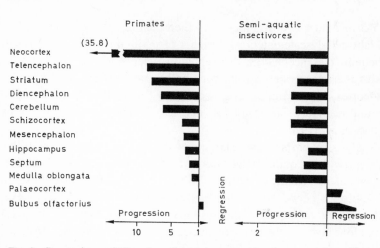

Fig. 9. Comparison of IP values for brain parts of primates and semiaquatic insectivores; from BAUCHOT and STEPHAN [1968, p. 268].

medulla that can be traced to the development of vibrissae and an enlarged trigeminal nerve). BAUCHOT and STEPHAN [1968, p. 267] ascribe these similarities to the need for better integration and organization in 'three dimensional environments' (trees and water).

4. Multivariate Analyses of the Data of Stephan, Bauchot and Andy

SACHER performed a factor analysis on all species (insectivores, prosimians, and anthropoids) and variables listed in STEPHAN *et al.* [1970]. His results are an elegant epitome of long verbal and bivariate arguments presented, for example, in STEPHAN and ANDY [1964]. Table II shows factor loadings on the first two varimax axes for logarithmically transformed data. Neocortex has the highest loading on the first axis. Of major brain divisions, cerebellum and diencephalon are next. STEPHAN and ANDY [1964] state that these two divisions follow the neocortex in degree of evolutionary advance on the 'ascending primate scale'. STEPHAN and ANDY [1964], previously discussed on p. 269 of this article, divided the parts of the telencephalon into three groups based upon the extent and consistency of their increase in the 'ascending primate scale'. Neocortex and striatum were in the first group; these have the highest loading on the first axis. Septum, schizocortex, and hippocampus were next; these load with intermediate values. Olfactory bulb and paleocortex occupied the lowest group; these have the smallest loadings on the first axis. The correspondence is complete. SACHER's second axis is

Table II. Loadings on varimax factor axes for 63 species of insectivores, prosimians, and anthropoids [from SACHER, 1970]

	Axis 1	Axis 2
Medulla oblongata	0.942	0.304
Cerebellum	0.969	0.235
Mesencephalon	0.956	0.262
Diencephalon	0.977	0.203
Olfactory bulb	0.183	0.912
Paleocortex and amygdala	0.864	0.464
Septum	0.937	0.340
Striatum	0.978	0.187
Schizocortex	0.924	0.358
Hippocampus	0.934	0.330
Neocortex	0.990	0.097
Brain weight	0.981	0.188
Body weight	0.889	0.380

almost an inverted image of the first: olfactory bulbs and paleocortex have highest loadings; septum, schizocortex, and hippocampus are intermediate; neocortex and striatum are low. SACHER [1970, p. 267] concludes: 'We have to do with a strong factor for neocorticalization and another, almost as strong, related to the paleocortical-olfactory-limbic structures.'

Factor scores of all species on these two axes are roughly arranged in three 'distinct strata' [SACHER, 1970, p. 269], corresponding to insectivores, prosimians, and anthropoids (though *Tarsius*, as previously mentioned, is closer to anthropoids than to prosimians). Since neocortex dominates axis 1 while olfactory bulb controls axis 2, SACHER [1970, p. 269] concludes: 'Insectivora, Prosimii and Anthropoidea form three distinct strata with respect to olfactory lobe-neocortex relationships. The transition from one group to the next may be approximately described as an increase of the ratio of neocortex to olfactory lobe, due primarily to increase in neocortical volume. The fact that this ratio has only two discrete increments suggests that a change in the ratio requires a major reorganization of the brain.' This represents a reincarnation, in modern guise, of the primal argument in allometric studies of the brain: that the evolution of the vertebrate brain proceeds by discrete shifts from one level to another. The empirical base of the original claim rests upon the observation, abundantly reaffirmed since its proposal by DUBOIS, that interspecific curves tend to have the same slope (0.66 in our best modern estimates), but different y-intercepts. (On log-log

Table III. Varimax factor loadings for insectivores, prosimians, and anthropoids

	Insectivores		Prosimians		Anthropoids	
	axis 1	axis 2	axis 1	axis 2	axis 1	axis 2
Body weight	0.542	0.717	0.888	0.397	0.741	0.599
Brain weight	0.768	0.587	0.841	0.444	0.873	0.463
Medulla oblongata	0.715	0.539	0.818	0.479	0.841	0.483
Cerebellum	0.790	0.539	0.891	0.423	0.858	0.484
Mesencephalon	0.789	0.543	0.823	0.441	0.866	0.465
Diencephalon	0.816	0.531	0.882	0.435	0.870	0.472
Telencephalon	0.760	0.613	0.894	0.441	0.874	0.463
Olfactory bulbs	0.458	0.872	0.356	0.927	0.449	0.886
Paleocortex and amygdala	0.542	0.823	0.650	0.738	0.849	0.483
Septum	0.730	0.612	0.808	0.546	0.810	0.522
Striatum	0.799	0.548	0.862	0.500	0.876	0.461
Schizocortex	0.820	0.518	0.835	0.494	0.797	0.504
Hippocampus	0.780	0.558	0.867	0.442	0.823	0.463
Neocortex	0.895	0.399	0.919	0.388	0.875	0.462

scales, two parallel lines have the same ratio of y-values [brain weights] at *any* x-value [body weight]. Thus, DUBOIS and his school believed that the ratio of y-intercepts [b-values of power functions] reflected the extent of this discrete advance in evolutionary organization.)

I have also analyzed the data of STEPHAN *et al.* [1970] in a different way. An appropriate multivariate technique depends upon the question asked. SACHER [1970] considered insectivores, prosimians and anthropoids as a single system representing evolutionary advance. Since factor analysis assumes that all samples are drawn from a single statistical universe, his procedure is appropriate for his perspective (though no one would claim that the sample is random – an impossible demand when samples are species with unique adaptations in taxonomic groups of relatively small total membership). The existence of three distinct strata may, however, suggest a fundamental inhomogeneity better treated by separate analyses of each of the three levels. Moreover, we cannot be sure from SACHER's analysis whether the strong distinction between neocortical and paleocortical-olfactory-limbic components of the brain arises only from variation among the three levels [where we already know it exists from verbal descriptions of STEPHAN and ANDY, 1964], or whether this among-level variation is an extension of patterns

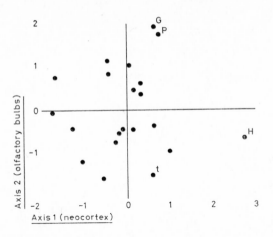

Fig. 10. Factor scores for 21 anthropoid species on the first two varimax axes (dominated by neocortex and olfactory bulbs respectively) of a factor analysis, performed by the author on data of STEPHAN *et al.* [1970]. G = Gorilla; P = Chimpanzee; H = *Homo sapiens*; t = *Cercopithecus (Miopithecus) talapoin*, a highly cephalized old-world monkey.

present within the individual groups. We cannot sort 'within' from 'among' variation in his analysis. The three levels might seem somewhat less distinct if their differences involved the same interrelationships of characters that control variation within groups. Consequently, I followed SACHER's procedure (factor analysis by BMD program O3M on log transformed data), but performed separate analyses for insectivores, prosimians and anthropoids. To assess differences among the three groups, I performed a stepwise discriminant analysis (BMDO7M) on the entire matrix (but placing *Tupaia* among the insectivores, rather than among prosimians).

Table III presents the first two varimax factor axes for the three analyses (98–99% of the total information in all three cases). In each case, the two factors that SACHER determined by analyzing the total sample are, again, clearly evident. Neocortex loads most strongly on the first axis in two of three analyses (second by a mere 0.0007 for anthropoids). With only one exception of a possible 15 (the schizocortex of insectivores), the five highest loadings for each analysis include only those variables identified by STEPHAN and ANDY [1964] as the strongest correlates of evolutionary 'progress' on the 'ascending primate scale' (i.e., brain weight, telencephalon, diencephalon, and cerebellum among major divisions; neocortex and striatum among parts

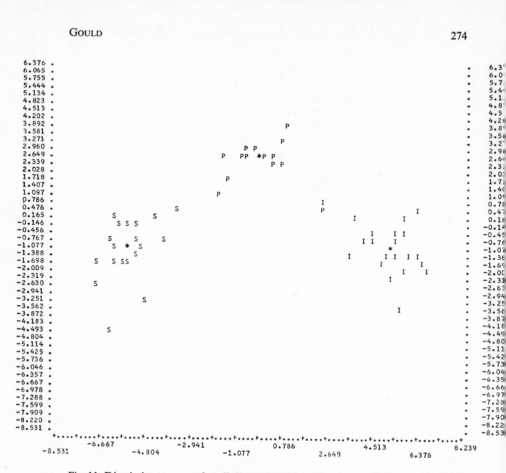

Fig. 11. Discriminant scores for all 63 species of insectivores, prosimians and anthropoids on the two canonical axes constructed to separate the three *a priori* groups; analysis by author from raw data of STEPHAN *et al.* [1970]. The separation is well axcomplished by very few variables; only a single species is misclassified. I = Insectivore; P = prosimian; S = anthropoid (simian); stars represent group centroids. Some points represent more than one species. Insectivores project strongly on the first canonical axis which includes telencephalon and neocortex with strong *negative* loadings.

of the telencephalon). For insectivores (in order): neocortex, schizocortex, diencephalon, striatum, cerebellum. For prosimians: neocortex, telencephalon, cerebellum, body weight, diencephalon. For anthropoids: striatum, neocortex, telencephalon, brain weight, diencephalon. In all three analyses, olfactory bulbs are, by far, the dominant element of factor 2. In two of three cases (not in anthropoids), paleocortex is a strong second. In the two primate groups, body size is strongly associated with the first axis; (this is not true

Table IV. F-values to enter step-up discriminant analysis for insectivores, prosimians, and anthropoids; first 5 cycles

	At outset	Neocortex removed	Schizo-cortex removed	Brain weight removed	Olfactory bulb removed
Body weight	28.21	22.32	5.43	1.79	1.16
Brain weight	67.04	39.35	22.25	–	–
Medulla oblongata	37.60	33.73	12.63	0.72	0.46
Cerebellum	54.41	46.46	8.12	6.24	5.35
Mesencephalon	49.67	21.67	3.31	0.07	0.95
Diencephalon	67.52	27.18	1.60	1.32	0.39
Telencephalon	75.96	24.93	17.67	4.13	4.25
Olfactory bulbs	3.79	44.91	5.76	16.37	–
Paleocortex and amygdala	22.84	27.34	8.86	0.73	7.21
Septum	37.30	44.80	7.35	0.91	0.10
Striatum	71.16	16.55	3.90	0.53	0.10
Schizocortex	33.70	68.72	–	–	–
Hippocampus	36.63	39.15	0.70	1.07	1.34
Neocortex	105.19	–	–	–	–

in insectivores in which several of the 'basal' species are large). This footnote for allometricians suggests that evolutionary 'advance' in the primate brain tends to occur in concert with increasing body size. In any event, within-group variation seems to be determined by the same interrelation of characters that analysis of the total matrix reveals.

Figure 10 presents, as an example of a revealing mode of depiction, factor scores for anthropoids plotted against the first two factor axes of table III. Man is clearly separated by his high score upon the neocorticalization axis of varimax factor 1. Chimpanzee, gorilla, and talapoin are high, as expected, but *Cebus albifrons* ranks second. Man is well separated from the great apes by the contrast of his weak, and their strong, projection on axis 2, dominated by the olfactory bulbs.

The stepwise discriminant analysis demonstrates that the three groups are well differentiated by a very small number of variables. Inclusion of all 14 variables results in a single misclassification among 63 species (fig. 11, the prosimian *Urogale everetti* is barely classified among the insectivores occupying a position adjacent to the expected outlier *Tupaia*). The use of only the four best discriminating variables – neocortex, schizocortex, brain weight,

Table V. Eigenvector for first canonical axis to separate insectivores, prosimians, and anthropoids

Variable	Axis
Body weight	− 1,41
Brain weight	57.21
Medulla oblongata	− 5.77
Cerebellum	− 2.71
Mesencephalon	− 3.47
Diencephalon	0.64
Telencephalon	−43.77
Olfactory bulbs	2.40
Paleocortex and amygdala	− 1.00
Septum	− 0.92
Striatum	− 0.30
Schizocortex	4.59
Hippocampus	1.74
Neocortex	− 7.14

and olfactory bulbs in that order – entails only two misclassifications. Table IV is a chart of changing F values (for entrance) as variables are sequentially included in the discriminant analysis. The initial ranking (column 1) shows the basic discriminating power of variables. It is difficult to make much sense of this chart alone because the effect of correlation is not removed. Neocortex is the best discriminator; olfactory bulbs by far the worst (since their variation in size among the three groups is so much smaller than that of any other character). However, when neocortex is entered, the revised values of F (column 2) now remove the effect of correlation between other variables and the neocortex; the olfactory bulb now has the third highest F value. With entrance of the schizocortex (column 3) and brain weight (column 4), very little discriminatory power remains in any variable but the olfactory bulbs which enter at the next pass. The remaining variables achieve their raw power of discrimination (column 1) primarily through their correlation with other variables previously entered. The step-up technique reveals a hierarchy of discrimination.

The first canonical variable neatly separates the three *a priori* groups and encompasses 88% of the total information of all samples. The raw eigenvector (table V) demonstrates that the separation of groups proceeds as we would have predicted from results of SACHER [1970] and STEPHAN and

ANDY [1964]. Brain weight is cast in strong negative association with telencephalon and neocortex – i.e., brains with a relatively small telencephalon and neocortex score most strongly on this axis. Hence (fig. 11), insectivores have the highest mean score, followed by prosimians and anthropoids.

IV. Scaling and Evolutionary Mechanisms

When we turn to the inference of evolutionary mechanisms from allometric data, we find that a fallacious, but deeply entrenched habit of thought has dominated past consideration. This is the almost automatic tendency to look towards the interspecific curve for evolutionary information. The famous 0.66 scaling of brain weight has prompted a multitude of speculations about evolutionary mechanisms for more than 80 years. Prominent among them, as detailed above (p. 250), is the claim that the brain 'improves' by discontinuous steps because the interspecific curves of 'lower' and 'higher' groups generally form a series of parallel lines on log-log scales.

Ironically, of all types of scaling, the interspecific curve is by far the worst to consult for information about evolutionary *mechanisms*. It represents the static scaling of contemporary adults within a group; it is not the result of ancestral-descendant sequences among the forms constituting it. It provides scaling criteria for the functional morphologist, but it does not express evolutionary variation within an evolving group. It can be used to predict blueprints for adaptation; it does not represent the path travelled by any organism towards that adaptation. To illustrate: a plot of leg width versus body length in any terrestrial group of bipeds or quadrupeds will show strong positive allometry. We may use the parameters of the power function as criteria for predicting the relative leg width of forms not yet discovered or evolved. If correlation is strong and the parameters represent well-understood physical principles (as in this case), we may even feel confident in extrapolating to sizes not represented by actual data. But the curve cannot tell us how our prospective phyletic giant will attain the predicted value; for it represents neither the ontogenetic path to adult values, nor the pattern of intraspecific variation within the ancestral population.

I suggest that we look to other types of scaling for ideas about evolutionary mechanisms. For example:

(1) Intraspecific scaling: The allometry of adults within a single population represents the correlated variability upon which evolution works. We can determine the rate of increase of organs and parts with body weight and

predict the proportions that might accompany larger size if size alone is the object of selection. If evolutionary sequences scale along an intraspecific curve, we may wish to conclude that ancestors and descendants are animals of the same basic design expressed at different sizes – i. e. descendants represent simple prolongations to different sizes of patterns in variation already present in ancestors.

(2) Geometric scaling: Proportional scaling with increasing or decreasing size is a violation of both ontogenetic and mechanical predictions that almost invariably include allometry. Yet it occurs often in evolution and is often under the control of simple hormonal and genetic mechanisms [GOULD, 1971b]. It can have rather surprising effects as an agent of evolutionary 'advance'. (I say surprising because another deeply-rooted, though fallacious, habit of thought indicates that no change of shape is no evolution.) The brain, for example, scales at 0.66 in a static series of contemporary adults. If it should increase in geometric similarity to body size within an evolving lineage, then a large descendant will have a relative larger brain than a large contemporary of its ancestor at the same body size.

(3) Ontogenetic scaling: Evolutionary size increase may occur by prolongation of ontogenetic trends. In complex curves with different parameters for juveniles and adults, prolongation or truncation of juvenile values yield the phenomena of paedomorphosis or recapitulation. These are of great significance in thousands of evolutionary sequences (including, probably, that of man), despite the general unpopularity of the subject these days.

I believe that each of these three types of scaling has played an important role in the evolution of brains in primates.

1. Intraspecific Scaling

Although prosimians, monkeys, and great apes generally form three interspecific regressions of slopes near 0.66 and increasing b-values, certain species occupy anomalous positions. Most outstanding are the high cephalization of the prosimian *Daubentonia madagascarensis* and the cercopithecid *Cercopithecus (Miopithecus) talapoin* and the low cephalization of the cebid howler monkey *Alouatta* and of *Gorilla* (the neocortical IP of *Gorilla* is only 32.1, even lower than that of the prosimian *Daubentonia*). BAUCHOT and STEPHAN [1969, pp. 267–268] offer the ingenious (and probably correct) proposal that the talapoin has such a relatively large brain because it is a phyletic dwarf, miniaturized (as are some small breeds of domestic dogs) along the intraspecific curve of ca. 0.23. This will provide it with a much larger brain than an average monkey of its body weight (which stands in a scaling relationship of 0.66 to the larger ancestor of the talapoin): 'It is probable that, in this case of miniaturization, the brain diminished in volume following the intraspecific curve of about 0.23... These miniaturized forms appear, in the value of their index of cephalization, as favored as juveniles of a species compared with adults. This last comparison shows that we cannot use indices of cephalization thus attained to infer that species have reached a superior level of brain evolution' [BAUCHOT and STEPHAN, 1969, pp. 267–268]. STEPHAN [1972, p. 162] offers the same explanation for *Daubentonia*.

So far so good. The obvious extension of the argument would mark *Gorilla* and *Alouatta* as phyletic giants, scaled up along the intraspecific curve [JERISON, 1973, has made a nonquantitative argument of this sort for *Megaladapis*, cited herein on p. 258]. But STEPHAN [1972], caught in a philosophical trap, shrinks from such a proposal. He is willing to allow for increase in cephalization by downward intraspecific scaling, but not decrease by upward scaling. He maintains his allegiance, as do many continental evolutionists, to a principle of idealistic morphology according to which adaptation to specific circumstances must be separated as an evolutionary event from general evolutionary direction: 'The size of the brain and its various divisions are influenced by two major factors: first the various changes inherent in the process of evolution and second, specialization such as presented by vision, olfaction, etc.' [STEPHAN, 1972, p. 72]. It is this belief in general direction that leads to such concepts as the 'ascending primate scale' (a notion foreign to most English and American evolutionists) despite the following proviso: 'It represents no direct evolutionary line, in which the respective higher forms have passed through the individual lower stages. However, it is very possible, and even likely, that one or the other of the lower forms is similar to those stages, through which the higher forms have evolved during phylogenesis' [STEPHAN and ANDY, 1970, p. 113]. The component of general direction simply will not allow for a decrease of cephalization during evolution: 'We do not believe that forms with lower levels of neocorticalization were derived from forms with higher levels ... neocortical development has never been retrograde during phylogeny' [STEPHAN, 1972, pp. 165, 174]. This leads to both an unusual assessment and to an elaborate proposal for the phylogeny of gorillas: 'Both genera *Gorilla* and *Alouatta* are obviously more primitive in their brain development than hitherto assumed' [STEPHAN and ANDY, 1970, p. 114]. 'The low position of the gorilla may be interpreted as follows: the Hominoidea probably radiated from forerunners with a level of neocorticalization that was less than that of the gorilla. After separation of the two lineages leading to recent gorillas and recent chimpanzees, the further development of higher centers of integration must have occurred at a much faster rate in the chimpanzee than in the gorilla' [STEPHAN, 1972, p. 165].

I see no reason why cephalization might not decrease in a variety of ecologic situations (other organs 'regress' after all). But more important, I do not see how scaling along an intraspecific curve can be interpreted as a deterioration of the brain. Large adults within a population seem to do as well as small ones, despite the scaling of 0.2–0.4 among them. When evolution follows an intrapopulational curve, allometric organs may be scaling as passive consequences of selection for size, i. e., ancestor and descendant may be the 'same' animal expressed at different sizes. The idea that *Gorilla* is a phyletic giant is not new [GILES, 1956]. Like any good primate, I will go out on a limb and predict that *Alouatta* is one also. SCHULTZ [1955] has shown that *Alouatta* exceeds all other ceboids in forward migration of the prosthion and backward movement of the occipital condyles during postnatal ontogeny; these are general trends in primate ontogeny. The large size of *Alouatta* may also be relevant in this context.

I am intrigued with the idea that intraspecific scaling may have played an important role in the evolution of man. If something like *Australopithecus africanus* represents a credible common ancestor for both the robust lineage of australopithecines and for the sequence leading eventually to us, then a brain-body curve for *A. africanus→A. robustus→A. boisei* scales strongly at 0.327 with r = 0.97 (based on brain and body estimates kindly supplied by D. PILBEAM). The static sequence *Pan paniscus→Pan troglodytes→Gorilla gorilla* scales equally well at 0.338 with r = 0.99. Could it be that the robust lineage merely represents the same 'creature' (or, better, the same level of adaptation) expressed at different sizes? Meanwhile, the *A. africanus→H. habilis→H. erectus→H. sapiens* lineage was scaling (with poorer correlation) at somewhere between 1.1 and 1.9 (estimates of THENIUS, TOBIAS and PILBEAM and spanning the extremes of conjecture for australopithecine body sizes). Scaling may, in fact, have been close to geometric; but, as argued previously (p. 278), this would represent a marked increase in brain size with respect to the 0.66 predictions of the interspecific curve of mechanical scaling.

HOLLOWAY [1972] has argued provisionally that the major reorganization of the human brain had occurred by the australopithecine level and that subsequent increase was tied to enlarging body size. (Previous statements that the *erectus→sapiens* sequence involved brain improvement with no increase in body size are being reassessed as various workers become convinced that the Trinl femur belongs to *Homo sapiens* rather than to '*Pithecanthropus*'. *Homo erectus* at Choukoutien was smaller than average *Homo sapiens*). But there is a marked difference between correlation and the parameters of regression. The correlation between brain weight and body weight in the sapiens lineage is very strong, but this does not imply that brain size is a passive consequence of increasing body size. Correlation does not imply determination. One must consider the parameters of regression; a slope greater than one can only denote a significant improvement in brain size. PILBEAM and I [in press] will explore more fully the consequences of allometric scaling in the evolution of man.

2. Geometric Scaling

Without doubt, the average brains of different phyletic groups vary greatly in internal arrangement. We cannot claim that they are related to each other by any type of proportional scaling. Nonetheless, the vast majority of evolutionary events are small changes within rather similar body plans; the potential scope for geometric scaling is very wide.

Many authors have commented (often with surprise) that brains of large and small animals within a related group are remarkably similar in the relative proportions of their parts. HARMAN [1957, pp. 14–15] wrote: 'Within orders of mammals, the increase in size of the neopallium is a remarkably constant phenomenon, the neocortical volume tending toward a constant percentage of the total brain volume in both Carnivora and primates ... Years ago the brain plan within groups was roughed out, and subsequent evolution took place without destroying the fundamental pattern'. JERISON [1970, pp. 240–241] commented upon the 'surprising fixity in relative size of the parts of the brain as a function of body size'. He notes that most logarithmic relations of parts of the brain to each other have an isometric slope of 1 (plots involving olfactory bulbs are generally exceptional): 'The implication of this argument is that it is, in principle, likely that given the total brain size one may predict with accuracy the size of any of the parts of the brain, and that, furthermore, the allometric functions for the total brain size and for the size of the parts as a function of body size may be considered to have equal slopes' [JERISON, 1970, p. 240].

These comments refer to interspecific relations. Geometric similarity of parts has also been recorded for ontogenetic and intraspecific variation. BRUMMELKAMP [1940a] divided the neocortex of growing sheep into five fields and showed that these retained strikingly constant proportionality throughout ontogeny: 'the neocortex develops equally in all its parts'

[BRUMMELKAMP, 1940a, p. 6]. For intrapopulational variation, STEPHAN *et al.* [1970, pp. 295–296] remark: 'According to our experience, extremely small or extremely large brains within one species have generally the same composition. There is no indication of directed change in composition related to the size of the brain.'

If, within a group, large brains are often geometrically similar to small brains, we obtain what might be called an 'enabling criterion' for evolution in true geometric similarity. If phyletic slopes for brain/body allometry scale with a slope of one, then large descendants will not only have the same brain /body ratio as small ancestors; the parts of their brain will also scale in constant proportion to each other. This is not a merely esoteric possibility. It could well represent a (if not the) major way to increase cephalization in evolution. Size increase is no rare event in evolution, and size increase in geometric similarity yields a marked increase in cephalization. It might even produce the famous set of interspecific parallel lines still generally interpreted as reorganization involving 'hidden' increases in the early ontogeny of descendants (fig. 12). Again, interspecific curves do not prescribe the evolutionary mechanisms that produced them.

WEIDENREICH [1941, p. 352] noted that dwarf wild dogs, unlike domestic breeds, are often close to geometrically similar with larger ancestors for brain/body relationships. PIRLOT and STEPHAN [1970, p. 436] calculated a high interspecific brain/body slope of 0.765 for bats. They ascribe the upward departure from 0.66 to a phyletic scaling that deviates from the interspecific curve in the direction of geometric scaling: 'Since the more primitive species are predominantly small, and the more developed species predominantly large, the slope tends to be high.'

Curves for insectivores, prosimians and anthropoids display the characteristic parallel pattern with increasing b-values. However, mean body size also increases from groups to group. In fact, if we connect points for mean body sizes on the three curves [from data of BAUCHOT and STEPHAN, 1969], the slope is indeed indistinguishable from 1, the criterion for geometric scaling (fig. 13). I have no desire to fall into my own trap and equate interspecific scaling of modern animals with phyletic events: an average modern insectivore is not a prosimian ancestor. Nonetheless, there has been both size increase and advancing cephalization in true lineages within the general sequence; the two events may not be unrelated. Elsewhere [GOULD, 1971b], I have presented a method for computing criteria of geometric similarity from sets of allometric curves with equal slopes.

BAUCHOT [1972] has recently tested my claim [GOULD, 1971b] that size

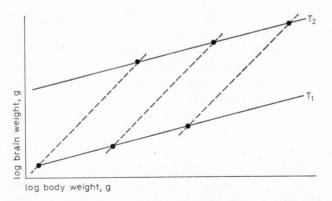

Fig. 12. An illustration of how two interspecific regressions with equal slopes might be produced by evolution in geometric similarity rather than via the 'standard' interpretation of reorganization of proportions at the outset of ontogeny; from GOULD [1971 b]. T_1 and T_2 are interspecific curves for ancestors and descendants respectively. The three species of T_1 evolve in geometric similarity (dotted lines have a slope of 1) as they increase in size; this yields the 'improved' cephalization of T_2.

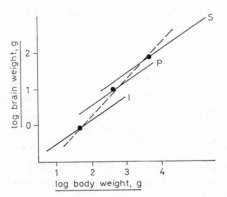

Fig. 13. Interspecific brain-body plots for adult insectivores (I), prosimians (P), and simians (S). Slopes are 0.67, 0.68, and 0.69 respectively. Dotted line of slope = 1 is an excellent fit to the three joint means. Thus, geometric similarity is preserved through the sequence (for gross weights). An average insectivore has the same brain/body ratio as an average anthropoid. Computed by author from data of BAUCHOT and STEPHAN [1969].

increase and cephalization are related in the insectivore-prosimian-anthropoid grouping. A brain/body plot for a massed sample of 124 species in all three groups yields a slope of 0.931. BAUCHOT [1972, p. 443] comments: 'The maintenance of geometric form, which entails an important increase of cephalization, can thus appear as a factor in evolution.' SACHER [1970, p. 255] computed a slope of 0.947 for a sample of 63 speices.

The most outrageous extension of this hypothesis would attempt to apply it to the evolution of man. Some estimates for the phyletic slope of the africanus-habilis-erectus-sapiens lineage are, indeed, close to 1. I shall pursue the matter no further beyond what may prove to be a strangely prophetic quote in an early paper by DUBOIS (also the perpetrator of most early work on brain/body allometry) describing his discovery of 'Pithecanthropus' [1896, p. 10]:

> We have only to double the length and breadth, both of the thigh-bone and of the skull of a *Hylobates syndactalus*, to have dimensions exactly corresponding to those of the Java form. By doubling all dimensions of a *Hylobates* we would obtain an imaginary product with a corresponding cranial capacity also. But certainly such an enlargement alone would not be adequate to explain in reality the large cranial capacity, as, with an enlargement of the size of the body in nearly allied species of mammals, we do not find a corresponding enlargement of the cranial capacity. The Anthropoid would therefore not only have grown in the size of the body, but his brain would have grown faster relatively to the body that we are accustomed to see in homogeneal mammal species of different size. This is actually what we find to be the case in the remains of *Pithecanthropus*.

TOBIAS [1967, p. 46] has written: 'Some of the earlier crania of *A. africanus* and of *Homo habilis* seem like perfect miniatures, pygmy versions, of modern human skulls.'

3. Ontogenetic Scaling

The complex ontogeny of the brain provides a rich store of evolutionary possibilities for its phyletic increase. It is an old observation, and a foundation of the paedomorphic theory of human origins (a notion of doubtful popularity that I regard as fundamentally correct): the cranium of modern man bears a remarkable similarity to common early ontogenetic stages of higher primates (and, consequently, man's skull changes less during his own ontogeny than other primates do in theirs). KUMMER [1951] and STARCK [1953] have applied D'ARCY THOMPSON's transformed coordinates to this problem and confirmed the hypothesis in general outline (fig. 14). There are exceptional points, of course: the relative positions of ear region and occiput change more during the ontogeny of man than during that of the great apes.)

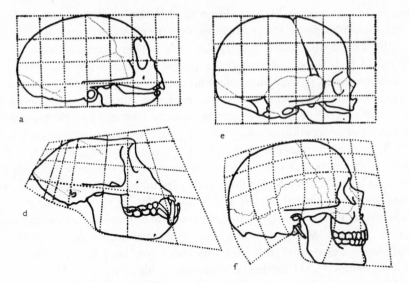

Fig. 14. Transformed coordinate analysis of *Macaca mulatta* and *Homo sapiens* to illustrate greater departure of the rhesus monkey from its original shape at birth and the corresponding retention in man of some important juvenile proportions as an adult. The basic trends are the same, but man's cranium undergoes a smaller relative decrease and his jaws a smaller relative increase, thus preserving original proportions. From KUMMER [1951, p. 143] a = Rhesus monkey a few days after birth; d = an old male; e = newborn *Homo sapiens*; f = adult *Homo sapiens*.

BOYCE [1964] performed a multivariate analysis on 99 skull characters for male, female and juvenile pongids and fossil and recent hominids. The first axis of a principal component analysis separated adult apes from adult men. But the second axis provided a primary expression of ontogenetic allometry, for it separated juveniles from adults. Juvenile apes, juvenile modern man and the Taung skull occupied extreme positions on this axis; adult apes and australopithecines grouped at the other extreme. But the skulls of adult human female and male, despite their large size, fell far closer to the juveniles. The unusual position of adult humans on an otherwise ontogenetic axis demonstrates our retention, in a multivariate sense, of a characteristically juvenile morphology for higher primates.

One can follow WEIDENREICH [1941] and ascribe all the peculiarities of man's skull to his large brain itself. In this case, the similarities between fetal skulls of great apes and adult skulls of humans might be accidental results of possessing large brains for different reasons. But by what ontogenetic mechanism did man achieve his large brain size? Here even WEIDENREICH affirms a paedomorphic postulate, one that involves the allometric curve of ontogeny and its scaling with ultimate body size. The two-part curve includes a fetal period of high slope and a postnatal period of much lower inclination. If, as size increased in evolution, the fetal curve were prolonged to occupy a larger and larger segment of total ontogeny, then the adult brain would increase in size (while the prolongation of early tendencies might promote a retention of proportions characteristic of early ancestral ontogeny). COUNT [1947] compared ontogenetic curves for man and 'monkey' and found two major differences: (a) man's fetal curve has about the same slope as the 'monkey's' but it has a higher y-intercept; (b) man's fetal curve extends over a much greater range of relative size. This is by far the more important of the two effects; it also reflects paedomorphosis by ontogenetic prolongation of juvenile trends: 'Man and monkey have subequal fetal slopes. Man climbs that slope … along a slightly higher elevation (per body weight). More notably, he climbs it much farther before turning aside to the right. After he turns, he travels but a short distance, and on a much more gradual rise than the monkey. As a result, he is bigger-bodied, bigger-brained, more highly cephalized' [COUNT, 1947, p. 1016].

I have demonstrated, I hope, that a distinction and characterization of the types of scaling can provide some insights into problems of the evolution of the brain.

V. Summary

Allometry should be defined broadly as the study of size and its consequences, not narrowly as the application of power functions to the data of growth. Variation in size may be ontogenetic, static or phyletic.

Errors of omission and treatment have plagued the study of allometry in primates. Standard texts often treat brain size as an independent measure, ignoring its allometric relation with body size – on this basis, gracile australopithecines have been accorded the mental status of gorillas. Intrinsic allometries of the brain/body are likewise neglected: many authors cite cerebral folding as evidence of man's mental superiority, but folding is a mechanical correlate of brain size itself. Confusion among types of scaling heads errors of treatment in both historical primacy [DUBOIS' ontogenetic inferences from interspecific curves] and current frequency. The predicted parameters of brain-body plots differ greatly for ontogenetic, intrapopulational, interspecific and phyletic allometries.

I then discuss basic trends in bivariate allometry at the ordinal level for internal organ weights, skeletal dimensions, lifespan and fetal weight. In considering the causes of basic bivariate allometries, I examine the reason for differences among types of scaling in brain-body relationships. The interspecific exponent of 0.66 strongly suggests a relationship to body surfaces, but we have no satisfactory explanation for why this should be so. The tripartite ontogenetic plot is a consequence of patterns in neuronal differentiation. We do not know why intraspecific exponents fall between 0.2 and 0.4; several partial explanations have been offered. Multivariate techniques have transcended the pictorial representation of transformed coordinates and offer new, powerful approaches to total allometric patterns.

Allometry is most often used as a 'criterion for subtraction". In order to assess the nature and purpose of an adaptation, we must be able to identify and isolate the aspect of its form that depends both upon its size and the size of the body within which it resides. Cranial indices and limb lengths are misinterpreted when authors apply no correction for body size. The search for a criterion of subtraction has been most diligently pursued in studies of the brain. Clearly, brain size must be assessed by comparison with a 'standard' animal of the same body size. But how shall size be measured, especially in fossils; and how shall a standard animal be construed. I discuss and criticize three methods recently used: RADINSKY's foramen magnum criterion; JERISON's minimum convex polygons and cephalization quotients; and the indices of progression in comparison with 'basal insectivores' of BAUCHOT, STEPHAN and their colleagues. With these criteria, we can affirm (for example) that: Eocene and Oligocene lemuroids were much more highly cephalized than their contemporary mammals but are, for the most part, slightly less cephalized than modern lemurs; all australopithecines are more highly cephalized than any nonhuman primate; above the prosimian level, primates are strongly and consistently more cephalized than average mammals [contrary to claims of STEPHAN]; the parts of the brain can be arranged in a 'hierarchy' according to their correlation with neocortical enlargement in the evolution of primates. SACHER and I have performed multivariate analyses, with slightly differing techniques and from different premises, on the data of BAUCHOT, STEPHAN and ANDY. We affirm the identification and separation of neocortical from paleocortical-limbic-olfactory complexes in both inter- and intragroup analyses. Insectivores, prosimians and anthropoids are well separated on discriminant axes by very few variables: neocortex, schizocortex, brain weight and olfactory bulbs; other parts of the brain add little to discrimination beyond their correlation to these variates.

Inferences about evolutionary mechanisms have been based most often upon interspecific curves. Of all types of scaling, these are the most inappropriate for such a purpose. They are static curves expressing, at best, mechanical criteria for adaptation. Intraspecific, phyletic and ontogenetic curves can reflect evolutionary mechanisms directly. The intraspecific curve expresses normal adult variation; evolutionary change in size may occur easily along its path. Size increase along this slope of ca 0.3 will lead to a decrease in cephalization relative to the interspecific curve of 0.66; I suggest that *Gorilla* and *Alouatta* have so evolved. Size decrease leads to improvement of cephalization [as BAUCHOT and STEPHAN suggest for *Miopithecus talapoin*]. The brain size of robust australopithecines may have scaled along an intraspecific curve. I argue that phyletic curves often scale with a slope near 1. With phyletic size increase, this maintenance of geometric similarity marks improvement in cephalization. Modern insectivores, prosimians and anthropoids are so arranged, but, of course, they provide no necessary insight into paleontological events. The tendency of parts of the brain to scale in geometric similarity with each other is an 'enabling criterion' for overall evolution of the brain in geometric similarity. The complex curve of ontogeny can be modified in a variety of ways. Prolongation of high fetal slopes to larger sizes and later ages may well be both an aspect of man's paedomorphosis and a cause of his high cephalization.

VI. References

ALCOBÉ, S. und PREVOSTI, A.: Versuch einer Analyse des allometrischen Wachstums beim Menschen. Homo 2: 15–19 (1951).

ASHTON, E. H. and SPENCE, T. F.: Age changes in the cranial capacity and foramen magnum of hominoids. Proc. zool. Soc. Lond. 130: 169–181 (1958).

BAUCHOT, R.: La densité en neurones des noyaux gris diencéphaliques: J. Hirnfors. 6: 327–330 (1964).

BAUCHOT, R.: Encéphalisation et phylogénie. C.R. Acad. Sci. 275: 441–443 (1972).

BAUCHOT, R. et STEPHAN, H.: Le poids encéphalique chez les insectivores malagaches Acta zool. 45: 63–75 (1964).

BAUCHOT, R. et STEPHAN, H.: Données nouvelles sur l'encéphalisation des insectivores et des prosimiens. Mammalia 30: 160–196 (1966).

BAUCHOT, R. et STEPHAN, H.: Etude des modifications encéphaliques observées chez les insectivores adaptés à la recherche de nourriture en milieu aquatique. Mammalia 32: 228–275 (1968).

BAUCHOT, R. et STEPHAN, H.: Encéphalisation et niveau évolutif chez les simiens. Mammalia 33: 225–275 (1969).

BIEGERT, J. und MAURER, R.: Rumpfskelettlänge, Allometrien und Körperproportionen bei catarrhinen Primaten. Folia primat. 17: 142–156 (1972).

BOAS, F.: The cephalic index. Amer. Anthrop. 1: 448–461 (1899).

BONIN, G. VON: Sidelights on cerebral evolution: brain size of lower vertebrates and degree of cortical folding. J. gen. et Psychol. 25: 273–282 (1941).

BOYCE, A. J.: The value of some methods of numerical taxonomy with reference to hominoid classification; in HEYWOOD and McNEELL Phenetic and phylogenetic classification, pp. 47–65 (Systematics Ass. 1964).

BRUMMELKAMP, R.: On the equal growth of the neocortex during the ontogenesis of the sheep. Verh. nederl. Akad. Wetensch. 43: 3–6 (1940a).

BRUMMELKAMP, R.: Brain weight and body size: a study of the cephalization problem. Verh. Kongr. nederl. Akad. Wetensch. 39: 1–57 (1940b).

CAMPBELL, B.: Human evolution, p. 425 (Aldine-Atherton, Chicago 1966).

CLARK, M. E.: Contemporary biology, p. 632 (Saunders, Philadelphia 1973).

CLARK, W. E. LeGROS: Deformation patterns in the cerebral cortex; in LE GROS CLARK and MEDAWAR Essays on growth and form presented to D'Arcy Thompson, pp. 1–22 (Clarenden Press, Oxford 1945).

CORRUCCINI, R. S.: Allometry correction in taximetrics. Syst. Zool. 21: 375–383 (1972).

COUNT, E. W.: Brain and body weight in man: their antecedents in growth and evolution. Ann. N.Y. Acad. Sci. 46: 993–1122 (1947).

DUBOIS, E.: On Pithecanthropus erectus: a transitional form between man and the apes. Sci. Trans. roy. Dublin Soc. 6: 1–18 (1896).

DUBOIS, E.: Über die Abhängigkeit des Hirngewichtes von der Körpergrösse bei den Säugetieren. Arch. Anthrop. 25: 1–28 (1897).

DUBOIS, E.: Über die Abhängigkeit des Hirngewichtes von der Körpergrösse beim Menschen. Arch. Anthrop. 25: 423–441 (1898).

EDINGER, T.: Brains from 40 million years of camelid history; in HASSLER and STEPHAN Evolution of the forebrain, pp. 153–161 (Thieme, Stuttgart 1966).

FREEDMAN, L.: Growth of muzzle length relative to calvaria length in *Papio*. Growth *26:* 117–128 (1962).

FRICK, H.: Das Herz der Primaten; in HOFER, SCHULTZ and STARCK Primatologia, vol. 3, part 2, pp. 163–272 (Karger, Basel 1960).

GEIST, V.: On the relationship of social evolution and ecology in ungulates. Proc. Annu. Meet. AAAS, 1972, p. 28 (1973).

GERVEN, D. P. VAN: The contribution of size and shape variation to patterns of sexual dimorphism of the human femur. Amer. J. phys. Anthrop. *37:* 49–60 (1972).

GILES, E.: Cranial allometry in the great apes. Human. Biol. *28:* 43–58 (1956).

GOULD, S. J.: Allometry and size in ontogeny and phylogeny. Biol. Rev. *41:* 587–640 (1966).

GOULD, S. J.: D'Arcy Thompson and the science of form. New Lit. Hist. *2:* 229–258 (1971a).

GOULD, S. J.: Geometric scaling in allometric growth: a contribution to the problem of scaling in the evolution of size. Amer. Nat. *105:* 113–136 (1971b).

GOULD, S. J.: Positive allometry of antlers in the 'Irish elk', *Megaloceros giganteus*. Nature, Lond. (1973).

GOULD, S. J.: The evolutionary significance of 'bizarre' structures: antler size and skull size in the 'Irish elk'. Evolution (in press).

HARMAN, P. S.: Paleoneurologic, neoneurologic and ontogenetic aspects of brain phylogeny. James Arthur Lecture. Amer. Mus. nat. Hist. (1957).

HEMMER, H.: Über allometrische Beziehungen zwischen Hirnschädelkapazität und Hirn-schädelwölbung im Genus Homo. Homo *15:* 218–224 (1964).

HEMMER, H.: Allometrische Untersuchungen am Schädel von *Homo sapiens* unter be-sonderer Berücksichtigung des Brachykephalisations-Problems. Homo *17:* 190–209 (1966).

HOLLOWAY, R. L.: jr.: Some questions on parameters of neural evolution in primates Ann. N.Y. Acad. Sci. *167:* 332–340 (1969).

HOLLOWAY, R. L., jr.: Neural parameters, hunting and the evolution of the human brain; in NOBACK and MONTAGNA Advances in primatology, vol. 1, pp. 209–224 (Appleton Century Crofts, New York 1970).

HOLLOWAY, R. L., jr.: Australopithecine endocasts, brain evolution in the Hominoidea, and a model of hominoid evolution; in TUTTLE The functional and evolutionary biology of primates, pp. 185–203 (Aldine-Atherton, Chicago 1972).

HOWELLS, W. W.: Analysis of patterns of variation in crania of recent man; in TUTTLE The functional and evolutionary biology of primates, pp. 123–151 (Aldine-Atherton, Chicago 1972).

HUXLEY, J. S.: Problems of relative growth (MacVeagh, London 1932).

IMBRIE, J.: Biometrical methods in the study of invertebrate fossils. Bull. Amer. mus. nat. Hist. *108:* 211–252 (1956).

JERISON, H. J.: Interpreting the evolution of the brain. Human. Biol. *35:* 263–291 (1963).

JERISON, H. J.: Gross brain indices and the analyses of fossil endocasts; in NOBACK and MONTAGNA Advances in primatology, vol. 1, pp. 225–244 (Appleton Century Crofts, New York 1970a).

JERISON, H. J.: Brain evolution: new light on old principles. Science *170:* 1224–1225 (1970b).

JERISON, H. J.: Evolution of the brain and intelligence (Academic Press, New York, 1973).

JOLICŒUR, P.: The multivariate generalization of the allometry equation. Biometrics *19:* 497–499 (1963).

KERMACK, K. A. and HALDANE, J. B. S.: Organic correlation and allometry. Biometrika *37:* 30–41 (1950).

KERR, G. R.; KENNAN, A. L.; WAISMAN, H. A., and ALLEN, J. R.: Growth and development of the fetal rhesus monkey. I. Physical growth. Growth *33:* 201–213 (1969).

KROGMAN, W. M.: Child growth, p. 231 (Univ. of Michigan Press, Ann Arbor 1972).

KUMMER, B.: Zur Entstehung der menschlichen Schädelform (ein Beitrag zum Fetalisationsproblem). Verh. anat. Ges. *49* (suppl. to Anat. Anz. vol. 98): 140–145 (1951).

KURTÉN, B.: On the variation and population dynamics of fossil and recent mammal populations. Acta zool. fenn. *76:* 1–122 (1953).

LEUTENEGGER, W.: Newborn size and pelvic dimensions of *Australopithecus.* Nature, Lond. *240:* 568–569 (1972).

LUMER, H. and SCHULTZ, A. H.: Relative growth of the limb segments and tail in macaques. Human Biol. *13:* 283–305 (1941).

LUMER, H. and SCHULTZ, A. H.: Relative growth of the limb segments and tail in *Ateles geoffroyi* and *Cebus capucinus.* Human Biol. *19:* 53–67 (1947).

MATSUDA, R.: Relative growth of Negro and white children in Philadelphia. Growth *27:* 271–284 (1963).

McMAHON, T.: Size and shape in biology. Science *179:* 1201–1204 (1973).

MEDAWAR, P. B.: Size, shape and age; in LE GROS CLARK and MEDAWAR Essays on growth and form presented to D'Arcy Thompson, pp. 157–187 (Clarenden, Oxford 1945).

MOLLISON, J.: Die Körperproportionen der Primaten. Morph. Jb. *42:* 79–304 (1910).

MOSIMANN, J. E.: Size allometry: size and shape variables with characterizations of the log-normal and generalized gamma distributions. J. amer. statist. Ass. *65:* 930–945 (1970).

NOBACK, C. R. and MOSS, M. L.: Differential growth of the human brain. J. comp. Neurol. *105:* 539–551 (1956).

PAKKENBERG, H. and VOIGT, J.: Brain weight of the Danes. Acta anat. *56:* 297–307 (1964).

PILBEAM, D. and GOULD, S. J.: Size and scaling in human evolution. Science (in press).

PIRLOT, P. and STEPHAN, H.: Encephalization in Chiroptera. Canad. J. Zool. *48:* 433–444 (1970).

PLATEL, R. and BAUCHOT, R.: L'encéphale de *Scincus scincus* (L.) (Reptilia, Sauria, Scincidae). Recherche d'une grandeur de référence pour des études quantitatives. Zool. Anz. *184:* 33–47 (1970).

RADINSKY, L.: Relative brain size: a new measure. Science *155:* 836–838 (1967).

RADINSKY, L.: The fossil evidence of prosimian brain evolution; in NOBACK and MONTAGNA (Advances in Primatology, vol. 1, pp. 209–224 (Appleton Century Crofts, 1970).

REEVE, E. C. R. and HUXLEY, J. S.: Some problems in the study of allometric growth; in LE GROS CLARK and MEDAWAR Essays on growth and form presented to D'Arcy Thompson, pp. 121–156 (Oxford Univ. Press, London 1945).

SACHER, G. A.: Dimensional analysis of factors governing longevity in mammals. Proc. Int. Congr. Gerontology, Vienna, p. 14 (1966).

SACHER, G. A.: Allometric and factorial analysis of brain structure in insectivores and primates; in NOBACK and MONTAGNA Advances in Primatology, vol. 1, pp. 245–287 (Appleton Century Crofts, New York 1970).

SCHAEFFER, U.: Gehirnschädelkapazität und Körpergrösse bei Vormenschenfunden in allometrischer Darstellung. Zool. Anz. 168: 149–164 (1962).

SCHALLER, G.: The year of the gorilla (Univ. of Chicago Press, Chicago 1964).

SCHEPERS, G. W. H.: The brain casts of the recently discovered Plesianthropus skulls. Transvaal Mus. Mem. 4: 89–117 (1949).

SCHULTZ, A. H.: Fetal growth in man and other primates. Quart. Rev. Biol. 1: 465–521 (1926).

SCHULTZ, A. H.: Die Körperproportionen der erwachsenen catarrhinen Primaten, mit spezieller Berücksichtigung der Menschenaffen. Anthrop. Anz. 10: 154–185 (1933).

SCHULTZ, A. H.: The size of the orbit and of the eye in primates. Amer. J. phys. Anthrop. 26: 389–408 (1940).

SCHULTZ, A. H.: The relative thickness of the long bones and the vertebrae in primates. Amer. J. phys. Anthrop. 11: 277–311 (1953).

SCHULTZ, A. H.: The position of the occipital condyles and of the face relative to the skull base in primates. Amer. J. phys. Anthrop. 13: 97–120 (1955).

SCHOLL, D.: The quantitative investigation of the vertebrate brain and the applicability of allometric formulae to its study. Proc. roy. Soc. Lond. 135: 243–258 (1948).

SNEATH, P. H. A.: Trend surface analysis of transformation grids. J. Zool. 151: 65–122 (1967).

SNELL, O.: Das Gewicht des Gehirns und des Hirnmantels der Säugetiere in Beziehung zu deren geistigen Fähigkeiten. Sitz. Ges. Morph. Physiol. München 7: 90–94 (1891).

SPRENT, P.: The mathematics of size and shape. Biometrics. 1972: 23–37.

STAHL, W. R.: Similarity and dimensional methods in biology. Science 137: 205–212 (1962).

STAHL, W. R.: Organ weights in primates and other mammals. Science 150: 1039–1041 (1965).

STAHL, W. R. and GUMMERSON, J. Y.: Systematic allometry in five species of adult primates. Growth 31: 21–34 (1967).

STARCK, D.: Morphologische Untersuchungen am Kopf der Säugetiere, besonders der Prosimier, ein Beitrag zum Problem des Formwandels des Säugetierschädels. Z. wiss. Zool. 157: 169–219 (1953).

STEPHAN, H.: Evolution of primate brains: a comparative anatomical investigation; in TUTTLE The functional and evolutionary biology of primates, pp. 155–174 (Aldine-Atherton, Chicago 1972).

STEPHAN, H. and ANDY, O. J.: Quantitative comparisons of brain structures from insectivores to primates. Amer. Zool. 4: 59–74 (1964).

STEPHAN, H. and ANDY, O. J.: The allocortex in primates; in NOBACK and MONTAGNA Advances in Primatology, vol. 1, pp. 109–135 (Appleton Century Crofts, New York 1970).

STEPHAN, H. und BAUCHOT, R.: Hirn-Körpergewichtsbeziehungen bei den Halbaffen (Prosimii). Acta Zool. 46: 209–231 (1965).

STEPHAN, H.; BAUCHOT, R., and ANDY, O. J.: Data on size of the brain and of various brain parts in insectivores and primates; in NOBACK and MONTAGNA Advances in Primatology, vol. 1, pp. 289–297 (Appleton Century Crofts, New York 1970).

STEPHAN, H. and PIRLOT, P.: Volumetric comparisons of brain structures in bats. Z. zool. Syst. EvolForsch. *3:* 200–236 (1970).

THOMPSON, D. W.: On growth and form, p. 1116 (Cambridge Univ. Press, London 1942).

TOBIAS, P. V.: General questions arising from some Lower and Middle Pleistocene hominids of the Olduvai Gorge, Tanzania. Sth Afr. J. Sci. *63:* 41–48 (1967).

TOBIAS, P. V.: Brain size, grey matter and race – fact or fiction? Amer J. phys. Anthrop. *32:* 3–26 (1970).

TOBIAS, P. V.: The brain in hominid evolution, p 170 (Columbia Univ. Press, New York 1971).

WANNER, J. A.: Relative brain size: a critique of a new measure. Amer. J. phys. Anthrop. *35:* 255–258 (1971).

WEIDENREICH, F.: The brain and its role in the phylogenetic transformation of the human skull. Trans. amer. Phil. Soc. *31:* 321–442 (1941).

WENT, F. W.: The size of man. Amer. Sci. *56:* 400–413 (1968).

Author's address: Dr. STEPHEN JAY GOULD, Museum of Comparative Zoology, Harvard University, *Cambridge, MA 02138* (USA)

In SZALAY: Approaches to Primate Paleobiology
Contrib. Primat., vol. 5, pp. 293–325 (Karger, Basel 1975)

Significance of Tooth Sharpness for Mammalian, Especially Primate, Evolution

R. G. EVERY

Centre for the Study of Conflict, and Department of Zoology, University of Canterbury, Christchurch, New Zealand

Contents

As LeGROS CLARK [1962] has aptly emphasized, '...the evidence which the dentition contributes towards the solution of ... problems of the inter-relationships of the Primates' is of prime importance, for '...no anatomical features have yielded more fruitful evidence in enquiries of this sort than the morphological detail of the teeth'. This well-recognized feature is further noteworthy in that the 'fruitful evidence' has come, almost exclusively, from the tooth-crown's unworn exterior surface. Despite recent interest in the 'wear-facet' [e. g. BUTLER, 1952; MILLS, 1955; CROMPTON, 1971], odontologists still tend to be preoccupied with the relatively unworn condition. With the detection of the tooth-sharpening phenomenon: thegosis [EVERY, 1960, 1972], a new dimension in the study of mammals is now available. All aspects of the tooth, not only the morphological detail of its unworn external structure, but also that exposed by wear, become equally important.

I. Sharpness

Giving further emphasis to the importance of teeth, LeGros Clark [1962] adds: 'It may be accounted a happy circumstance, also, that, because of the durability of the enamel and dentine of which they are composed, in fossil material teeth are more commonly preserved than any other part of the skeleton; indeed, in many instances they are the only remains available for study'. Despite the general awareness of this remarkable characteristic, general appreciation of its dominant evolutionary significance seems remarkably slow. For it is this very characteristic (hardness and durability – resistance to wear and fracture) that allows the tooth's prime evolutionary advantage: its sharpness. It is the sharpness that gives the tooth its capacity to penetrate and divide exogenous material.

The most primitive tooth, the dermal denticle [Tomes, 1923], functioned by sharpness: the apex of the simple cone was simply sharp. Overwhelmingly, the greater part of the subsequent phylogenesis of dentitions is dominated by the evolutionary advantage of the sharp apex to the recurved conical crown. The sharp edge of a blade running from the apex is the prime secondary advantage. Furthermore, the incusive surface [Every, 1972] (the 'basin'), a relatively recent evolutionary event, although giving significant additional advantage, does not (even in primates) take precedence to sharpness as the dominant functional feature [Every, 1972, in press]. Its dominant function is to confine a resistant exogenous material to the action of a sharp cusp or blade.

Such features, therefore, require careful observation and analysis, especially in view of the prevailing concept of 'the lessening importance of the teeth' in the most advanced forms of primate evolution: the hominids leading to man, himself. In this area, the thinking has been overwhelmingly preoccupied with 'flat and blunt', 'regressed', and 'mortar-and-pestle' 'grinding' functions for the teeth [e.g. LeGros Clark, 1962; Butler, 1952; Mills, 1955; Zuckerman, 1958; Leakey, 1960; Washburn, 1960].

II. Continuity of Sharpness

A sharp tooth when constantly used nonetheless deteriorates, just as occurs with man's artificial instruments; penetrating spikes and cutting edges eventually become blunted. Biological evolution on the one hand and cultural on the other, have, where useful, compensated for this disadvantage by roughly comparable processes (except that one is unlearnt and the other learnt).

This is achieved in two principal ways. Sharpness is maintained, either by a process of (repeated) discard and replacement of the entire instrument, or by a (repeated) restoration of the sharpness of the original instrument itself.

The first process is the well-understood characteristic condition in reptiles, generally. The second, apart from replacement of the deciduous by the permanent dentition, is the characteristic general condition in mammals. This fundamental mammalian feature has until recently remained obscured. Yet the phenomenon of thegosis appears to have been a crucial factor, not only in the reptilian-mammalian transition, but also in the subsequent mammalian radiation. For the argument is being developed that without this 'instinctive' behavioral phenomenon the precision scissorial blade unit [EVERY, 1972], a dominant morphological characteristic of the mammalian (as distinct from the reptilian) dentition, could not have evolved. Essential to the scissorial blade system is the effective maintenance of its closely appositioned sharp-cutting edges [EVERY, 1972].

Although thegosis appears in a few isolated reptiles, e.g., *Sphenodon* [EVERY, 1965], *Iguanodon* (certain birds, e.g., *Diomedea* thegose the tomia [TUNNICLIFFE, 1973; TUNNICLIFFE *et al.*, in press], and a few isolated invertebrates, e.g., *Jasus*, the 'mandibles' [TUNNICLIFFE, 1973; SCALLY *et al.*, in press], and *Evechinus*, the 'teeth' [TUNNICLIFFE, 1973; SCALLY, 1973]), thegosis is not a reptilian characteristic. Reptilian teeth characteristically alternate and by-pass each other when jaws are closed. There is seldom an instance of tooth opposition and thus the possibility of tooth contacting tooth. Moreover, apart from the thegotic reptiles, the reptilian jaw (especially limited by the pterygoid flange) cannot move in such a way as to bring teeth into contact. When the simple, recurved, conical tooth progressed to the tricuspid condition with sharp-edged blades running from the apices, as in the mammal-like reptile *Thrinaxodon*, for example, such teeth did not function scissorially, as do the carnassials of carnivorous mammals, but by deep 'over-bite' of relatively closely appositioned blades. This is comparable to the action of the premolars immediately mesial to the carnassials of most mammalian carnivores. Limited cutting does occur but it is by no means as effective as with blades subject to thegosis.

III. Jaw Movements

Concurrent with the emergence of the phenomenon of thegosis is a complex of changes in the bones of the jaws and skull and the articulation of the jaw joints. These are mainly: the displacement of the articular, quad-rate, and hyomandibular bones to the ear; further reduction of the remaining bones in the lower jaw to leave one, the dentary, on either side; a reversal of the convex-concave arrangement of the jaw articulation with the skull (i.e., the rounded condylar head becomes a feature of the lower jaw, and the articular fossa, of the skull); and the removal of the limiting action of the pterygoid flange on the lower jaw.

This complex of changes allows the advantage of greater rigidity and precision of movement under more varied, as well as more forceful, muscular action.

Two features basic to the new mammalian action are involved:

(1) A horizontal component in the characteristic mammalian action is added to that characteristic of reptiles, where there is simple vertical (orthal) closure. Part of the mandibular movement, therefore, is at an angle (some-times considerable) to the vertical. Moreover, the initial mandibular displace-ment (geared for the scissorial action) may, in some mammals, be exaggerated (sometimes considerably), which gives further advantage: it facilitates collec-tion of food from the buccal sulcus for recycling on the masticatory table. This is a particular feature of many herbivores and omnivores.

Other fundamental displacements occur in preparatory phases of mandibular cycles. These are geared to locate the functional action at the appropriate area where a tooth or a group of teeth are specialized for that function (mastication, incision, grasping, crushing, splitting, mixing, slashing, etc.). The main instance of this is when segments of posterior teeth on one side of the mouth are put into function, the spatial relationships are such that the corresponding (mirror-image) segments on the other side are automati-cally disengaged. Similarly, all posterior segments are automatically dis-engaged when anterior segments are brought into function.

Characteristically, in none of these actions do the tooth-substances, themselves, collide. The cuspal planes (i.e., the faceted surfaces) do not, as is commonly thought, guide the mandibular action. Proprioceptive mech-anisms are presumably adapted to control both the course and the termina-tion of mandibular action [EVERY, 1970].

(2) In marked contrast to any action involved in tooth function, masti-catory or otherwise, thegotic action involves forceful grinding contact of one tooth against another. Masticatory (or other) action and thegotic action, there-fore, occur on separate occasions. Although the horizontal component of the thegotic action may occur in the same direction as the masticatory (or other) action, in many mammals it is discretely variant, even diametrically opposite.

IV. Scissorial Action

The biological machinery of the scissorial blade system is detailed and precise [EVERY, 1972]. Its main features involve two sharp-edged blades, each shaped and orientated to the other so as to confine exogenous material to their mutual point-cutting [EVERY, 1972] action. Essential to this action is the close apposition of the leading edges when approximated. This, together with a number of other precisely adapted features [EVERY, 1972], allows balanced reactions of force acting on exogenous material interposed between the blades, thus to ensure its division. There is minimal tendency, also, for a resistant material to become wedged in between the overlapping sides of the blades, uncut.

In the masticatory cycle, scissorial blades are, of course, applied to a bolus of food, and thus to cutting action, well before the scissorial action, itself, occurs, and are liable to damage. Such pre-scissorial action is, as already suggested, comparable to the molar action of the premammalian reptile *Thrinaxodon* and that of the non-scissorial-bladed premolars of most mammalian carnivores. Whenever the damage occurs, thegotic servicing of the scissorial leading edges is essential.

Precise morphological features are adapted to facilitate this servicing. For example, the tooth surface on which thegosis occurs is hollowed out, leaving a narrow strip of thegosed enamel trailing the leading edge. Hollowing may be primary (developmental), or secondary (by functional abrasion of differentially hard tooth-substances) [EVERY, 1972].

Morphological adaptations for high nutrient (low abrasive) and low nutrient (high abrasive) foods are marked, and are categorized into three main types [EVERY, 1972]; briefly:

(1) *α-dentition* (low abrasive food), where the thegosis occurs on the ends of the enamel prisms (e. g., carnivores, and insectivores, generally).

(2) *β-dentition* (high abrasive food), where the thegosis occurs (predominantly) along the sides of the enamel prisms to include the amelodentinal junction (e. g., herbivores, generally).

(3) *α-β-dentition* (moderately abrasive food), where α-thegosis occurs for a considerable part of the tooth's lifetime before β-thegosis takes over (e. g., omnivores, generally).

Primate dentitions are characteristically α-β.

V. Thegosis Facet

When one tooth is rubbed violently against another, clusters of enamel prisms tend to be dislodged, dragged across the surface of both teeth, gouging out further prisms. The process eventually produces (relatively) flat surfaces on each tooth, covered by striations parallel to the line of action. When this action is consistently in the one direction, all the striations are parallel. The line of action, however, may vary (sometimes considerably) and cross-hatch the surface striations. Cross-hatching ensures a uniformly flat facet and thus a uniformly straight-sided blade and is particularly advantageous in circumstances of some β-dentitions (e. g., *Cavia*) where both the thegosis and the functional actions are protrusive.

The morphology of a thegotic tooth is so adapted that the produced facet characteristically presents an acute angle to another (adjacent) tooth surface. This acute-angled boundary is characteristically situated on that part of the tooth that first enters the food in the masticatory (or other) process. It is thus adapted as the leading edge in the cutting action.

Inevitably, with masticatory (or other) function, abrasive scratches, pits, and gouges become superimposed on and eventually obliterate the thegotic marks, and the leading edge becomes rounded and blunted. Correction of this disadvantage is precise and limited; evolutionary processes have ensured that the crucial (sometimes irreplaceable) tooth substances are, within the limits of their functions, effectively conserved. Thegotic servicing, therefore, is precisely adapted both in its occasion and its frequency.

The thegosis facet provides a fertile source of evidence in the study of tooth function; its features in the characteristic primate molar warrant some emphasis:

(1) The facet, characteristically, is a discrete area with circumscribed boundaries clearly defining it from adjacent tooth surfaces.

(2) Superficially, the facet is smooth and shiny.

(3) Detailed examination, however, shows parallel striations which are limited entirely to the facet area, itself, i.e., they do not extend beyond the facet boundary on to any adjacent surface.

(4) These striations are matched by correspondingly parallel striations on the corresponding (thegosis-) facet of an opposing tooth.

(5) They are, therefore, evidence of the wear of tooth substance forced directly against tooth substance.

(6) They are, therefore, evidence of surface wear in the absence of an interposed exogenous material.

(7) A facet is frequently found clouded by surface markings super-imposed on the parallel striations. These appear as curved (sometimes straight) scratches, pits, and gouges that are generally aligned in the direction of the masticatory movement. The marks on the teeth, therefore, run from lingual to buccal in a mandibular molar and (relatively) buccal to lingual in a maxillary.

(8) On a transverse facet, therefore, although tending to diverge, their general alignment is parallel to the original thegotic striation.

(9) On an oblique facet, however, they cross diagonally.

(10) Abrasive marks, furthermore, tend to diverge from each other, particularly on an incusive surface where their divergence is often conspicuous; the divergence, moreover, occurs equally on either side of the (general) transverse direction [EVERY, in press].

(11) These (superimposed) abrasive markings also appear on the corresponding (thegosis-) facet of an opposing tooth.

(12) In contrast to the thegotic striations, however, they are not correspondingly matched.

(13) Furthermore, they extend beyond the boundaries of the facet into foveae and on to adjacent tooth surfaces.

(14) Their characteristics, therefore, give evidence of the functional forces that produce them. They are the markings of abrasion, the results of friction of exogenous material forced over the surface of the tooth's crown.

(15) The characteristics of abrasion, particularly the consistent orientation of the markings in the general line of the action that produced them, shows the fallaciousness of the concept that facets, especially oblique facets, i.e., MILLS' 'lingual phase of occlusion' [MILLS, 1955], originate from masticatory or other such action.

(16) An abrasion area made quasi-discrete by peripheral enamel edges is frequently mistaken for a facet.

(17) Four levels of differential wear markings can be differentiated: (a) a facet with clearly defined margins, matching thegotic, parallel striations, and no abrasion; (b) a facet with somewhat less clearly defined margins and matching, thegotic, parallel striations freshly superimposed on old abrasive marks; (c) a facet with less clearly defined margins and abrasive marks freshly superimposed on old, matching, thegotic, parallel striations; (d) an area with the margins of the old thegosis facet ill-defined and rounded, and all signs of matching, thegotic, parallel striations obliterated; that is, the facet is replaced by an abrasive area, only its former location is recognizable.

(18) The evidence of differential wear markings found on molar teeth,

therefore, suggests differential causal actions [EVERY, 1960, 1970; EVERY and KÜHNE, 1971].

VI. Occlusal Relationships

The spatial relationships of dynamic occlusion are not easy to appreciate, even with the concrete objects in one's hands and manipulated manually. Cinefluorography is currently attracting much attention. Yet the enthusiasm seems more proportional to the cost of the equipment than any elucidation from the results. This may seem unduly harsh, but that cinefluorography should fail to be impressive was predictable. Granted, opening and closing movements may be generally analysed. But these are concerned more with grasping and, together with the direct action of lip, cheek, and tongue muscles, the manipulation of material specific to the scissorial or incusive characteristics of the teeth.

These movements, apart from collecting partly processed material from the buccal and lingual sulci for recycling on the occlusal table, sort out various components for specialized treatment on appropriate areas of the occlusal surfaces, reject other components, and generally mix with saliva. For example, bone is channelled to the talonid on the distal segment of the dog's carnassial, to the specialized incusive premolar mesial to the hyaena's carnassial, and rejected from all the teeth of the cat. Despite the apparent rigidity of the carnivore's jaw joints, there are nonetheless discrete variations of the jaw movements.

When, as in lagomorphs, such preparatory movements occur perpendicular to the, often short, action of the teeth in terminal occlusion, it is hardly surprising that they are mistaken as the fundamentally significant occlusal action itself.

Sorting and manipulating commonly occurs at the minute level of a pinhead-sized seed or spicule of bone. But it is with the analysis of such differential movement that cinefluorography cannot be expected to cope.

A radiographic picture, furthermore, merely gives a differentially radioopaque profile of an undefined thickness of material. It gives scant information of surface contour or detail. Radiography in dentistry is notoriously inadequate as a source of accurate information of occlusal detail or, for example, the contours of a restorative preparation. Even when locating the course of the preparation for root-canal therapy, disturbing error is frequently encountered.

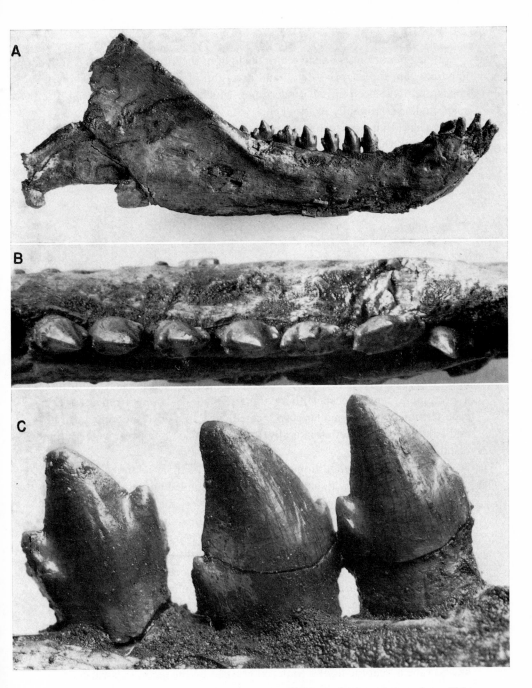

Fig. 1. Thrinaxodon sp., early Jurassic mammal-like reptile; lateral view of mandible (A), occlusal view of the cheek teeth (B), and lateral view of three cheek teeth (C).

When dentitions are analysed in the light of thegotics, it becomes readily apparent that the adaptive detail and precision of the biological machinery concerned have been grossly underrated. Much of this evidence, being unsuspected, is continually overlooked.

It seems that, apart from the impressive accumulation of specimens and their reports, the elucidation of the selective forces in the evolution of mammalian dentitions has remained little advanced beyond the efforts of COPE and OSBORN of the last century, and those of SIMPSON [1936].

VII. Discussion

The following attempt to elucidate some of these forces in the light of the phenomenon of thegosis directs the attention to a narrow selection of specific illustrations. Understanding the primate dentition begins, at least, with the reptilian antecedents of mammals.

In figure 1, the right mandible and teeth of *Thrinaxodon*, an early Jurassic mammal-like reptile, are shown. In occlusal view, *Thrinaxodon's* laterally compressed, tricuspid, cheek teeth present a mesiodistally orientated row of sharp penetrating spikes and cutting blades. Each tooth has a prominent central cusp with two sharp blades running from its apex, each terminating in a smaller cusp located close to a corresponding cusp in and adjacent tooth. Viewed laterally, the mesial blade is convex, whereas the distal is concave.

In the ancestral condition where there is a simple recurved cone, the apex of the crown only is sharp, the walls are rounded and blunt. The tooth is wholly adapted as a 'grasping hook'. Moreover, the wide interproximal space separating it from neighbouring teeth allows the entire length of its crown to be fully functional. The features allow deep penetration of grasped material with minimal tendency to tear or cut, thus minimal tendency to fragmentation and, significantly, to impaction between the teeth. The relatively intact material is 'hooked' into the confines of the oral cavity and swallowed whole. (The homodont dentition of the common dolphin, for example, shows a clear instance of a mammalian readaptation of this condition.)

The situation is contrasted in *Thrinaxodon* where the 'grasping hook' has become a 'reaping hook'. Material grasped by the teeth is sliced, cut and segmented rather than held in one piece. Even a slippery struggling prey has little chance of escape without serious injury. Advantages additional to immobilizing the prey are the reduction of larger prey to segments and the facilitation of deglutition and digestion.

A system of cusps-and-blades-in-a-row functions as a comminuting agent even more advantageously when each tooth's sharpness is made relatively continuous with that of the adjacent tooth. *Thrinaxodon*'s laterally compressed, mesiodistally elongate teeth arranged closely adjacent to each other show a considerable adaptation of this mammal-like feature. Yet, in this respect, i.e., the contiguity of sharpness in proximal teeth, the characteristic reptilian mechanism of tooth-replacement presents a major disadvantage. ROMER [1967] describes the reptilian condition: 'In a great variety of lower vertebrates tooth replacement (in contrast to mammals) continues throughout life. Seldom do we see in such forms (unless arbitrarily restored in a figure) an even tooth row; instead, we generally see an uneven arrangement, with some teeth large and fully developed, others small and obviously recently erupted, and still other tooth positions represented by toothless sockets in which replacement is taking place'.

The disadvantage of the situation where there is repeated discontiguity of closely adjacent teeth is not only the discontinuity of the cutting action, but, and more importantly, the tendency to impaction. Especially is this so as the system is geared to process exogenous material into fragments hazardously liable to impaction. The advantage of *Thrinaxodon*'s small secondary cusp situated at the base of each mesial and distal blade is, therefore, seen as a stop that doubly functions to confine the material to the action of the blade and to protect the vulnerable interproximal tissue from the destructive pathology that rapidly results from unresolved impaction.

Thrinaxodon's cheek teeth show another important feature: the surfaces alongside the cutting edges, mesially and distally, both buccally and lingually, are hollowed out. This allows an extra fineness to the cutting edges [EVERY, 1972].

Because the characteristic reptilian jaw is rigidly limited in transverse jaw action, the teeth are seldom arranged to allow their cutting blades to be brought into close apposition as in a pair of scissors. When jaws are closed, there is, therefore, some appreciable gap between the blades as they pass, and cutting efficiency is correspondingly limited. The capacity to penetrate exogenous material interposed between the opposing teeth of opposing jaws is then achieved by a markedly deep over-bite. When jaws are fully closed, mandibular teeth are entirely overlapped by maxillary teeth and their buccal surfaces are wholly concealed.

The emergence of mammals, however, is characterized in molar teeth

by the close apposition of their cutting edges. It is an adaptation made possible by the introduction of three major features: an ontogenetic continuity
of the tooth row through single replacement of the dentition; a horizontal
component in the mandibular action; and, significantly, an ontogenetic continuity of the sharp leading edges through the agency of a discrete process
separate from any other function, masticatory or otherwise, i.e., thegosis.

Figure 2 shows the left mandibular molar of a late Jurassic mammal,
Priacodon (NMNH 2696), a triconodont (distal cusp lost in preparation).
Although not in the mammalian ancestral main stream, *Priacodon's* primitive, laterally compressed, cusps-and-blades-in-a-row molars show a number
of fundamental aspects of the mammalian elaboration.

Since the interproximal area in the mammalian (single replacement)
dentition has become relatively stabilized, mammalian molars can be brought
into close interproximal contact. In fact, by extending a blade to overlap a
corresponding blade in an adjacent tooth, continuity of the cutting action is
further ensured and impaction from wedging more effectively precluded.

In the primitive triconodont, this overlapping appears as a 'tongue-and-
groove' formation; the 'tongue' situated at the distal, and the 'groove' at the
mesial extremity of each tooth whether maxillary or mandibular. In later
mammals, particularly eutherians, this overlapping, or 'stegoid formation'
[EVERY, 1972], is further developed and is differentiated in maxillary and
mandibular teeth in a more balanced system. As in *Thrinaxodon*, the hollowed-out surface is featured. *Priacodon*, however, shows the characteristic
mammalian thegosis facet: the discrete narrow strip of worn surface separating the hollowing from the leading edges of the blades. In this mammal, the
horizontal component of the scissorial action is transverse, i.e., from buccal
to lingual. Facets are, therefore, located on buccal surfaces of mandibular
teeth, palatal of maxillary, and, at the leading edges, are appropriately be-
velled (i.e., are acutely angled) to the adjacent tooth surfaces. The freshly
thegosed surface shows parallel striations (faintly seen in fig. 2) which match
corresponding striations in the opposing tooth and are discretely exclusive
to the facet itself.

Not only do the teeth in opposing jaws alternate, but also the cusps
themselves, i.e., they interdigitate in a cusp-to-valley relationship with the
blades arranged in a zig-zag fashion somewhat comparable to tailor's pink-
ing shears.

Blades running from the cusp apices of mandibular molars are steeper
than those of the opposing maxillary molars. When approximated, the
cutting edges, therefore, cross diagonally. It is effectively a combination of

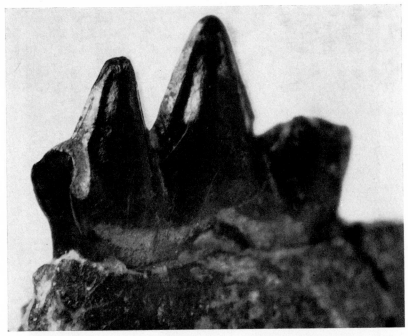

Fig. 2. Priacodon sp. (NMNH 2696), late Jurassic triconodont; left mandibular molar.

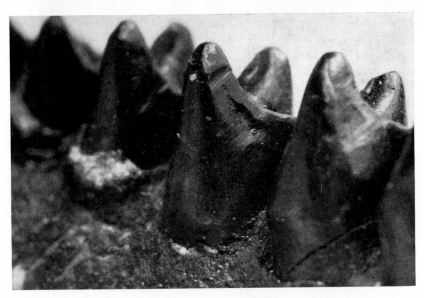

Fig. 3. Late Jurassic, dryolestid; right mandibular molars, mesiobuccal view.

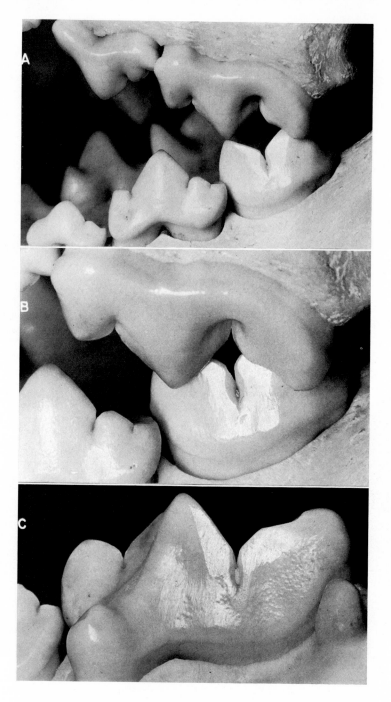

scissor-like and guillotine-like actions, which features the important advantage of *point-cutting* (i.e., where the force of the action is concentrated at a point [EVERY, 1972] in contrast to *line-cutting* (i.e., where the force is dispersed along a line [EVERY, 1972]).

In the symmetrodontoid 'reversed triangles' condition, point-cutting action is further developed by extending the occlusal surfaces transversely, i.e., buccopalatally in maxillary teeth and linguobuccally in mandibular teeth. The basic cusp-to-valley zig-zag arrangement is maintained, but a further zig-zagging, this time of tooth with tooth, is superimposed, thus giving considerable extension to the length of the blades.

In the Triassic mammal *Kuehneotherium* (the earliest known ancestor of the line leading to the 'tribosphenic' molars of the Theria), mandibular molars are virtually cusp-and-blades-in-a-row, whereas maxillary molars already show a degree of the symmetrodontoid expansion. This symmetrodontoid characteristic is developed and persists in both maxillary and mandibular molars through the Peramuridae in the Jurassic and the Aegialodontidae in the Cretaceous. There is little modification other than a more transverse (buccolingual) orientation of the distal blades of mandibular teeth, mesial of maxillary; also various elaborations of the stegoid formations, especially in the 'talonid' which, however, show no evidence of incusive functions as in molars SIMPSON calls 'tribosphenic' [SIMPSON, 1936].

Incusive function is first known in *Pappotherium* in the early to middle Cretaceous. It is suggested that this basic feature occurred concurrently with a fundamental change in the symmetrodontoid system [EVERY, 1972]. Instead of cusp-to-valley interdigitation there is cusp-to-cusp apposition with blades vertically concave and horizontally convex.

This system, but devoid of incusive characteristics, appears in the Jurassic Dryolestidae and Paurodontidae, mammals presently grouped with Amphitheriidae and Peramuridae as pantotheres. Yet, even without the incusive feature, their morphology and functional characteristics are so fundamentally contrasting they cannot validly be considered to occupy the same ordinal grade.

Unelaborated by incusive characteristics, however, the typical dryolestid molar (particularly mandibular) clearly illustrates many fundamental principles of the scissorial blade system basic to modern therians.

Figure 3 shows the right mandibular teeth of a late Jurassic dryolestid.

Fig. 4. Part of left cheek dentition of a leopard; lateral view of occluding cheek teeth (A, B) and lingual view of left P4 (C).

Although the dryolestid, as *Priacodon*, is not ancestral to modern therians, the molars conveniently show further aspects of the mammalian advance. These features can best be discussed using the terminology for mammalian teeth devised to connotate integral aspects of morphology and function [EVERY, 1972]. Terms used here are briefly explained for teeth in the maxilla; the features in teeth in the mandible are denoted by the suffix -id.

The pointed cusp at the terminal end of a sharp-edged blade is an *akis* (Greek *akis*, small pointed object, splinter, arrow-head, point of a chisel). When the blade edge running from the akis is concaved, thus functioning to confine exogenous material to the blade's cutting action, it is a *drepanon* (Greek *drepane*, sickle). A drepanon with a second akis, i.e., at the other end of the blade, is a *diakidrepanon*. When two diakidrepana are combined in a unit (sharing an akis) the feature is a *triakididrepanon*. When viewed in the occlusal aspect, a triakididrepanon appears as an asymmetrical Gothic Arch (the base on the buccal boundary of the tooth) with the mesial diakidrepanon oriented transversely, and the distal obliquely across the dental arch. The oblique blade is, therefore, longer than the transverse. It is also more (horizontally) convexed. The relationships in a triakididrepanid are reversed. In the row of dryolestid's triakididrepanoid molars illustrated, the molar in focus shows the main akid (shared by the two diakidrepanids) and part of the oblique blade, the akid at the lingual extremity of which is hidden. Behind, slightly out of focus, is the akid at the lingual extremity of the transverse blade. These three akids, therefore, are: (1) *protoakid* (pr^d) (the shared akid), (2) *parakid* (pa^d) and (3) *metakid* (me^d). The blades are: (1) Proto-diakidrepanid-obliqua, abbreviated to *Protobliquid* (Pro^d) (situated on the mesial face of the tooth); (2) Proto-diakidrepanid-transversa, abbreviated to *Prototransversid* (Prt^d) (situated on the distal face).

This terminology, not only accurately defines the morphological characteristics, but also the functional. It also facilitates easier conception of that most difficult of all aspects of odontology, i.e., the spatial relationships involved in dynamic occlusion. For example, the oblique blade of a mandibular tooth articulates scissorially with an oblique blade of one opposing maxillary tooth. Similarly, the transverse blade articulates with a transverse blade, but in another tooth distal to the first.

In the mono-triakididrepanoid teeth of the dryolestid, this relationship is relatively easy to appreciate. But when further scissorial units, together with incusive units, appear (as in *Pappotherium*), the spatial relationships of the occlusion become extremely confusing.

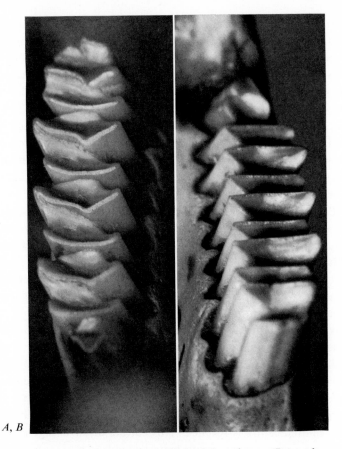

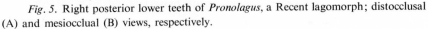

Fig. 5. Right posterior lower teeth of *Pronolagus*, a Recent lagomorph; distocclusal (A) and mesiocclual (B) views, respectively.

This also can be simplified if the features in the mandibles are regarded as the basic, dynamic, male units that articulate with corresponding, but static, female units in the maxilla.

This is illustrated in the uncomplicated dryolestid where the combined pair of diakidrepanids of the male unit (the 'trigonid') functions with the female unit made up of two discrete diakidrepana of two teeth which proximate at the stegoid formation. The male unit is the Proto-Triakididrepanid, which is abbreviated to *Proto-Triakid* (PrT^d). The female unit is the Stego-Tetrakididrepanon, which is abbreviated to *Stego-Tetrakis* (Stego-T).

The focused tooth in figure 3 shows clearly (despite the concealed parakid) the vertical drepanoid feature of this dominant blade of the tooth, the Protobliquid. It also shows the discrete narrow strip of thegosis-faceting trailing the leading edge, also the hollowed-out surface trailing the facet. Some parallel thegotic striations are visible, especially in the deep groove near the apex of the protoakid that seem to have been caused by an irregularity in the maxillary tooth.

In the same general line of the thegotic striations are numbers of slightly divergent pit and gouge scratches of abrasion which extend into the hollowed-out area and over the rounded crown wall. They show clearly the line of action of the tooth in mastication. It is precisely in line with the orientation of the drepanoid features.

When the Protobliquid approximates the Paraobliquis in the scissorial action, each akid/akis pair is lined up together. That is, the protoakid approximates the paraobliquakis and the parakid the parakis. The intervening vertically drepanoid blades, therefore, oppose and approximate each other and, at each end, their cutting edges are diagonally related. Exogenous material is, therefore, trapped in the open area between the blades and confined to their mutual point-cutting actions.

The protoakid/paraobliquakis pair is the first to approximate and point-cutting begins here. The parakid/parakis pair approximates soon after, and point-cutting follows from this end of the blade opposite to, i.e., towards the original pair. Eventually, the two actions meet at the blades' mutual centre.

Were dryolestid's Paraobliquis shaped in a simple drepanoid curve as shown in the Protobliquid, point-cutting would give place to line-cutting when the blade's mutual centre is reached. This cutting disadvantage is the more significant since a resistant exogenous material tends to be concentrated at this very area.

The disadvantage is obviated by the morphological feature where the central concavity of one or both blades is fissured [Every, 1972].

In Dryolestidae the transverse blade of the mandibular molar is fissured whereas the oblique is not. The oblique blade of the maxillary molar, however, shows a feature that is virtually an eversion of fissuring and appears as a small cusp [EVERY, 1972]; this cusp (called 'cusp C') has long confused odontologists, particularly its homology with the 'metacone'. Fissuring has proven to be more efficient, and it is this feature that appears especially in α-dentitions of modern therians. (It is discussed more fully below.)

Another important characteristic of the triakididrepanoid condition is the horizontal convexity of the scissorial blades. This is especially evident in the oblique blade. In the illustration of dryolestid's Protobliquid, the convexity faces the viewer and is only slightly apparent by the shading. The striking aspect of the convexity of this blade surface is that it opposes a corresponding convexity of the blade surface of the Paraobliquis. Admittedly, the complicated spatial relationships of this system are not easy to appreciate but, along with the vertical (diakidrepanoid), opposing, blade concavities, the whole system is adapted as an automatic confining device. Exogenous material is restricted to the cutting action and wedging between the blades is obviated. (The phenomenon is discussed more fully below.)

Figure 4 depicts the left P^3, P^4 and P_3, P_4, M_1 of a leopard. The cutting blades of the leopard's cheek teeth show a striking combination of reptilian and mammalian characteristics. Only the mandibular first molar and the distal one-half of the maxillary fourth premolar are thegotic. Despite the small but distinct (transverse) horizontal component, characteristic of the mammalian scissorial action, all other blades remain characteristically reptilian, i.e., their cutting blades, as in *Thrinaxodon*, cannot contact each other to wear by direct contact of tooth substance against tooth substance. Although morphologically sharp on eruption, when deteriorated through use, their cutting edges and cusp apices cannot be resharpened. These teeth penetrate and cut to a limited degree and virtually function as added grasping agents of material held in the large diastematic space immediately posterior to the grasping canine teeth.

The carnassial teeth, on the other hand, show fundamental characteristics of the α-dentition of modern therians. Thegosis occurs on the ends of the enamel prisms, is confined to a relatively narrow strip of faceting trailing the sharp leading edge of the blade, and bears the characteristic parallel striations of tooth-to-tooth contact wear. Trailing the faceting is the accentuated hollowing-out of the crown surface. The hollowed-out area on the distal segment of the palatal surface of P^4 (fig. 4C) shows the original morphological detail relatively intact. The mesial segment, on the other hand, shows this detail partially obliterated by the scratchings of abrasion, which marks, although generally lined in the direction of the scissorial action, are in many instances noticeably divergent.

Cusp-to-cusp apposition, contrasting cusp-to-valley interdigitation, is illustrated in figure 4A. The mesial akid/akis pair is just approximating, whereas the distal pair is overlapped and point-cutting action already established. The carnassial blades in felids are limited to oblique diakidrepana, the transverse components of the original triakididrepanoid units having disappeared. The obliquity of the blades, moreover, has become almost longitudinal (mesiodistal), and with fissuring in both blades their drepanoid characteristic has become somewhat distracted. In some carnivores, the mandibular blade, only, is fissured; the maxillary remains a simple diakidrepanon. With the double fissuring, point-cutting action is doubly ensured, as is illustrated in figure 4B.

Although the facets of both carnassial teeth illustrated in figure 4A and C face the viewer, their shadings give some indication of their horizontal convexities. In the articulated specimen illustrated in figure 4A, the convexity of the mandibular blade, therefore, faces the viewer, whereas that of the maxillary blade faces away. The spatial relationships of the leading edges of the blades are, therefore, such that the central area of each overlaps the escapement surface (the off-side surface) of the other. Exogenous material interposed between the blades is, therefore, forced into the escapement area rather than become wedged in between the blades' approximating (faceted) surfaces. It is a device that automatically ensures close alignment of the scissorial edges from the force of the biting action alone.

Figure 5 shows the right posterior lower teeth of *Pronolagus*, a lagomorph; distocclusal view (5A) looking directly at the facet surfaces, and mesiocclusal view (5B) looking directly down the facet surfaces, i.e., in line with the scissorial action.

Pronolagus is chosen as a sample of a β-dentition because of the discrete 'simplicity' of many of its features. The hypsodonty is complete; all teeth grow from perpetual pulps. There are two scissorial blades to each molar, neither is a primary blade, i.e., neither has the leading cutting edge situated on an exterior enamel surface. Being secondary blades, their leading cutting edges are situated on an interior enamel surface, i.e., at the amelo-dentinal junction, indicating that they become functional only when secondary hollowing-out occurs.

In contrast to the leopard's molar where all scissorial blades have disappeared except one oblique blade, in this lagomorph all except two transverse blades have lost their scissorial functions. (Other lagomorphs, however, like the leopard, are reduced to the single blade.)

In figure 5A, both blades are seen as discretely diakidrepanoid. The main blade, the Prototransversid, is fissured (markedly so for the β-dentition), whereas the Hypotransversid is plain.

Thick enamel is restricted to the scissorial blades alone, all other enamel is vestigially thin.

The line of scissorial action is clearly evidenced by two main features: (1) the orientation (seen at a glance) of the diakidrepanoid leading edges, i.e., they are orientated to a mesiodorsal mandibular action; (2) the general line (seen at close scrutiny) of the abrasive scratches, found on both the facet and the escapement surfaces of the blades.

The line of thegotic action is also evidenced by two main features: (1) the (seen at a glance) parallel, transversely planned row of blades when viewed down the line of scissorial action (fig. 5B); (2) the (seen at close scrutiny) transverse, parallel striations of a freshly thegosed specimen (mostly obliterated by abrasion in this one), found on the facet areas only.

Figure 6 shows a mesiopalatal view of the occlusal surfaces of the left M^1 and M^2 of *Absarokius*, an early Eocene primate (YPM 17483 in fig. 6A and YPM 18686 in fig. 6B).

α-Thegosis stage of wear of the transverse blades is shown in figure 6A and the beginning of β-thegosis in shown in figure 6B.

Failing an appreciation of the phenomenon of thegosis, odontologists could be expected to misinterpret the primate molar. The overt enlargement of the basins would, in terms of generally accepted theory, inevitably suggest an enhancement of the 'mortar-and-pestle', 'grinding', 'milling' functions and a depreciation of any other functions the ancestral tooth may have had. But this view overlooks fundamental aspects of the α/β-dentition adapted specifically for the mixed, i.e., omnivorous, diet, an adaptation, moreover, in terms of prolonged ontogenetic continuity of both the scissorial and the scissorio-incusive triakididrepanoid advantages.

Although the enlarged basins clearly have adaptive significance (otherwise they would not be here), their advantage as agents of comminution is less the action of bluntness confining exogenous material to bluntness than that of bluntness confining to sharpness.

This fundamental error seems to have permeated odontologists' efforts to understand the evolution of the mammalian, especially the primate, dentition. It is inherent in SIMPSON's [1936] basic term 'tribosphenic' (Greek tribein, to rub, grind, or pound), perpetuated by concepts of 'mortar-and-pestle', 'surface-shearing' 'rotary' actions, and now exacerbated by MILLS' [1955] fallacious [EVERY, in press] concept of a 'lingual phase of occlusion' in addition to the buccal. And overlooked is the concomitant of an enlarged basin: the lengthening of its peripheral blades [EVERY, in press].

By definition, the new term *incusion* (Latin *incuso*, to hammer; *incus*, anvil) connotates the reality of the dominant functional significance of the primate dentition, i.e., its sharpness. The blunt surface of the anvil, not only confines a material to the action of the blunt surface of a hammer, but also to that of the sharp edge of a (hammered) chisel or sharp point of a (hammered) punch. The axe and chopping block is perhaps even more appropri-

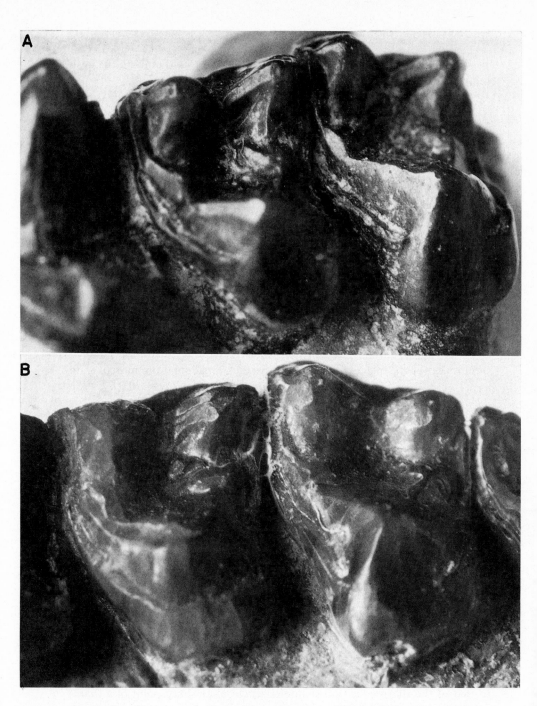

Fig. 6. Absarokius, sp., early Eocene omomyid primate (YPM 17483); mesiopalatal view.

ately analogous.

The facility of the new terminology [EVERY, 1972, and in press] in allowing appreciation of occlusal relationships has already been instanced with scissorial blades, i.e., transverse blades articulate with transverse blades and oblique with oblique. There is further facilitation in the appreciation of incusion. For incusive blades cross-diagonally, i.e., transverse blades occlude with oblique blades and oblique with transverse.

It is this relationship of interlocking scissorio-incusive triakididrepanoid features that allows the mutual actions of bluntness with bluntness and bluntness with sharpness.

Concomitant with the enlargement and heightening of incusive features characteristic of the α/β-dentition is the progressive orientation of the purely scissorial triakididrepana (i.e., Para-Triakis and Meta-Triakis) on the periphery of the crown. The angle between the blades progressively becomes obtuse and the orientation of the blades themselves progressively longitudinal, i.e., mesiodistal. Despite these changes (which affect the transverse blades more than the oblique) phylogenetic relationships remain consistent. The basic terminology is, therefore, still appropriate.

The illustration of α-thegosis in *Absarokius* (fig. 6A) highlights the transverse blades. Peripheral orientation inevitably shortens the scissorial blades, but the disadvantage is compensated for by increasing their number. Small triakididrepana at the base of the Para-Triakis and Meta-Triakis appear. They are: *proto-Para-Triakis* and *proto-Meta-Triakis*. And a new blade, the *Metastylotransversid*, running distolingually from the metakid, has developed to articulate scissorially with the Prototransversis.

Cingular blades have also appeared (or been elaborated) orientated precisely as ontogenetic reserve blades in the gerontic condition.

Features basically characteristic of the α-dentition are evident. Diakidrepanoid blades with primary (developmental) hollowed-out surfaces trail the narrow strip of thegosis- faceting at the blade's leading edge. With the shortened blades, however, fissuring has disappeared.

The early stages of β-thegosis in *Absarokius* (more particularly of the transverse blades of the Meta-Triakis of M[1]) is illustrated in figure 6B. With the breaching of the amelodentinal junction and establishment of secondary (abrasive) hollowing out, a second blade appears, trailing the primary blade (itself now β-thegosed).

When β-thegosis becomes established at the incusive basin, the secondary blade becomes the dominant blade and, in the gerontic condition, the primary blade eventually disappears.

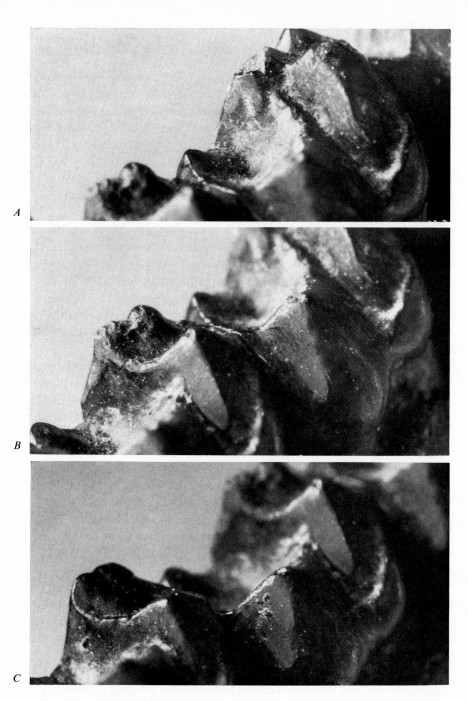

Fig. 7. Plesiadapis sp., late Paleocene archaic primate (PU 88108); distobuccal views.

Fig. 8. Plesiadapis sp., late Paleocene archaic primate (PU 88108); buccocclusal view.

In figures 7 and 8 is a series of distobuccal (7A–C) and buccocclusal (fig. 8) views, respectively, of the right M_1, M_2, and M_3 of *Plesiadapis* (PU 88108), a late Paleocene primate. (There is some *post mortem* and preparation damage to the metakid of M_1 and M_2, and M_3 is extruded so that the Proto-Triakid of M_3, where it proximates the Hypo-Triakid of M_2, is not level).

M_3 shows some of the characteristics of a lower molar on eruption. Yet unaffected by wear, the sharp leading edge and diakidrepanoid characteristics of the Prototransversid are apparent; also the primary (developmental) hollowing-out of the distal blade surface. Thegosis is well established in M_1 and M_2.

In the ancestral condition with the high Proto-Triakid and low Hypo-Triakid, the spatial relationships were such that the Prototransversid articulated primarily with the Paratransversis and secondarily with the Prototransversis.

With the raising of the Hypo-Triakid so that its proximal contact with the Proto-Triakid of the distal molar becomes level (fig. 7B), the spatial relationships of the occlusion were then such that the (secondary) articulation of the Prototransversid with the Prototransversis was lost. Instead, secondary articulation was established with a new feature on the maxillary molar: the *proto-Paratransversis* (seen in *Absarokius*, fig. 6A). And secondary articulation with the Prototransversis was established with a new feature in the mandibular molar: the *Metastylotransversid* (seen in *Plesiadapis*, fig. 7 B). In some primates, e.g., *Palaeopropithecus* and *Lepilemur*, this feature has become so developed that the metastyloakid has replaced the entoakid entirely, and the escapement valley that previously existed between these two akids is now established between the metastyloakid and the free end of the Hypotransversid, i.e., the hypotransversakid [EVERY, in press].

With the raising of the Hypo-Triakid to meet the level of the Proto-Triakid at the proximal contact point, the Protobliquid and the Hypotransversid, also, are affected. The free ends of their scissorial blades are now terminated at the contact point. This shortening affects the Protobliquid more markedly than the Hypotransversid which normally is a shorter blade because of the presence of the entoakid. The scissorial characteristics of blades lingual to the contact point are, therefore, lost. In *Plesiadapis*, the original parakid is strongly evident; in most other primates, it is vestigial or entirely lost. Its scissorial functions are replaced by a new akid: the *protobliquakid* [EVERY, in press].

The Protobliquid in the ancestral α-dentition is the dominant scissorial blade of the mandibular molar. That it is still a significant blade in the α/β-dentition is suggested by the common appearance of an ontogenetic reserve blade trailing it at the cingulum (as in *Plesiadapis*). Contrasting the α/β-dentition's shortening of the Protobliquid is the lengthening of the Hypobliquid. This is markedly evident in M_1 of *Plesiadapis*, where the blade extends virtually to the metakid.

In figure 9, aspects of the mandibular teeth of 'Taungs child', a South African australopithecine, are shown. Figure 10 shows the right M^2 and M^3 (MLD28) of an australopithecine from Makapansgat. At the hominid stage of evolution, Hypo-Triakid/Proto-Triakid levelling is complete. There is, moreover, a marked thickening of enamel over the whole of the crown, clearly an advantage to the α-mode of occlusion. In the unworn condition, however, this enamel thickening often appears as blunt ridges and mounds, approaching that of the bunodont dentition. And when worn so that saucer-shaped depressions appear in the exposed dentine this seems only to add (mistakenly) to the general effect of bluntness. It is hardly surprising, therefore, that, especially when preoccupied with tool-making theory, theorists believe there is a lessening of importance of teeth in the evolution of hominoids [LEGROS CLARK, 1962], that the teeth wear down from their tips to become flat and blunt [ZUCKERMAN, 1958; LEAKEY, 1960], and, ultimately, in apes and man, chewing is now entirely a cusp-in-fossa grinding action [MILLS, 1972].

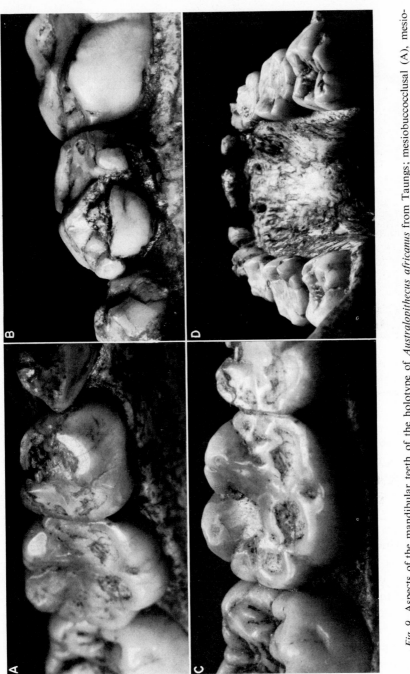

Fig. 9. Aspects of the mandibular teeth of the holotype of *Australopithecus africanus* from Taungs; mesiobuccocclusal (A), mesio-linguocclusal (B), buccocclusal (C), and distocclusal (D) views, respectively.

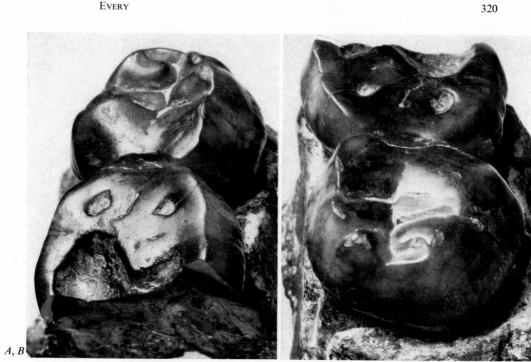

A, B

Fig. 10. Aspects of the right maxillary teeth of M²⁻³ of a Plio-Pleistocene hominid (MLD 28), Makapansgat; mesiocclusal (A), distocclusal (B), buccal (C), and mesial (D) views, respectively.

The bunodont condition is markedly apparent in the scarcely worn first molars of the Taungs child. Also apparent (despite the focusing in the illustration being concentrated on another feature, discussed below) is the multiple secondary basining in the first and second deciduous molars.

Since theorists are much concerned with the 'blunted' worn condition, it is this stage, i.e., the β-stage, that is discussed here.

The most striking instance of a bunodont condition is seen in *Phacochaerus*, the wart-hog. On eruption, the last molar shows some 30 enamel-coated cusps each completely surrounded by cementum. Both masticatory and thegotic wear soon expose the dentine so that 30 or so islands of dentine surrounded by rings of projecting enamel and surrounded, again, by cementum, form the functional occlusal surface. The tooth is implanted in the jaw to a great depth with little tendency to form roots, thus approaching the perpetual pulp condition.

C

D

Fig. 10

Clearly, in this system, there is no cusp to penetrate into any basin; there is no mortar-and-pestle analogy for the occlusal action. Yet, when analyzed, it is apparent that the occlusion presents a multitude of convex and concave enamel blades, virtually all interrelated for β-scissorial point-cutting action.

Thegosis faceting in the second deciduous molar (i.e., dm_2) of the Taung child is still in the α-stage for the leading blades of the tooth (i.e., the blades of the lingual cusps). This contrasts the trailing blades (i.e., the blades of the buccal cusps) where secondary (β-) blades are well established. Diakidrepanoid β-thegosis is clearly apparent in the first deciduous molar, dm_1 (fig. 9A). (Unfortunately, the secondary hollowing-out in this specimen is confused by the conglomerate of material misplaced in preparation. The whole lingual enamel plate of the metakid is, in fact, cemented in place upside-down: fig. 9B.)

Especially when viewed posteriorly (fig. 9D), the functional significance of the leading blades (of the functional teeth) is apparent. The sharp, primary, scissorial edges surmount the relatively flat, vertical, lingual surfaces (fig. 9B). (The relative flatness of this surface could have facilitated the preparator's error in capsizing the metakid of dm_1.)

The orientation of the blades in maxillary molars is, of course, reversed, i.e., the sharp, primary, scissorial edges surmount the relatively flat, vertical, buccal surfaces. Transverse (buccolingual) mandibular action, therefore, articulates the (in this specimen) primary, α-thegosed, leading blades with the secondary, β-thegosed, trailing blades in adaptively efficient, scissorial, point-cutting action. There is no analogy, whatsoever, with the action of a mortar-and-pestle.

When the characteristics of the two discrete wears of thegosis and abrasion are understood, the analysis of dentitions is markedly facilitated. The facet that forms the long graceful bevel of the leading cutting edge of Taung child's entoakid, for example (highlighted in fig. 9C), shows, on close scrutiny, crucial evidence of the adaptive precision of thegotic servicing. The base of this facet shows two discrete sets of parallel striations. One formed by lingual-to-buccal, transverse, mandibular action, i.e., rotating about a condylar axis on the ipsilateral side), and the other by distobuccal-to-mesiolingual, oblique, mandibular action (i.e., about an axis on the contralateral side).

The evidence that, in the hominids, these are the mandibular actions concerned, i.e., fundamentally opposite to the masticatory action, emerged from observations and experiments on man during some 20 years of a psychiatrically orientated dental practice. Little of this is yet documented. The field, briefly, covers a study of the differential diagnosis of loss of tooth substances; mainly: violent rubbing, tooth-to-tooth contact; abrasion from friction of exogenous material; fracture and bruising not from external violence; chemical erosion from acid of, for example, citrus fruits, preservatives, regurgitation of gastric juice; caries [EVERY, 1960]. It involved an intensive study of facet chains (aided by devised techniques for metal casts and mechanical articulations) and correlated with what emerged as the syndrome of extreme mandibular movements [EVERY, 1960]. Other evidence came from the study of actions of temporal, masseter, and internal pterygoid muscles contrasted with that of the external pterygoid [EVERY, 1970]; directions of pressure flaking and flow of maleable metal restorations [EVERY, 1970]; and experimental apparatus worn only when awake or only when asleep.

Thegotic servicing of the limited and irreplaceable substances of brachyodont teeth is necessarily conservative. Since there are extreme variations of emotional motivations and environmental circumstances, dentitions may be found, therefore, that vary from complete faceting of virtually all occlusal surfaces to complete absence of faceting where abrasion, alone, is present.

For both anatomical and motivational reasons, it is to be expected that thegosis in the gerontic condition ontogenetically tapers off.

Some of the factors that maintain functional sharpness in this condition are illustrated in the specimen MLD 28 (fig. 10), where thegosis faceting has disappeared, only its former orientation being apparent. Conveniently, one tooth is fractured showing a cross-section through both leading and trailing blades as well as the intervening (secondary) hollowing-out of dentine. Figure 10A shows a mesiobuccocclusal view, and figure 10B a distocclusal view of the still sharp primary leading edges which surmount the relatively flat vertical buccal surfaces (seen directly in fig. 10C) of both teeth. They also show, in M^2, the (secondary) leading edge of the (secondary) trailing blade, i.e., situated at the amelodentinal junction, and the manner in which this feature is supported on the tooth crown, i.e., surmounting a distinctly rounded palatal surface.

VIII. Conclusion

It seems evident that an understanding of primate dentitions involves an analysis of evolutionary events that took place very much earlier than the earliest known primate itself. Among other aspects of dental adaptations, it involves a fundamental appreciation of the phenomenon of thegosis, an adaptation that seems to have been crucial to the emergence of Mammalia, itself.

IX. Acknowledgments

For allowing me to study and photograph specimens, I thank the Bernard Price Institute *(Thrinaxodon)*; Smithsonian Institution *(Priacodon)*; Freie Universität Berlin (the dryolestid); University of Auckland *(Felis)*; National Museum of Rhodesia *(Pronolagus)*; Yale Peabody Museum *(Absarokius)*; Princeton University *(Plesiadapis)*; Witwatersrand Medical School (the australopithecines).

X. References

BUTLER, P. M.: The milk molars of the Perissodactyla, with remarks on molar occlusion. Proc. zool. Soc., Lond. *121:* 777–817 (1952).

CLARK, W. E. LeGros: The antecedents of man; 2nd ed., pp. 388 (Edinburgh Univ. Press, Edinburgh 1962).

CROMPTON, A. W.: The origin of the tribosphenic molar; in KERMACK and KERMACK Early mammals. Zool. J. Linn. Soc. *50:* suppl. 1, pp. 65–87 (1971).

EVERY, R. G.: The significance of extreme mandibular movements. Lancet *ii:* 37–39 (1960).

EVERY, R. G.: The teeth as weapons: their influence on behaviour. Lancet *i:* 685–688 (1965).

EVERY, R. G.: Sharpness of teeth in man and other primates. Postilla *143:* 1–30 (1970).

EVERY, R. G.: A new terminology for mammalian teeth: founded on the phenomenon of thegosis, pp. 64 (Pegasus Press, Christchurch, N.Z., 1972).

EVERY, R. G.: Thegosis in prosimians; in MARTIN, WALKER and DOYLE Prosimian biology (Duckworth, London, in press).

EVERY, R. G. und KÜHNE, W. G.: Funktion und Form der Säugerzähne. I. Thegosis, Usur und Druckusur. Z. Säugetierk. *35:* 247–252 (1970).

EVERY, R. G. and KÜHNE, W. G.: Biomodal wear of mammalian teeth; in KERMACK and KERMACK Early mammals. Zool. J. Linn. Soc. *50:* suppl. 1, pp. 23–27 (1971).

LEAKEY, L. S.: Finding the world's earliest man. Nat. Geogr. *118:* 240–435 (1960).

MILLS, J. R. E.: Ideal dental occlusion in the Primates. Dent. Pract. *6:* 47–61 (1955).

MILLS, J. R. E.: Evolution of mastication. Proc. roy. Soc. Med. *65:* 392–396 (1972).

ROMER, A. S.: Notes and comments on vertebrate paleontology, pp. 384 (Univ. of Chicago Press, Chicago 1968).

SCALLY, K. B.: Thegosis in the sea urchin *Evechinus chloroticus* (Val.) (Echinodermata: Echinoidea). J. dent. Res. *52:* 583 (1973).

SCALLY, K. B.; EVERY, R. G., and TUNNICLIFFE, G. A.: Thegosis in the red spiny lobster *Jasus edwardsii* (Crustacea: Decapoda: Palinuridae) J. dent. Res. (in press).

SIMPSON, G. G.: Studies of the earliest mammalian dentitions. Dent. Cosmos *78:* 791–800 (1936).

TOMES, C. S.: A manual of dental anatomy: human and comparative; 8th ed., pp. 547 (Churchill London 1923).

TUNNICLIFFE, G. A.: Preliminary observations of the phylogenesis of thegosis. J. dent. Res. *52:* 583 (1973).

Tunnicliffe, G. A.; Every, R. G., and Scally, K. B.: Audiospectrographic analysis and behavioural significance of thegotic sounds. J. dent. Res. (in press).

Washburn, S. L.: Tools and human evolution. Sci. Amer. *203:* 3–15 (1960).

Zuckerman, S.: Correlation of change in the evolution of higher primates; in Huxley, Hardy and Ford Evolution as a process, pp. 300–350 (Allen & Unwin, London 1958).

Author's address: Dr. R. G. Every, Centre for the Study of Conflict, 25 Clifton Ter., Sumner, *Christchurch 8* (New Zealand)

Subject Index